Friction and Wear of Ceramics

Friction and Wear of Ceramics

Principles and Case Studies

Bikramjit Basu
Indian Institute of Science, Bangalore, India

Mitjan Kalin
University of Ljubljana, Ljubljana, Slovenia

B. Venkata Manoj Kumar
Indian Institute of Technology Roorkee, Roorkee, India

The American Ceramic Society

WILEY

Published by John Wiley & Sons, Inc., Hoboken, New Jersey.
Published simultaneously in Canada.

Registered Office
John Wiley & Sons, Inc., 111 River Street, Hoboken, NJ 07030, USA

Editorial Office
111 River Street, Hoboken, NJ 07030, USA

For details of our global editorial offices, customer services, and more information about Wiley products visit us at www.wiley.com.

Wiley also publishes its books in a variety of electronic formats and by print-on-demand. Some content that appears in standard print versions of this book may not be available in other formats.

Library of Congress Cataloging-in-Publication data

Names: Basu, Bikramjit, author. | Kalin, Mitjan, author. | Kumar, B. Venkata Manoj, author.
Title: Friction and wear of ceramics : principles and case studies / Bikramjit Basu, Indian Institute of Science, Bangalore, India, Mitjan Kalin, University of Ljubljana, Slovenia, B. Venkata Manoj Kumar, Indian Institute of Technology Roorkee, India.
Description: First edition. | Hoboken, NJ : Wiley-American Ceramic Society/John Wiley & Sons, Inc., [2020] | Includes bibliographical references and index.
Identifiers: LCCN 2020001871 (print) | LCCN 2020001872 (ebook) | ISBN 9781119538387 (case) | ISBN 9781119538837 (adobe pdf) | ISBN 9781119538400 (epub)
Subjects: LCSH: Ceramic materials–Mechanical properties. | Friction. | Mechanical wear. | Tribology.
Classification: LCC TA455.C43 B377 2020 (print) | LCC TA455.C43 (ebook) | DDC 620.1/404292–dc23
LC record available at https://lccn.loc.gov/2020001871
LC ebook record available at https://lccn.loc.gov/2020001872

Cover Design: Wiley
Cover Images: (top and bottom left) Courtesy of B. V. Manoj Kumar, (top and bottom right) Reproduced with permission of Elsevier

Set in 9.5/12.5pt STIXTwoText by SPi Global, Pondicherry, India

MIX
Paper from responsible sources
FSC® C013604

Contents

About the Authors *xiii*
Foreword by Dr. Sanak Mishra *xvii*
Foreword by Prof. Koji Kato *xviii*
Preface *xix*

Section I Fundamentals of Ceramics: Processing and Properties *1*

1 Introduction: Ceramics and Tribology *3*
1.1 Introduction *3*
1.2 Classification of Engineering Materials *6*
1.3 Engineering Ceramics *8*
1.4 Structural Ceramics: Typical Properties and Tribological Applications *9*
1.5 Structure of the Book *14*
1.6 Closure *17*
References *17*

2 Processing of Bulk Ceramics and Coatings *21*
2.1 Introduction *21*
2.2 Conventional Processing of Ceramics *21*
2.2.1 Sintering Mechanism *22*
2.2.2 Conventional Processing of Ceramics *24*
2.2.2.1 Powder Processing and Compaction *24*
2.2.2.2 Pressureless Sintering *27*
2.2.3 Advanced Processing of Ceramics *28*
2.2.3.1 Hot Pressing *28*
2.2.3.2 Microwave Sintering *28*

2.2.3.3 Spark Plasma Sintering *29*
2.3 Thermal Spray-Based Coating Deposition *30*
2.3.1 Basics of Thermal Spray Deposition *33*
2.3.1.1 Plasma Spray Deposition *33*
2.3.1.2 Flame Spray Deposition *34*
2.3.1.3 Wire Arc Spray Deposition *35*
2.3.1.4 High-Velocity Oxy-Fuel Spray Deposition *36*
2.3.1.5 Detonation Spray Coating *36*
2.3.2 Bond Strength of Thermal Spray Coatings *39*
2.3.2.1 Bond Mechanism *39*
2.3.2.2 Test Methods *40*
2.3.3 Coating Structure *42*
2.3.3.1 Particle and Substrate Material Properties *42*
2.3.3.2 Particle Temperature and Velocity *42*
2.3.4 Case Study: WC-Co Coatings *42*
2.4 Closure *49*
References *49*

3 Conventional and Advanced Machining Processes *53*
3.1 Introduction *53*
3.2 Conventional Machining *54*
3.3 Advanced Machining Processes *57*
3.3.1 Electro-Discharge Machining *57*
3.3.1.1 Working Principle *59*
3.3.1.2 EDM Process Variables *61*
3.3.1.3 EDM Parameters *62*
3.3.1.4 Surface Analysis *63*
3.3.1.5 EDM of Ceramic-Based Composites *64*
3.4 Closure *66*
References *66*

4 Mechanical Properties of Ceramics *71*
4.1 Defining Stress and Strain *71*
4.2 Comparison of Tensile Behavior *78*
4.3 Brittle Fracture of Ceramics *80*
4.4 Cracking in Brittle Materials *84*
4.5 Experimental Assessment of Mechanical Properties *87*
4.5.1 Hardness *87*
4.5.2 Compressive Strength *88*
4.5.3 Flexural Strength *89*
4.5.4 Tensile Strength *91*
4.5.5 Elastic Modulus *91*

4.5.6 Fracture Toughness *95*
4.5.6.1 Notched Beam Test *96*
4.5.6.2 Indentation Microfracture Method *97*
4.5.7 Practical Guidelines for Reliable Measurements *98*
4.6 Closure *99*
References *100*

Section II Fundamentals of Tribology *103*

5 Contact Surface Characteristics *105*
5.1 Nature and Roughness of Contact Surfaces *105*
5.2 Surface Roughness Measurement *108*
5.2.1 Stylus Method *108*
5.2.2 Atomic Force Microscopy *109*
5.2.3 Optical Interferometry *110*
5.2.4 Laser Surface Profilometry *111*
5.2.5 Scanning Electron Microscopy *111*
5.3 Bearing Area Curve and Cumulative Distribution Function *111*
5.4 Nominal Versus Real Contact Area *112*
5.5 Hertzian Contact Stress *115*
5.6 Closure *116*
References *118*

6 Friction and Interface Temperature *119*
6.1 Theory of Friction *119*
6.1.1 Friction Laws and Mechanisms *120*
6.2 Types of Friction *125*
6.2.1 Static and Kinetic Friction *125*
6.2.2 Slip-Stick Friction *126*
6.2.3 Rolling Friction *126*
6.3 Friction of Engineering Material Classes *127*
6.4 Frictional Heating and Temperature at the Interface *139*
6.4.1 Heating Due to Friction *140*
6.4.2 Understanding the Temperature in the Contact: The Bulk and Flash Temperatures *141*
6.5 Analytical Models Used to Predict the Temperatures in the Contact *145*
6.6 Implications of the Important Contact Temperature Models *146*
6.6.1 Archard Model *147*

6.6.2 Kong–Ashby Model *148*
6.7 Closure *149*
References *150*

7 Wear of Ceramics and Lubrication *155*
7.1 Introduction *155*
7.2 Testing Methods and Quantification of Wear of Materials *157*
7.3 Classification of Wear Mechanisms *158*
7.3.1 Tribomechanical Wear *159*
7.3.1.1 Adhesive Wear *159*
7.3.1.2 Abrasive Wear *161*
7.3.1.3 Fatigue Wear *164*
7.3.1.4 Fretting Wear *165*
7.3.1.5 Erosive Wear *167*
7.3.2 Tribochemical Wear *171*
7.3.2.1 Oxidative Wear *174*
7.4 Lubrication *175*
7.4.1 Regimes of Lubrication and the Stribeck Curve *175*
7.4.2 Influence of Lubricant Composition, Contact Pressure, and Temperature on Lubrication *178*
7.5 Closure *181*
References *182*

Section III Case Study: Sliding Wear of Ceramics 185

8 Sliding Wear of SiC Ceramics *187*
8.1 Introduction *187*
8.2 Materials and Experiments *188*
8.3 Friction and Wear Behavior of SiC Ceramics Sintered with a Small Amount of Yttria Additive *189*
8.4 Influence of Mechanical Properties on Sliding Wear of SiC Ceramics *191*
8.5 Wear Mechanisms *191*
8.6 Closure *192*
References *193*

9 Sliding Wear of SiC-WC Composites *195*
9.1 Introduction *195*
9.2 Microstructure and Mechanical Properties of SiC-WC Composites *196*
9.3 Influence of Mating Material and WC Content on Tribological Properties *197*

9.3.1 Friction and Wear Behavior *197*
9.3.2 Mechanisms of Material Removal *198*
9.3.3 Friction and Wear of SiC-WC Composites: System-Dependent Properties *202*
9.3.4 Wear Mechanisms *202*
9.4 Reciprocated Sliding Wear Behavior of SiC-WC Composites *203*
9.4.1 Frictional and Wear Behavior *205*
9.4.2 Critical Analysis of Wear Mechanisms *206*
9.4.2.1 Wear Debris Analysis *206*
9.4.2.2 Effect of Temperature *208*
9.4.2.3 Effect of Test Configuration on Wear Behavior *208*
9.5 Closure *209*
References *210*

10 Sliding Wear of Zirconia-Toughened Alumina *215*
10.1 Introduction *215*
10.2 Mechanical Properties of ZTA *216*
10.3 Sliding Wear Properties of ZTA *219*
10.4 Correlation with Theoretical Analysis *222*
10.5 Closure *224*
References *225*

11 Abrasive Wear of Detonation Sprayed WC-12Co Coatings *227*
11.1 Introduction *227*
11.2 Coatings and Abrasive Wear *228*
11.3 Abrasive Wear Results *230*
11.4 Surface and Subsurface Damage Mechanisms *231*
11.5 Closure *233*
References *234*

12 Solid–Lubricant Interaction and Friction at Lubricated Contacts *237*
12.1 Introduction *237*
12.2 Materials and Sliding Wear Experiments *239*
12.3 Wetting and Spreading Properties *240*
12.4 Surface Energies of Different Classes of Materials *242*
12.5 Wetting Evaluation of Engineering Surfaces *242*
12.6 Effect of Wetting on EHL Friction *246*
12.7 Correlation Between Spreading Parameter and Friction *247*
12.8 Closure *249*
References *250*

Section IV Case Study: Erosive Wear of Ceramics 253

13 Erosive Wear of SiC-WC Composites 255
13.1 Introduction 255
13.2 Materials and Erosion Tests 256
13.3 Influence of Type of Erodent on Erosive Wear Behavior 256
13.4 Influence of Impingement Angle and WC Content on Erosive Wear Behavior 258
13.5 Correlating Erosive Wear Behavior with Microstructural Characteristics 259
13.6 Correlating Erosive Wear Behavior with Mechanical Properties 259
13.7 Erosive Wear Behavior at High Temperature 260
13.8 Closure 262
References 263

14 Thermo-Erosive Behavior of ZrB_2-SiC Composites 265
14.1 Introduction 265
14.2 High-Temperature Erosion Tests and Computational Modeling 267
14.3 Computational Modeling of Thermo-Erosive Behavior 269
14.4 High-Temperature Erosion Test Results 270
14.5 Transient Thermal Studies Using FE Analysis 271
14.6 Coupled Thermo-Structural Analysis 271
14.7 Thermo-Erosive Behavior 273
14.8 Closure 274
References 275

15 Erosive Wear of WC-Co Coating 279
15.1 Introduction 279
15.2 Materials and Erosion Experiments 280
15.3 Erosive Wear Mechanisms (Surface Damage) 282
15.4 Erosive Wear Mechanisms (Subsurface Damage) 287
15.5 Correlating Wear Mechanism with Erodent and Coating Properties 290
15.6 Closure 293
References 293

Section V Case Study: Machining-Induced Wear of Cermets 295

16 Crater Wear of TiCN Cermets in Conventional Machining 297
16.1 Introduction 297

16.2 TiCN Cermets and Machining Conditions *298*
16.3 Wear Mechanisms of TiCN-WC-Ni Cermets *299*
16.4 Machining with TiCN-WC-TaC-Ni-Co Cermet Tools *300*
16.5 Correlating Cermet Composition, Microstructure, and Wear During Machining *304*
16.6 Closure *306*
References *306*

17 Wear of TiCN-Based Cermets in Electrodischarge Machining *309*
17.1 Introduction *309*
17.2 Materials and EDM Tests *310*
17.3 Wear of TiCN-Cermets During EDM *310*
17.4 Mechanisms of Material Removal During EDM *311*
17.5 Closure *313*
References *313*

Section VI Future Scope *317*

18 Perspective *319*
18.1 Innovation Cycle for Wear-Resistant Materials *319*
18.2 *In Situ* Diagnosis of Tribological Interaction *321*
18.3 High-Temperature Wear Testing *321*
18.4 Modeling and Simulation in Tribology *322*
18.5 Tribomaterialomics – A New Concept *323*
18.6 Education and Mentoring of Next-Generation Researchers *327*
References *328*

Appendix: Appraisal *329*
A.I Multiple Choice Questions *329*
A.II Select the Appropriate Combination *350*
A.III Fill in the Blanks with the Most Appropriate Response *352*
A.IV Mention the Appropriate Material/Equipment in the Blank *353*
A.V Identify Whether the Following Statements are True/False *354*
A.VI Short Review Questions and Descriptive Questions *354*
A.VII Analytical Questions *363*
A.VIII Model Answers *366*

Index *369*

About the Authors

Bikramjit Basu
Professor, Materials Research Center
Indian Institute of Science, Bangalore,
Karnataka, India

Prof. Bikramjit Basu is currently a Professor at the Materials Research Center and holds Associate Faculty position at Center for Biosystems Science and Engineering, Indian Institute of Science (IISc), Bangalore. He is currently Visiting Professor at University of Manchester, UK and Guest Professor at Wuhan University of Technology, China. After his undergraduate and postgraduate degree in Metallurgical Engineering from NIT Durgapur and IISc respectively, he earned his PhD in the area of Engineering Ceramics at Katholieke Universiteit Leuven, Belgium in March, 2001. Following a brief post-doctoral stint at University of California, Santa Barbara; he served as a faculty of Indian Institute of Technology Kanpur during 2001–2011. Bikramjit's international standing and impact on the field are illustrated by his prolific publication record (>300 peer-reviewed research papers, including 28 papers in Journal of American Ceramic Society and 35 papers in journals of high impact factor, more than 20 invited review papers, total citation >10,500 and H-index: 54). He has authored 7 textbooks, 2 edited books and one research monograph in the interdisciplinary areas of Biomaterials, Ceramics, Tribology and Energy. In addition to writing textbooks, he has created two online course curricula, including one on Tribology, which are watched over 9000 times worldwide through National Programme on Technology Enhanced Learning (NPTEL).

Prof. Basu's contributions in Engineering Science have been widely recognised. He is the first biomaterials scientist to receive in 2013, India's most coveted science and technology award, **Shanti Swarup Bhatnagar Prize**, which was first awarded in 1958. A Chartered Engineer of UK, he is an elected Fellow of the Indian Academy of Sciences (2020) and of the International Union of Societies for Biomaterials Science and Engineering (2020). He is also Fellow of the American Ceramic Society (2019), American Institute of Medical and Biological Engineering (2017), Institute of Materials, Minerals & Mining, UK (2017), National Academy of Medical Sciences, India (2017), Indian National Academy of Engineering (2015), Society for Biomaterials and Artificial Organs (2014) and National Academy of Sciences, India (2013). He remains the only Indian from India to receive the prestigious 'Coble Award for Young Scholars' (2008) from the American Ceramic Society. He is also a recipient of the Abdul Kalam Technology Innovation National Fellowship from Indian National Academy of Engineering (2020). The nominee has established independent research programs, first at IIT Kanpur (2001–2011) and lately at Indian Institute of Science, Bangalore (2011–till date).

Bikramjit has made truly outstanding contributions at the frontiers of Engineering Ceramics, biomedical engineering and lately, solar energy. A common thread that runs through these research programs is his exceptional understanding in manipulating processing approaches (spark plasma sintering, 3D inkjet powder printing, etc.) and in combining computational analysis with performance-qualifying property measurements or prototype testing; while establishing process-structure-property linkages for a wide spectrum of oxide and non-oxide ceramics. A major research theme has been to enhance the fracture resistance through tailoring of process parameters and manipulation of microstructure at multiple length scales. Another major contribution has been to develop microstructure-wear resistance relationship for ceramics/cermets in unlubricated fretting/sliding condition and to establish analytical framework for tribochemical/tribomechanical wear. Many of the tribology studies in his research group involved cryogenic sliding wear of materials for space technology applications and erosive wear of ultra-high temperature ceramics at elevated temperature. Over last two decades, he has groomed a large number of PhD students, who have now established independent research programs on structural ceramics and tribology as faculty members in IITs/NITs in India.

Dr. Mitjan Kalin
Professor,
Faculty of Mechanical Engineering
University of Ljubljana
Ljubljana, Slovenia

Dr. Kalin's areas of research are the wear and friction mechanisms of advanced materials, nanoscale interface phenomena, and boundary films for novel green-lubrication technologies, including his widely recognized contribution to the lubrication of DLC coatings. He has delivered over 50 invited lectures worldwide, including 10 keynote and plenary lectures at the most renowned conferences in the field of tribology. He has published over 150 peer-reviewed journal papers with 3200 citations and an h-index of 31. He also published 10 book chapters, 2 books, and holds 11 patents, including US and EU patents. He acted as a member of the editorial boards of eight international journals, Associate Editor of the ASME *Journal of Tribology* and *Frontiers in Mechanical Engineering* – Tribology Section. Since 2012, he has served as the Editor-in-Chief of *Lubrication Science* (Wiley). In his career, he has led over 35 large, three-year projects, most of them international. He also collaborated in industrial projects with renowned companies in Europe, Japan, and the United States in over 140 R&D projects. He has received several awards, including a prestigious ASME Burt L. Newkirk Award (2006), Fellow of STLE (2012), the two highest Slovenian state awards, namely Zois Prize (2006) and Zois Award (2015), and the Top 10 scientific achievements at the University of Ljubljana (2014). Since 2010, he has been a Full Professor at the Faculty of Mechanical Engineering, University of Ljubljana, where he is the Head of the Laboratory for Tribology and Interface Nanotechnology and the Chair for Tribology and Maintenance Technology. He is currently also a visiting professor at Luleå University of Technology, Sweden. Since 2013, he has been a coordinator of a Joint European Master Programme on the Tribology of Surfaces and Interfaces (TRIBOS) under the Erasmus Mundus umbrella. In 2019, he also became a coordinator of the European Joint Doctorate programme GreenTRIBOS, an EU Marie Skłodowska–Curie Actions initiative. In 2007–2011, he was elected as a Vice-Dean for Research; in 2013–2017, as a Vice-Dean for Master and Doctoral Studies; and in 2017–2021, as Dean of the Faculty of Mechanical Engineering. He was also elected as Executive Board Member and Deputy President of the International Tribology Council (ITC) for the period 2017–2021.

B. Venkata Manoj Kumar
Associate Professor
Department of Metallurgical and Materials Engineering
Indian Institute of Technology (IIT) Roorkee
Roorkee
Uttarakhand, India

Dr. B. Venkata Manoj Kumar is currently working as Associate Professor at Department of Metallurgical and Materials Engineering, Indian Institute of Technology (IIT) Roorkee. Dr. Kumar obtained his PhD degree from IIT Kanpur in November 2007. Subsequently, he worked as a post-doctoral researcher at Seoul National University from January 2008 to January 2009, Research Assistant Professor at the University of Seoul from February 2009 to February 2011, and Assistant Professor at IIT Roorkee from March 2011 to April 2016. With the primary theme of understanding microstructure–property relations, Dr. Kumar has been actively involved in processing advanced ceramic systems like SiC, ZrB_2, B_4C, Si_3N_4, TiCN-Ni cermets, etc., and studying the influence of microstructural characteristics on their material removal mechanisms when subjected to sliding, fretting, erosion, or machining conditions. He is also co-instructor for a National Programme on Technology Enhanced Learning (NPTEL) online course on friction and war of materials. His research work is being supported by both government agencies and private industries in India. His research ideas are also supported by foreign funding agencies through bilateral projects. He has so far supervised four PhD theses. He has published around 55 peer-reviewed research articles with more than 1300 citations and h-index of 20. He has also published three review articles and three book chapters. He is presently a member of the editorial board for the *Tribology and Lubricants* journal published by the Korean Tribology Society. He is co-inventor of the filed two Indian patents and delivered more than 40 lectures in reputed conferences.

Foreword

The continuing growth of manufacturing and energy sectors demands the development of materials with superior performance in harsh tribological conditions over a longer lifetime. Among advanced materials, structural ceramics are increasingly used in tribological applications, because of their unique combination of superior properties. It is therefore important to understand the phenomena of friction and wear, from the viewpoint of ceramic science and engineering. However, most of the available books on tribology primarily provide an understanding of the mechanics and design of components, with less focus on materials aspects.

In a significant departure from conventional "mechanical engineering" viewpoint, this book has dealt with the essential concepts of tribology as well as basics of processing and properties of ceramics and ceramic coatings. Also, I recognize the authors have put their best efforts to analyze their own published results, related to sliding, erosion, and machining of ceramic components and systems. This part is particularly useful for the readers of varied background in recognizing microstructure–mechanical–tribological property correlation. Besides, the grand collection of conceptual and analytical questions in the Appendix section will be greatly useful in assessing the knowledge of the undergraduate and postgraduate students, wishing to pursue the field of tribology in their academic careers. This feature distinguishes the present book from currently available other tribology books.

I strongly feel the tribology community across the Materials Science and Mechanical Engineering domain will be greatly benefited from this new source of knowledge in the rapidly growing multidisciplinary field of tribology of ceramics.

Dr. Sanak Mishra
President, Indian National Academy of Engineering
www.inae.in
New Delhi, India

Foreword

Among all the engineering materials, advanced ceramics are being widely investigated in the last few decades for their potential use at tribological contacts, particularly under extreme conditions. Ceramic cutting tools, bearings, seals, hip joints, and coatings are some of the important advanced ceramic components developed in the past 30–40 years. The tribological performance of the ceramic components can be improved through innovative process design and microstructural engineering. While most of the presently available books on tribology are written by mechanical engineers or researchers from the mechanics background, there is an urgent need for understanding the ceramics tribology from materials science perspective.

In this context, I find this book highlights the importance of the microstructural characteristics–material property–tribological performance relationship for a range of advanced ceramic materials. The fundamental concepts of the processing and mechanical properties of ceramics are introduced for this purpose. This is particularly important for the researchers from the mechanical engineering discipline as they would acquire knowledge as how to tune the processing conditions to tailor the mechanical properties. This book also reviews some of the fundamental mechanisms of friction, wear, and lubrication. A range of case studies highlights how to design ceramic composites for better wear resistance.

I appreciate the authors' efforts in compiling seminal studies from their own research to fit into the broader framework of the book and helping the readers to realize the translation of basic tribology science to applications. Particularly, the description of the wear micromechanisms is expected to provide a strong background to the readers toward designing/developing ceramic composites for tribological components. This book has important pedagogical feature as a textbook with the rich collection of objective and subjective problems on important issues of ceramic tribology and will be very useful for students and practicing scientists.

Prof. Dr. Koji Kato
Nihon University, Japan

Preface

The Most Influential Science that Needs to Be Better Understood

Tribology is widely defined as the science and technology of interacting surfaces in relative motion. Of enormous practical value to key engineering industries – from energy, manufacturing, nuclear, thermal to biomedical – the field of tribology is drawing ever-increasing attention in recent decades across the multiple engineering disciplines.

The Paradox: How to Solve It

Yet the tribology field suffers from a paradox. As distinguished tribologist, Professor John Tichy of Rensselaer Polytechnic Institute (RPI), New York, once said, "Tribology is the most common area of science in the world that nobody knows about." The field has grown into one of the most influential sciences encountered in everyday life, drawing the most influential researchers from multidisciplinary and cross-disciplinary backgrounds.

Ceramics and their composites, despite their unique combination of hardness, elastic modulus, compressive strength, resistance against oxidation, and creep, find limited tribological application in the market on a commercial scale. Therefore, the development of advanced (bulk or coating) ceramic materials and the progress in research for superior tribological performance requires knowledge on microstructural engineering via innovative process design, and understanding the physics of material degradation. The improved understanding of the contribution of microstructural features and micromechanisms on tribological performance of ceramic components is essential. This helps in extending component lifetime and achieving higher energy efficiency.

The tribology of ceramics and ceramic composites – which comprises most of the applications in the area – is still not considered a focus area in many industries, or in academia, in spite of its remarkable potential and contribution. Most engineering students receive few hours of instruction in tribology of ceramics, most engineers and researchers do not have adequate grounding in this area, and most textbooks do not deal with real-world problems worth solving.

The present book highlights the importance of the correlation among microstructure, mechanical properties, friction, and wear resistance for a range of selected ceramic systems. The development of bulk ceramics and ceramic coatings, and understanding on their mechanical properties, has also been discussed in detail. The exploration of friction and wear mechanisms of several engineering ceramic systems provides a thorough background to the readers on how to design and to develop novel materials for tribological applications.

The Transition: Classroom to Industry

The perceived need to meet stringent needs of superior performance of ceramic components in tribological conditions has necessitated a book of this length and depth. The present book is the result of authors' several years of teaching and research in materials tribology, engineering ceramics, and composites at the Indian Institute of Technology (IIT) Kanpur, Indian Institute of Science (IISc), Bangalore, the Indian Institute of Technology (IIT) Roorkee, and the University of Ljubljana, Slovenia. Overall, the book is designed to enrich the knowledge on basic concepts on ceramic processing and tribology, and to upgrade the understanding of state-of-the-art research findings of advanced ceramic composites. Thus, it will be highly useful for beginners and students pursuing the tribology of ceramics, and for experienced researchers and practitioners engaged in understanding the science of tribology.

The Approach: Decoding the Chapters

The book is essentially structured into six sections, spanning 18 chapters. For those uninitiated in basic ceramics and tribology, the first section is a run-through of the fundamentals: processing, microstructure, and mechanical properties of bulk ceramics and ceramic coatings. The second section targets the more advanced complexities: contact surface characteristics, friction, interface temperature, wear, wear mechanisms, and lubrication. Keeping in view of the significance of advanced ceramic systems in improving wear resistance and product lifetime, significant results from the authors' own research groups are explained in three subsequent chapters. In the concluding section, the major issues that demand urgent attention, and future directions of tribology toward an improved and new generation of structural ceramic systems, are provided. In addition, a large collection of different types of qualitative and quantitative questions on tribology of ceramic materials is presented as Appendix, both for the novice and experienced tribologists, to assess their knowledge on the subject.

To Conclude: A Word of Thanks

Recognizing the collaboration over the last two decades, Professor Bikramjit Basu specifically acknowledges some of his students, including Debasish Sarkar, Subhadip Bodhak, Rohit Khanna, Manisha, G. B. Raju, T. Venkateswaran, Amartya Mukhopadhyay, and Shekhar Nath. Professor M. Kalin acknowledges cooperation and support from all his co-workers at Laboratory for Triblogy and Interface Nanotechnology (TINT) at University of Ljubljana, and numerous co-workers and friends worldwide in the field of ceramics tribology that shared the knowledge, experience and discussions in many occasions over more than two decades of engagement in this filed. Professor B. Venkata Manoj Kumar would like to sincerely thank all his students, particularly Vipin, Sandan, Vikas, Ashish Selokar, Suneel, Rajavel, Yashpal, Sonali, Smita, and Ashish Nayan, and his collaborators Professors Young-Wook Kim, Sinhoo Kang, S. K. Nath, and Sai Ramudu Meka, and Dr. T. Venkateswaran. The unconditional help from Surya, Nilesh, Rea, Nihal, Nandita and Prerana in each stage of preparation of the book is acknowledged. Professor Basu would like to convey heartfelt thanks to his father Mr. Manoj Mohan Basu for critically checking some part of the manuscript. Professor Basu is particularly grateful to Dr. Damayanti Datta for her help in shaping the Preface of this book. Professors Basu and Kumar would like to express their gratitude to Mr. Amit Ganguli and Mr. N. M. Dube for exciting discussion and suggestions, during the writing of this book. If someone who has helped in completing the book is not acknowledged here, authors sincerely admit that it is purely unintentional, but not due to lack of appreciation.

Further, the authors, Basu and Kumar would like to gratefully acknowledge their respective institutes, IISc Bangalore and IIT Roorkee. They would also like to express their sincere gratitude for the financial support received from various Government of India agencies, including the Indian Space Research Organization (ISRO), the Department of Biotechnology (DBT), the Department of Atomic Energy (DAE), the Defense Research and Development Organization (DRDO), the Department of Science & Technology (DST), and the Council of Scientific & Industrial Research (CSIR), which facilitated research in the field of tribology of advanced materials at IIT Kanpur, IISc Bangalore, and IIT Roorkee. One of the authors (Kalin) acknowledges the steady financial support from the Ministry of Higher Education, Science and Technology, and the Slovenian Research Agency over the years. The authors are grateful to Professor Koji Kato and Dr. Sanak Mishra for writing the foreword of this book. They are particularly thankful to Professor Kato for critical comments on some aspects of this book. Most importantly, the authors would like to convey heartfelt appreciation to their respective parents and other members of the families: Pritha and Prithvijit; Janja and Matija; and Lakshmi, Sugatri, and Surag, for their encouragement and support, while writing this book.

Bikramjit Basu
Indian Institute of Science, Bangalore, India

Mitjan Kalin
University of Ljubljana, Ljubljana, Slovenia

B. Venkata Manoj Kumar
Indian Institute of Technology Roorkee, Roorkee, India

Section I

Fundamentals of Ceramics: Processing and Properties

1

Introduction

Ceramics and Tribology

This foundational chapter introduces the readers to the multidisciplinary facets of tribology, viz. friction, wear, and lubrication. The technological significance of tribology is discussed and an overview of classification of engineering materials is provided. As this book largely discusses the tribological behavior of ceramics and their composites, typical properties and tribological applications of structural ceramic materials are emphasized. The overall structure of the book is presented toward the last part of the chapter.

1.1 Introduction

The word "tribology" originated from the Greek word "tribos" means rubbing [1, 2]. Tribology is described as "the science of interacting surfaces in relative motion and practices related there to." The science of tribology primarily embraces the study of three components: friction, wear, and lubrication, as illustrated in Figure 1.1.

While a committee of UK government coined the word "Tribology" in 1966 [3], the interest in the tribology field is much older than the documented history. It is worth finding the footprints of tribology in early human age when fire was invented by friction between stones and/or woods [4]. Archives also show the knowledge of ancestors in reducing friction during translatory motion by studded wheels of a harvest car or by lubricating the path for transporting heavy Egyptian statues. Other records on tribology concepts in Paleolithic age include drill bits for hole drilling, stone or wood wheels for grinding cereals, etc. It is to note that systematic scientific investigations of friction and lubrication date back to few centuries, whereas the concept of wear is much younger, almost five years. This delayed attention on wear concept is probably due to the unavailability of electron microscopy and other instrumentation tools in the past. However, with the recent development of

Friction and Wear of Ceramics: Principles and Case Studies, First Edition.
Bikramjit Basu, Mitjan Kalin, and B. Venkata Manoj Kumar.

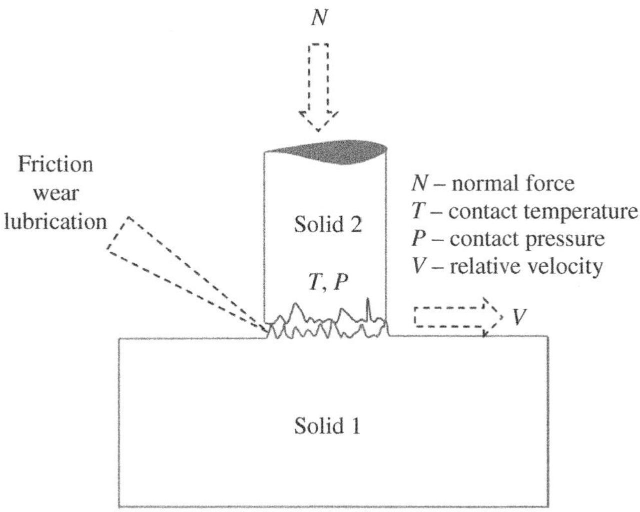

Figure 1.1 Schematic illustration of basic concept in tribology.

advanced microanalysis and spectroscopy tools, wear is now considered to be one of the potential components of tribology that helps in assessing the material loss and understanding the physics of material removal at surface and subsurface regions.

It is widely accepted in modern times that the friction and wear are major concerns in achieving sustainable growth in various industrial sectors like automotive, aerospace, construction, biomedical, optical, and microelectronics. In fact, a better understanding of tribology concepts is perceived to increase energy efficiency, reduce fossil fuels, and even improve health and lifestyle. As the performance of friction/wear components is strongly influenced by materials' behavior, novel material systems with superior properties and their processing technologies significantly influence the product efficiency and thereby cause impact on the economic sustainability.

> Friction is the resistance to motion that arises from the interactions of two solid surfaces at real contact area. It should be noted that the frictionless movement is needed in some applications, whereas some applications need friction.

For example, applications like bridge supports, hinges on doors, bearings, rivets, human knee or hip joints, etc., need less friction at the contacts. However, applications like clutches, brake pads, etc., demand high and controllable friction at the contact. Nevertheless, the performance and life of engineering components for a particular application are generally based on the efficient control of friction and surface properties.

> The progressive material damage from the surface due to relative velocity at the contact with other material is referred to as wear.

The phenomenon of wear can occur on surfaces of either or both mating materials, under different conditions like sliding, erosion, fretting, rolling, etc. It is important to note that mild wear in initial contact can progress rapidly during later stages to severe wear. Such transition can cause vibration and heating that may further lead to reduced product efficiency and loss. In fact, a majority of engineering disasters can be traced back to the instances of severe wear, which initiated in a mild scale. On the other hand, high and controllable wear rates are required in cases like polishing, grinding, machining, etc. Like friction, wear has to be controlled with early identification and design of suitable material systems for a better performance. In fact, the extent of friction and wear depends on the mating materials, contact conditions, and the surrounding environment. Thus, friction and wear are to be treated as system properties, not material intrinsic properties [5].

The extent of friction and material damage can be minimized by properly lubricating the contact. Several examples, where lubrication is useful, include metal cutting, gears, brakes, bearings, seals, orthopedic joints, etc. The role of the lubricant is to separate the contacting surface by forming viscous low-shear films. In fact, understanding the viscous flow characteristics of the lubricant, and interactions of solid surface and lubricant is necessary to control the friction [6–8]. Even in the absence of an external lubricant, it is quite possible that the contact is filled with a lubricating product, formed as a result of reaction between the mating materials and/or wear debris with the surrounding atmosphere. The later part of the book covers case studies where such surface interactions notably influence the wear and friction properties of ceramic composites.

> While friction is the temporary and immediate response of the body in contact, wear accounts for the history of contact conditions [9].

A thorough understanding on the behavior of materials with respect to varying contact conditions in terms of contact stress, contact temperature, and environment is strictly needed to achieve desirable tribological performance.

The science of tribology is unarguably an interdisciplinary field that can be explained by synergetic interaction among concepts drawn from fundamentals of Materials Engineering, Chemical Engineering, Mechanical Engineering, Physics, and Chemistry (see Figure 1.2).

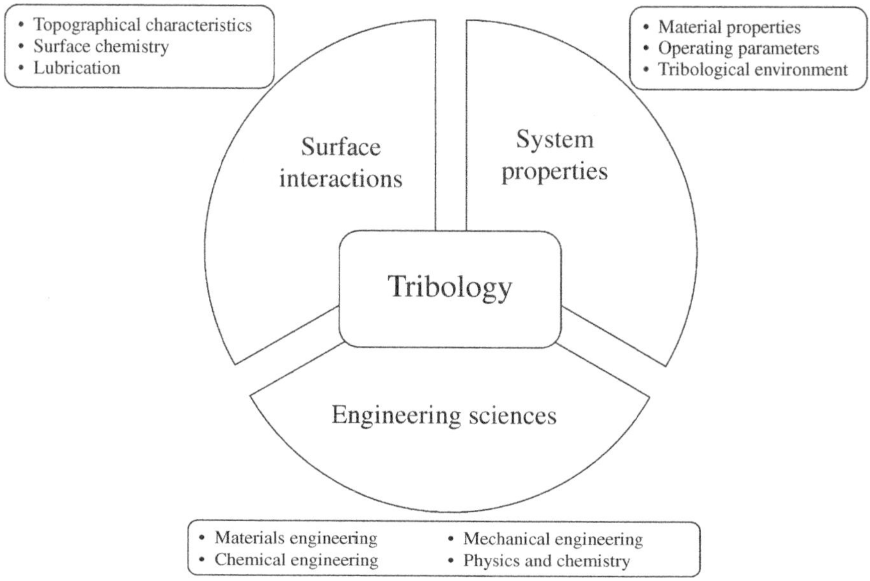

Figure 1.2 The conceptual illustration of contribution from disciplines of basic engineering and, system properties and surface interaction to tribology science.

While Mechanical Engineering concepts are greatly helpful to design the component, to understand contact mechanics and lubrication regimes; the Metallurgical Engineering concepts are useful in improving wear resistance by strengthening the bulk material or hardening the surface. On the other hand, the major aim of the material scientists is to control friction and wear by appropriately designing the process and tailoring the microstructure so that the performance in the given tribological conditions can be maximized [10]. Several technological, socioeconomic, and environmental requirements such as improved integrity, reliability, and performance of systems; higher durability of products; etc., drive the development of advanced material systems. In particular, the ever-increasing need for the development of advanced materials for extreme tribological applications necessitates comprehensive understanding on microstructure–properties–tribological performance relationship.

1.2 Classification of Engineering Materials

In general, engineering materials are classified into three primary classes: metals and alloys, ceramics and glasses, and polymers and elastomers [11–16] Among these primary classes, ceramics and glasses are widely being investigated for several engineering applications. The industrial-scale applications requiring moderate wear resistances are still dominated by steels and other metallic alloys. However, ceramics are potentially considered for use in extreme tribological environments.

While the widespread use of polymers and elastomers is driven by distinct advantages in terms of availability in different shapes or sizes, high flexibility, and low density; metallic materials have advantages of high toughness, high tensile strength, and manufacturability.

However, polymers and elastomers have a low melting point and very low elastic modulus and strength, whereas metals and alloys have much lower melting points, lower strength, hardness, and elastic modulus compared with ceramics and glasses. In particular, ceramics and glasses have superior properties: high melting points, high hardness, high compressive strength, high elastic modulus, and can withstand high temperatures without significant degradation of strength. Accordingly, ceramics are preferred for various elevated temperature structural or tribological applications.

On the other hand, another class of engineering materials, composites, is being developed to combine beneficial properties of the three primary classes of materials.

The composites are generally defined as "a class of materials that comprise of at least two intimately bonded microstructural phases aimed to tailor properties (e.g. elastic modulus, hardness, and strength) for specific applications."

The microstructure of a composite consists of three important constituents: matrix, reinforcement, and their interface. While equiaxed or an elongated grain structure is generally found in crystalline matrix phase, the reinforcement phases can have different morphology: particulates, fibers, and whiskers. The composites with fibers exhibit anisotropy in properties. Based on the major constituent type, composites are further classified as polymer matrix composites (PMCs), metal matrix composites (MMCs), and ceramic matrix composites (CMCs) [17, 18]. While several resin-bonded PMCs are largely studied for their use in aerospace applications, aluminum (Al)–silicon carbide (SiC) particulate composites and magnesium (Mg)–SiC particulate composites are widely investigated MMCs for their use as automotive and structural parts. Among CMCs, Al_2O_3–SiC whisker composites and Al_2O_3–zirconia (ZrO_2) particulate composites are largely studied for cutting tool inserts and other wear parts.

As the present book largely focuses on friction and wear behavior of ceramics and ceramic composites, a brief note on engineering ceramics followed by typical properties and important tribological applications of structural ceramics is provided in the following sections.

1.3 Engineering Ceramics

> Ceramics are defined as "a class of inorganic non-metallic materials that can be either processed or used at high temperatures and have an ionic and/or a covalent bonding."

Although the major use of ceramic materials in the last few decades was focused on traditional applications such as construction materials, kitchen wares, and sanitary wares, the progress of ceramic science and technology since the early 1990s enabled this important material class to extend applications to engineering fields like aerospace, electronics, nuclear, biomedical, etc. [11–16].

It is widely agreed that ceramics can be classified as traditional ceramics and engineering ceramics. Less expensive silica-based ceramics prepared for daily-life application are named as traditional ceramics. In contrast, engineering ceramics are prepared using highly pure and expensive ceramic powders and strategic process design to tailor properties for use in advanced applications. In this regard, the present book discusses the applicability of structural ceramics for tribological applications. From application view point, engineering ceramics are broadly categorized as functional ceramics and structural ceramics.

> The perfomance of functional ceramics is determined by magnetic, electric, dielectric, optical, and other properties; while the development and performance of structural ceramics is driven by the optimisation of mechanical properties, such as hardness, strength, and toughness.

Structural ceramics can be further classified as oxide ceramics (e.g. SiO_2, ZrO_2, and Al_2O_3) and non-oxide ceramics (e.g. TiN, Si_3N_4, TiC, SiC, TiB_2, and B_4C). As friction and wear properties largely influence performance of structural ceramics, it is imperative to understand the properties and tribological applications of structural ceramics.

1.4 Structural Ceramics: Typical Properties and Tribological Applications

Among advanced materials, ceramics are candidates for several wear-resistance applications owing to the unique set of their attractive properties like low density, high compressive strength, high hardness, high elastic modulus, and superior resistance against oxidation and creep [19]. The moderate fracture toughness can be improved by microstructural engineering via suitable process design.

The relevance of important properties of structural ceramics for tribological applications is presented in Figure 1.3. Table 1.1 summarizes typical properties of important metallic, polymeric, and ceramic materials. Detailed discussion on mechanical behavior of ceramics, fracture mechanics, and the assessment of different mechanical properties is provided in a later chapter. Using multiple case studies in this book, it has been emphasized that higher elastic modulus and higher hardness are, respectively, needed for superior resistance against Hertzian contact damage and abrasive wear. On the other hand, the shift of maximum Hertzian shear stress from bulk to the surface, and higher modulus-dependent contact pressures are some of the issues of concern with ceramics in tribological contacts. Further, few nitride and boride ceramics can be used for components exposed to temperatures close to 1500 °C, while Ni-based superalloys can be used only up to 1000 °C. In addition, ceramic composite (e.g. Si_3N_4-based or SiC-based composites) are demonstrated to retain high strength at temperatures as high as 1000 °C.

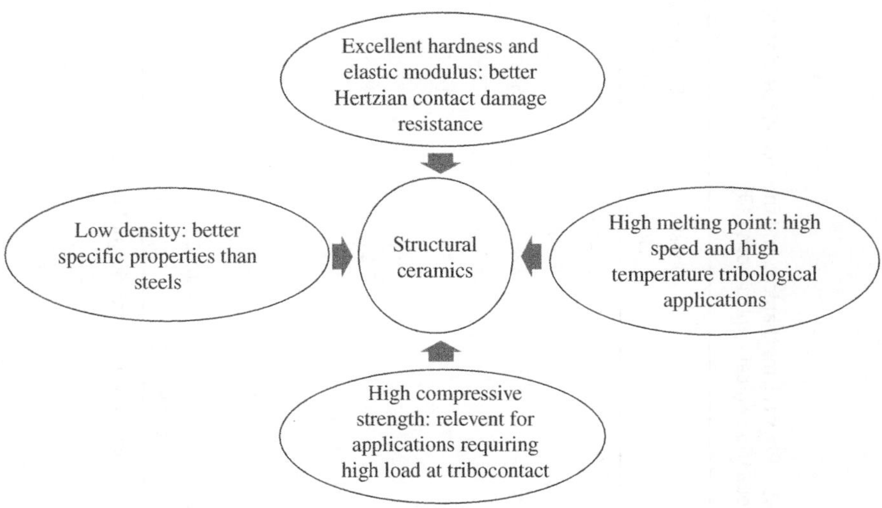

Figure 1.3 Important relevance of structural ceramics properties for tribological applications.

Table 1.1 Physical, mechanical, and thermal properties of some important metallic, polymers, and ceramic materials, which are relevant for various tribological applications [20].

Material	Density, ρ (g/cm^3)	Elastic modulus, E (GPa)	Fracture toughness, K_{IC} (MPa·m$^{1/2}$)	Vickers hardness, H_V (GPa)	Thermal conductivity, K (W/m·K)
Steel	7.8–7.9	21	5–214	1–9	3–6
Cast iron	7.1–7.4	64–181	2–6	1–8.5	3–6
Al-alloy	2.6–2.9	6 8	23–45	0.25–1.4	121–237
Al_2O_3	3.9	21–39	3–5	14–19	25–35
ZrO_2	5.6–6.25	14–21	1–8	12	2
Si_3N_4	3.2	17	4–7	16–18	5–25
SiC	3.2	45	4.5	25	9–125
Polyamide (PA)	1.1–1.14	2–4	3	0.8–1	0.25–0.35
Polyimide (PI)	1.3	3–5	—	—	0.37–0.52
Polytetrafluoroethylene (PTFE)	2.1–2.3	0.4	—	0.12	0.25
High-density polyethylene (HDPE)	0.92	0.2	1–2	0.13	0.33–0.57

Source: Reproduced with permission of Elsevier.

With respect to frictional heating, a large increase in friction-induced temperature is a concern with oxide ceramics having low thermal conductivity, whereas boride and carbide ceramics with relatively good thermal conductivity easily dissipate the heat from the contact. Further, density of ceramics is lower compared with many metals (see Table 1.1). This allows, high-speed machining possible with ceramic tool inserts.

The compressive strength of structural ceramics is almost eight times larger than the tensile strength and therefore, ceramics will be useful for tribological applications where contacts experience compressive loading conditions. It is to note that the compressive strength of ceramics is mostly superior to that of metals.

> Ceramics are identified as potential materials to replace existing materials for several tribological components like cutting tools, seals, valves, bearings, cylinder liners, etc.

However, the fundamental understanding of the relationship between material characteristics like microstructure, phases, etc., and wear behavior shall be understood for optimal use in tribological applications. For example, the wear behavior of non-oxide ceramics like SiC is largely reported to be influenced by test parameters, material, or environmental parameters [21]. As illustrated in Figure 1.4 [22], compared with SiC ceramics with equiaxed grains, the wear resistance was better for SiC ceramics having elongated grain morphology due to the *in situ* toughening by hard interlocking network of grains.

In general, the tribological materials development is focused mainly in two directions: ceramic coatings on metallic substrates and ceramics or ceramic composites. Coatings are mostly fabricated using nitrides, carbides, or borides with recent development of diamond or diamond-like (C–H) films at the higher end of the hardness-cost scale [23]. As the thickness of the coating is normally between 1 and 5 μm, property or performance of the relatively soft substrate is limited. In recent times, processing technology for diamond-like carbon (DLC) coatings has been improved to achieve low friction and wear for several lubricated and non-lubricated applications [24–31]. Thermal spraying can be applied for thicker coatings (in the millimeter range), but compatibility with substrate properties (thermal expansion, etc.) and cohesion are rather limited. This aspect is particularly discussed in later chapters of the book.

Monolithic ceramics, i.e. ceramics without any addition of second phase, particularly those with improved toughness and strength are being developed in several research labs and industries [32]. However, ceramic composites approach is now established to improve the overall properties of monolithic ceramics [33, 34].

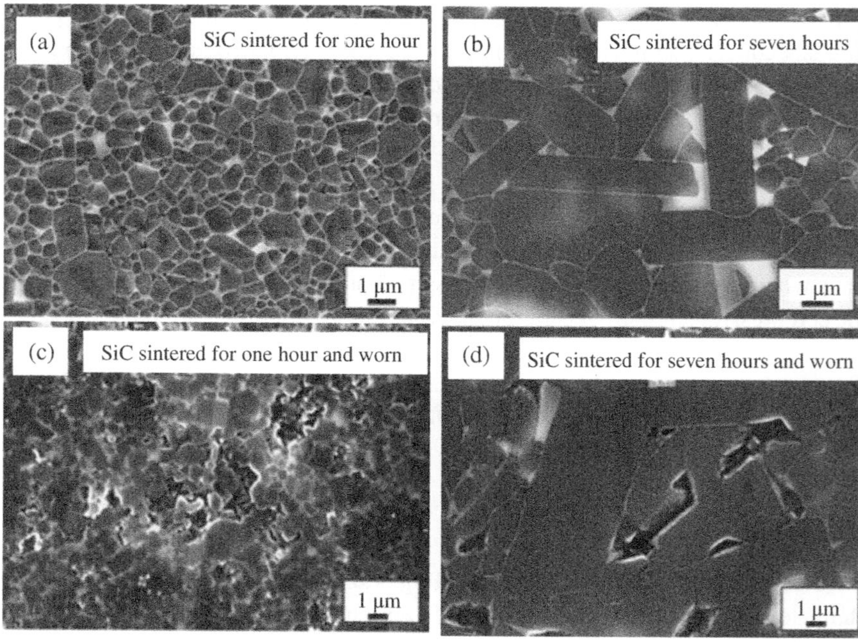

Figure 1.4 Typical SEM images of SiC ceramics (a, b) unworn and (c, d) after sliding against Si_3N_4 ball. Note the presence of elongated grains in (b) and equiaxed grains in (a), and comparatively less wear in (d) than in (c). Other conditions of sintering and sliding wear can be found in [22]. *Source:* Reproduced with permission of John Wiley & Sons.

Two major classes of ceramic composites are fiber-reinforced composites and particle/whisker-reinforced composites. Expensive ceramic fibers, composite production route, chemical compatibility between matrix and fiber, the fibers' oxidation at high temperatures are major drawbacks that restrict the widespread use of fiber-reinforced composites. To this end, particle-reinforced ceramic composites offer a feasible and a relatively cost-effective choice toward developing materials with improved properties. Tribological properties of particulate-reinforced CMCs are discussed in many case studies in this book.

The applications of wear-resistant ceramic materials range from bearings to leading edges of space shuttles, cutting tool inserts to orthopedic implants, automotive brake pads to dental crowns, and digital memory disks to turbine blades.

Few examples of engineering ceramic components, where wear resistance is of major concern are illustrated in Figure 1.5. Zirconia gears (see Figure 1.5a) are commonly used in heat engine applications. Bearings (see Figure 1.5b) with low-density silicon nitride balls are preferred in turbo pumps. Silicon carbide ceramics are used for bulletproof vest assembly units (see Figure 1.5c) because of higher hardness. Hot-pressed boron carbide nozzles (see Figure 1.5d) are preferred for nozzle applications, where abrasive resistance against silicon carbide and alumina abrasives is needed. TiCN-based cermets or their coating on steel are used for cutting tool inserts (see Figure 1.5e) in high-speed machining. In biomedical field, hydroxyapatite (HAp)-based ceramics are used for bone spacers, and Al_2O_3 or ZrO_2 ceramics are preferred for femoral ball heads (see Figure 1.5f) because of their hardness, abrasion resistance, and bicompatibility.

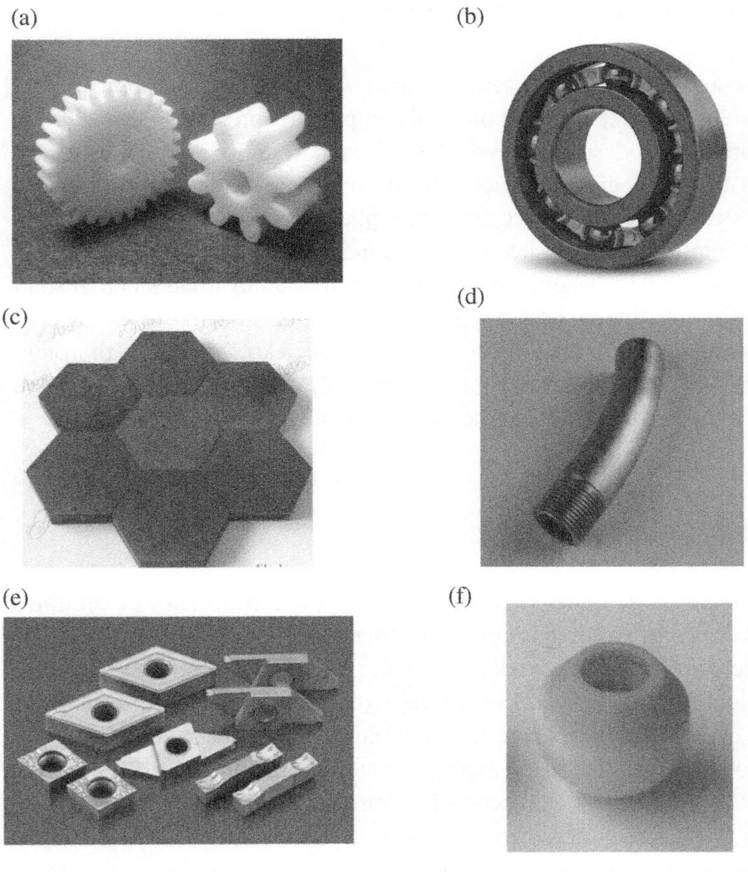

Figure 1.5 Few representative ceramic components requiring good tribological resistance: (a) zirconia gears [35], (b) Si_3N_4-based ball bearings [36], (c) SiC-based bulletproof vest assembly units [37], (d) hot-pressed boron carbide nozzle [38], (e) TiCN-coated cutting tool inserts [39], and (f) alumina femoral balls in total hip joint replacement (THR) application [40]. *Source:* Reproduced with permission of Springer Nature.

Furthermore, the use of advanced ceramics for critical transport engine components like fuel injector nozzles, needle valves, plungers, turborotors, etc., has been recently identified as a revolutionary approach for efficient and green transportation [41] in terms of reduction of carbon emissions, improved fuel efficiency, and superior vehicle performance, where material loss by sliding or erosive wear phenomenon is a serious concern.

> It is important to note that despite possessing superior properties and tremendous potential, the commercialization of ceramic and ceramic composites for tribological applications is rather limited, due to difficulties in processing and machining, and limited understanding on wear behavior.

For given conditions of design, environment, and lubrication, engineers and scientists involved in preparing tribological components prefer possible ceramic materials with required set of physical, mechanical, and thermal properties. Nevertheless, the tribological behavior of the ceramic component is not controlled unless microstructural features like grain size, phase assemblage, defect distribution, etc., are tailored and micromechanisms of material removal in different wear contacts like abrasion, adhesion, erosion, etc., are thoroughly understood. It shall be noted that innovative process design can be adopted to achieve engineered microstructures and a combination of superior properties, whereas a thorough understanding on the wear behavior is extremely essential to enhance the applicability of advanced ceramic materials for tribological contacts.

1.5 Structure of the Book

The present book covers basic issues of processing and properties of ceramics and ceramic coatings, and major concepts of friction, wear, and lubrication. In order to understand possible correlations among the microstructure, properties, and wear behavior, a collection of recent research on wear studies of important ceramic and ceramic coatings is presented in the later part of the book. The content of the book is organized in six thematic sections spanning over 18 chapters.

The first section of this book, Section I: "Fundamentals of Ceramics," is very important as this has been designed to facilitate readers who do not have a background in the area of ceramics. Considering the impact of processing or manufacturing approaches on the microstructural-development and resultant properties/performance of ceramics in tribological conditions, Chapter 2 is dedicated to explain fundamentals in ceramics processing wherein salient techniques for processing bulk ceramic materials via conventional and advanced sintering techniques. As ceramic coatings are preferred for several wear-resistant applications, a conceptual understanding on thermal spray-based ceramic coating techniques is also provided. Considering the particular preference of ceramics over steels or WC-Co hard materials in high-speed machining applications, basic concepts of conventional and advanced machining techniques are explained in Chapter 3. The advanced technique of electro-discharge machining is emphasized.

The mechanical behavior of ceramic materials is distinctly different than that of metallic and polymeric materials. It is widely reported that the elastic modulus, fracture toughness, and hardness strongly influence the wear behavior of ceramics in contact loading conditions. Therefore, brief discussion on stress and strain response, and details of fracture mechanics and toughening mechanisms are provided in Chapter 4. This chapter also discusses the experimental assessment for various mechanical properties of ceramics.

In addition to mechanical properties, wear test parameters (like load, speed, lubrication, etc., in sliding contacts; erodent type, size, velocity, angle of impingement, etc., in erosion contacts; depth of cut, speed, workpiece type, etc., in machining contacts) and environmental parameters (like temperature, relative humidity, inertness, etc.) also influence overall tribological behavior of the material. In fact, the interaction of mating materials with surrounding environment may influence contact conditions by forming a third body in contact. The friction and wear are largely influenced by the nature and stability of the third body. The application of lubricants can reduce the wear and friction by establishing a viscous and low shear strength film in contact. In the aforementioned perspective, Section II: "Fundamentals of Tribology" is dedicated to enlighten the reader with the fundamental concepts of tribology. The physical characteristics of contacting surfaces are explained in Chapter 5 and fundamentals of friction are provided in Chapter 6. As friction at the contact induces heat, basic understanding on the dissipation of frictional heating and contact temperatures is necessary in order to protect mating material surfaces from degradation. Therefore, key concepts of frictional heating and flash temperatures and the implications of important analytical models in flash temperature calculations are also provided in Chapter 6. Furthermore, as understanding on degradation behavior in wear contact provides a strong background to the readers as how to design and develop new tribological ceramic material systems, detailed explanation of wear mechanisms in terms of physicochemical changes at tribocontacts, leading to tribomechanical wear and tribochemical wear, is provided in Chapter 7. The importance of lubrication in reducing the friction *and* wear, and basics of lubrication regimes are also discussed in Chapter 7.

The later part of the book is important in realizing the implications of tribology concepts and material properties on friction/wear resistance of ceramic systems. For this, salient results obtained from studies of the author's own research groups on friction and wear behavior of ceramic or ceramics coatings subjected to sliding wear/erosive wear/machining induced wear are described. The selection of these wear contacts is justified in view of increasing contribution of ceramic systems toward energy and product efficiency and manufacturing ability. Based on the strong influence of materials on tribological behavior of systems in unlubricated (dry) operating conditions, majority of the results obtained in dry sliding wear studies on SiC ceramics, SiC–WC composites, ZrO_2–Al_2O_3 composites, and WC-Co coatings are discussed in detail in Chapters 8–11 (Section III: "Sliding Wear of Ceramics"). In addition, a comparative study on steels, coatings, polymers, and ceramics in lubricated sliding contacts is provided as Chapter 12 to explain the role of interface of solid surface with lubricant in controlling the friction.

Section IV: "Erosive Wear of Ceramics" deals with the solid particle erosive wear behavior of selected ceramic composites and ceramic coatings. The influence of material and operating parameters on erosive wear behavior of SiC–WC composites are discussed in Chapter 13, whereas the necessity and role of computational analysis in understanding the experimental erosive wear results for ZrB_2–SiC-based multiphase composites are highlighted in Chapter 14. The correlation of erosive wear performance with mechanical properties and erodent type for WC-Co coating is discussed in Chapter 15.

Further, considering the growing demand for light weight, hard, wear resistant, and chemically inert materials for cutting tool inserts, the performance and wear behavior of advanced cermets in machining of steel workpiece are discussed in Section V: "Machining-Induced Wear of Cermets." The results obtained in conventional machining and nonconventional electro-discharge machining conditions are elaborately explained in Chapters 16 and 17. It is to state that emphasis is given in each case study to explain dominant material removal mechanisms in selected sliding, erosive, or machining-induced wear conditions.

In the concluding section, Section VI: "Future Scope," authors' perspective for designing and developing next-generation ceramic materials with superior tribological properties is discussed in Chapter 18. The directions for the progress of tribology field are provided with an emphasis on the immediate need for developing advanced ceramics for self-lubricating films, robust testing equipment and in situ tribological interaction diagnostics facility, and simulation and modeling tools for improved understanding of experimental results. The necessity for data science and data mining concepts for better predictive capability of wear resistance of advanced materials is also highlighted. Also, a new conceptual approach "Tribomaterialomics" is suggested toward developing new generation wear-resistant materials.

Last but not the least, the most distinguishing part of this book is the Appendix: "Appraisal," wherein a rich collection of several types of questions is provided.

One can assess the understanding on fundamental concepts of tribology by attempting the numerous questions in 'appraisal' section. As per the knowledge of authors, such a large collection of qualitative and quantitative questions along with model answers is not available in any other tribology-related book, and this makes the present book unique. It is further expected that the 'appraisal' section serves as quick-reference and makes the book student-friendly and text-book type.

1.6 Closure

> The influence of materials on tribological performance shall be considered in a larger context as these materials are components of "tribological systems."

Materials for tribological applications include all major classes of materials like polymers, metals, ceramics, and their respective composites. Moreover, interfacial properties are affected by environment and tribochemical reactions. Being system dependent, these properties, including both mating materials, influence the tribological properties of any contact and component considerably.

Ceramics and ceramic composites are preferred for many components in friction and wear-resistant applications, because of the unique combination of their superior properties such as low mass forces, better abrasion resistance, and moderate tribochemical behavior. However, a better understanding on the assessment and mechanisms of material loss in friction and wear contacts is essential to increase the applicability of advanced structural ceramics for tribological components. In particular, the relationship among processing, microstructure, properties, and tribological performance is to be thoroughly understood.

In this regard, the present book is structured to enrich the fundamental knowledge on processing, microstructure, and mechanical properties of bulk ceramics and ceramic coatings, and friction, contact temperature, wear, wear mechanisms, and lubrication. In order to realize the implications of ceramic and tribology concepts, state-of-the-art research findings of tribological behavior of selected ceramic systems is presented in later part of the book. Further, a rich collection of conceptual, numerical, and analytical-type questions and model answers is provided in appraisal section to help readers in assessing their knowledge on the subject of tribology of ceramics.

References

1 Bhushan, B. (1999). *Principles and Applications of Tribology*. New York: Wiley.
2 Hutchings, I.M. (1992). *Tribology: Friction and Wear of Engineering Materials*. Guernsey, UK: Butterworth-Heinemann.

3 Jost, P. (1966). *Lubrication (Tribology): A Report on the Present Position and Industry's Needs*. London: Department of Education and Science, Her Majesty's Stationery Office.

4 Czichos, H. (1978). *Tribology*. Amsterdam: Elsevier.

5 Stachowaik, G.W. and Batchelor, A.W. (1993). *Engineering Tribology*, Tribology Series, vol. 24. Amsterdam: Elsevier.

6 Archard, J.F. (1953). Contact and rubbing of flat surfaces. *J. Appl. Phys.* 24: 981–988.

7 Johnson, K.L. (1985). *Contact Mechanics*, vol. 26, 23. London, New York, Sydney: Cambridge University Press.

8 Suh, N.P. (1986). *Tribophysics*. Englewood Cliffs, NJ: Prentice-Hall.

9 Rabinowicz, E. (1995). *Friction and Wear of Materials*, 2e. New York: Wiley.

10 Zum Gahr, K.H. (1987). *Microstructure and Wear of Materials*. Oxford: Elsevier.

11 Basu, B. and Balani, K. (2011). *Advanced Structural Ceramics*. Wiley and American Ceramic Society.

12 Barsoum, M.W. (2003). *Fundamentals of Ceramics*. Boca Raton, FL: Taylor & Francis.

13 Carter, C.B. and Norton, M.G. (2007). *Ceramic Materials*. New York: Springer.

14 Chiang, Y.M., Birnie, D.P., and Kingery, W.D. (1997). *Physical Ceramics*. New York: Wiley.

15 Zum Gahr, K.H. (1989). Sliding wear of ceramic--ceramic, ceramic–steel and steel–steel pairs in lubricated and unlubricated contact. *Wear* 133: 1–22.

16 Kingery, W.D., Bowen, C.R., and Uhlman, A. (1976). *Introduction to Ceramics*, 2e. New York: Wiley.

17 Gay, D., Hoa, S.V., and Tsai, S.W. (2003). *Composite Materials-Design and Applications*. Boca Raton, FL: CRC Press LLC.

18 Kaw, A.K. (2006). *Mechanics of Composite Materials*, 2e. Boca Raton, FL: CRC Press.

19 Czichos, H. and Habig, K.-H. (1992). *Tribology Handbook Friction and Wear*, 56. Braunschiweig: Vieweg Verlag.

20 Czichos, H., Klaffke, D., Santner, E., and Woydt, M. (1995). Advances in tribology: the materials point of view. *Wear* 190 (2): 155–161.

21 Sharma, S.K., Kumar, B.V.M., and Kim, Y.-W. (2016). Tribological behavior of silicon carbide ceramics – a review. *J. Korean Ceram. Soc.* 53: 581–596.

22 Lopez, O.B., Ortiz, A.L., Guiberteau, F., and Padture, N.P. (2007). Sliding-wear-resistant liquid-phase-sintered SiC processed using α-SiC starting powders. *J. Am. Ceram. Soc.* 9 (2): 541–545.

23 Subramonian, B., Kato, K., Adachi, K., and Basu, B. (2005). Experimental evaluation of friction and wear properties of solid lubricant coatings on SUS44C steel in liquid nitrogen. *Tribol. Lett.* 2 (3–4): 263–272.

24 Matthews, A., Leyland, A., Holmberg, K., and Ronkainen, H. (1998). Design aspects for advanced tribological surface coatings. *Surf. Coat. Technol.* 100–101: 1–6.

25 Grill, A. (1999). Diamond-like carbon: state of the art. *Diam. Relat. Mater.* 8: 428–434.

26 Fontaine, J., Donnet, C., Grill, A., and Mogne, T.L. (2001). Tribochemistry between hydrogen and diamond-like carbon films. *Surf. Coat. Technol.* 146–147: 286–291.

27 Erdemir, A. and Donnet, C. (2006). Tribology of diamond-like carbon films: recent progress and future prospects. *J. Phys. D: Appl. Phys.* 39 (18): 311–327.

28 Neville, A., Morina, A., Haque, T., and Voong, M. (2007). Compatibility between tribological surfaces and lubricant additives – how friction and wear reduction can be controlled by surface/lubes synergies. *Tribol. Int.* 4 (1–12): 168–195.

29 Kalin, M., Velkavrh, I., and Vizintin, J. (2008). Review of boundary lubrication mechanisms of DLC coatings used in mechanical applications. *Meccanica* 43: 623–637.

30 Velkavrh, I., Kalin, M., and Vizintin, J. (2008). The performance and mechanisms of DLC – coated surfaces in contact with steel in boundary-lubrication conditions – a review. *Strojniski Vestn. – J. Mech. Eng.* 54 (3): 189–206.

31 Kalin, M., Velkavrh, I., and Vizintin, J. (2009). The Stribeck curve and lubrication design for non-fully wetted surfaces. *Wear* 267: 1232–1240.

32 Cawley, J.D. and Lee, W.E. (1994). Oxide ceramics. In: *Structure and Properties of Ceramics*, Materials Science and Technology, vol. 11 (eds. R.W. Cahn, P. Haasen and E.J. Kramer), 101–114. Weinheim, Germany: Wiley-VCH.

33 Rühle, M. and Evans, A.G. (1989). High toughness ceramics and ceramic composites. *Prog. Mater. Sci.* 33: 85.

34 Basu, B. (2001). Zirconia-titanium boride composites for tribological applications. PhD thesis. Katholieke Universiteit Leuven, Belgium.

35 http://pktol.com/goods/131.html.

36 https://www.kugellager-express.de/full-ceramic-deep-groove-ball-bearing-ce-6203-si3n4-peek-open-17x40x12-mm

37 https://image.ec21.com/global/multiImage/popMultiImgNew.jsp.

38 http://www.hmt-cera.com/hmt/abrasive.html.

39 http://www.ntkcuttingtools.com/product/ss_grades_carbide.html#QM3.

40 Novikov, N.V., Rozenberg, O.A., Sokhan, S.V., and Mamalis, A.G. (2005). Finish diamond machining of ceramic femoral heads. *Int. J. Adv. Manuf. Technol.* 25: 244–247.

41 http://ceramics.org/learn-about-ceramics/ceramics-in-transportation.

2

Processing of Bulk Ceramics and Coatings

Considering the impact of processing approaches on the microstructure and resultant properties and tribological performance, this chapter is dedicated to explain the fundamentals of conventional and advanced consolidation techniques for bulk ceramics. As ceramic coatings are used for several wear-resistant applications, basic concepts of thermal spray-based ceramic coating techniques are also provided.

2.1 Introduction

The consolidation of ceramics has widely been considered as a great challenge in the materials community. Similarly, challenging is the deposition of adherent ceramic coatings on an engineering material. This demands both the fundamental understanding of the process science and knowledge over different competing processing approaches. Such knowledge would allow one to select appropriate processing schemes and to fine-tune the process parameters. In the above perspective, this chapter discusses conventional and advanced processing approaches for bulk ceramics. This is followed by the description of various coating deposition techniques. This chapter is expected to provide a platform for the reader to understand and identify specific processing route for ceramics, both in bulk or coating form.

2.2 Conventional Processing of Ceramics

Since this book focuses on the application of ceramics in the field of tribology, it is important for a reader to know how to make ceramics. This is particularly important for researchers of mechanical engineering discipline without formal knowledge of

Friction and Wear of Ceramics: Principles and Case Studies, First Edition.
Bikramjit Basu, Mitjan Kalin, and B. Venkata Manoj Kumar.

materials science. To start with, ceramics cannot be manufactured by melting–solidification route, as widely used for metals. High melting point restricts such an approach. Also, bulk deformation processes, like forging, rolling, and extrusion, cannot be easily adapted to shape ceramics, due to inherent brittleness. As described in this chapter, sintering of a powder compact is used to obtain dense ceramic. It is worthwhile to mention that sintering is also used for many refractory metals. Ideally, sintering should allow one to obtain more than 99% theoretical density, i.e. solid compact with little closed porosity. In some technological applications, like tissue engineering and porous bearings, it is allowed to obtain less than 95% theoretical density.

> As a manufacturing process, sintering refers to the consolidation of powders to a dense solid at temperature, $T > 0.5T_m$ (T_m ~ melting point) with or without external pressure application and is assisted by diffusional mass transport.

Sintering can be described as a phase transformation from a porous state to a dense state, involving neck formation among the powder particles.

2.2.1 Sintering Mechanism

The process of sintering at global scale can be best understood using a schematic diagram, as shown in Figure 2.1. Ideally, all the void spaces in a powder compact are removed by mass transport from solid particles to the available pores. This process can be kinetically faster under external pressure application and thermodynamically feasible at lower temperature with external pressure application (hot pressing (HP), hot isostatic pressing (HIP), etc.) compared with pressureless sintering.

> During sintering, solid/vapor (s/v) interfaces are replaced by solid/solid (s/s) interfaces leading to the reduction in surface energy of the system.

Thermodynamically, any irreversible process is preferred, provided it reduces the total energy of the system,

$$\mathrm{d}(\gamma A) = A\mathrm{d}\gamma + \gamma\mathrm{d}A < 0 \tag{2.1}$$

where A is the surface area and γ is the interface/surface energy.

As also shown in Figure 2.1, grain coarsening ($\gamma\mathrm{d}A$) term dominates at the final stage. This also indicates that one would expect a dense compact with larger grain

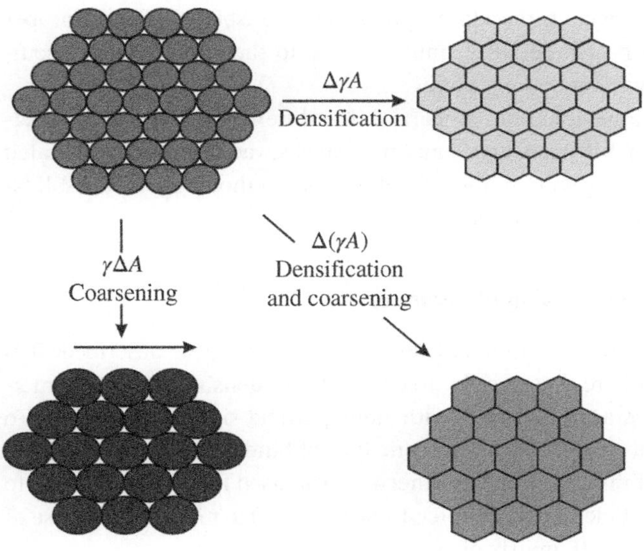

Figure 2.1 Schematic demonstration of the sintering of a porous powder compact to a dense ceramic with equiaxed grain morphology.

size, if sintering parameters are not optimised. In an ideal scenario, sintering should allow one to obtain dense compact with finer grains (see Figure 2.1). Three distinguishable stages can be identified in sintering of a powder compact:

1) **Initial stage**: Initially, the particles (of diameter D) come in contact and form a weak cohesive bond. Vapor-phase transport involving the surface diffusion leads to neck formation at the interparticle region. As shown in Figure 2.1, coarsening instead of shrinkage takes place at the initial stage for a longer holding time.
2) **Intermediate stage**: During continuous heating, neck growth effectively takes place due to grain boundary and lattice diffusion. In parallel to neck growth, the pore coarsening occurs, which results in the evolution of the interconnected pore channel. Importantly, the breakdown of continuous pore channel takes place toward the end of this sintering stage, leading to achieving around 70–92% theoretical density.
3) **Final stage**: In the intermediate stage, most of the open pores are removed. In the last or final stage of sintering, the closed pores undergo shape changes primarily by grain boundary/lattice diffusion-based mass transport mechanism. While being attached to grain boundaries, the isolated pores will be eventually removed. This would result in the attainment of 98–99% theoretical density. If the sintering temperature and holding time (at the final stage) are

not appropriately selected, the closed pores become isolated. The improper selection of sintering temperature/time can lead to the simultaneous occurrence of grain growth and pore growth.

As far as sintering atmosphere is concerned, the different atmospheres generally used are hydrogen, nitrogen, argon and natural gas, vacuum, and dissociated ammonia. In the following section, the typical process methodology of consolidation of a powder compact is described.

2.2.2 Conventional Processing of Ceramics

Various processing approaches to make dense ceramics are now described. It is important to mention here that in order to ensure faster densification, agglomerate-free high-purity ceramic powders with finer particle size are preferred. To enhance sintering, often a desired amount and type of binder or sinter-aid is used. This requires efficient milling, which is otherwise also used for two reasons: (i) to reduce the starting particle size powder feed stock, and (ii) to mix second phase of a different composition with matrix phase.

2.2.2.1 Powder Processing and Compaction

Ball milling is a common first step in the powder-based processing approach. This is used to mix the elemental/prealloyed powders using grinding balls (milling media) to cause size reduction or mixing of powders, as shown in Figure 2.2. Usually, stainless steel, WC, or other ceramic (Al_2O_3, ZrO_2, and agate) balls are used. Normally, the vials are made of agate or steel.

> The ball milling process leads to a reduction in particle size by comminution due to repeated crushing of powder particles by the balls, moving at high speeds.

As far as milling parameters are concerned, the ball size, milling time, and milling speed are all important factors. The critical speed for effective milling is inversely proportional to the internal diameter of the mill. At lower speed, the ball-impact to cause milling is minimal. On the other hand, any speed larger than the critical speed causes balls to revolve due to centrifugal force. An ideal movement of balls and powder is shown in Figure 2.2b. It should also be mentioned that ball to powder ratio (BPR) can vary in the range of 4 : 1 to 10 : 1 for effective milling. Increasing the ball diameter or the number density of balls also leads to an increase in BPR. Typically, the smaller the ball size, the more is the ball surface to induce comminution of powders. In general, ball size of 10 mm is largely used.

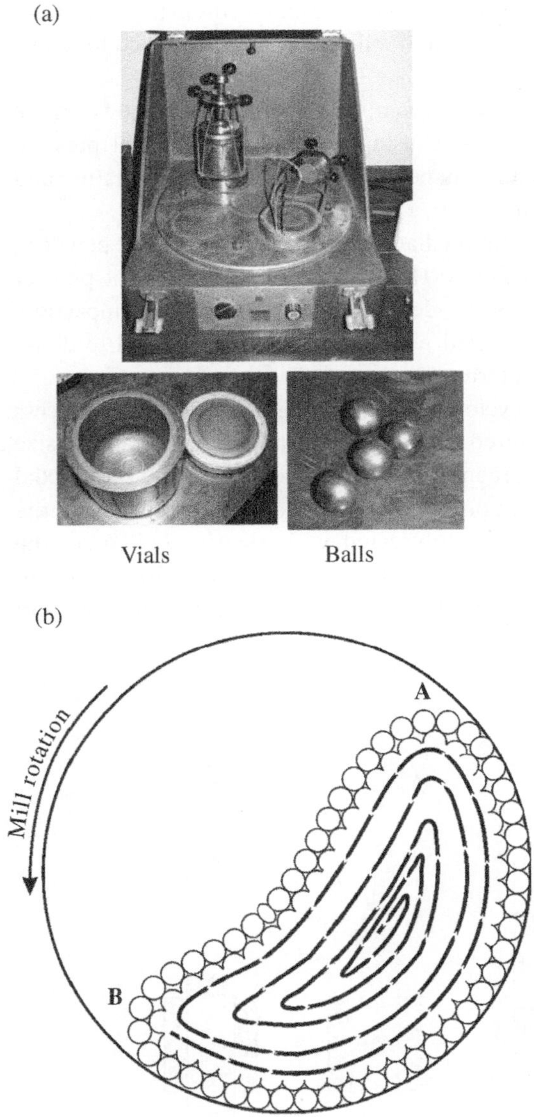

Figure 2.2 (a) Experimental setup of a planetary ball-mill and (b) schematic illustration showing the movement of balls and powders in a ball mill.

As mentioned earlier, effective ball milling can refine the powder particle size. Depending on the type of the material, maximum particle size reduction is attained at longer milling duration (24 or 48 hours). Longer milling can lead to undesired agglomeration of fine powders. To sum up, the ball milling variables including BPR, type of balls (harder than powders), liquid medium (acetone, toluene) or dry medium (air/vacuum/inert), and mill speed are to be tuned for effective milling.

The second step after the ball milling is compaction, wherein the powders are pressed into the desired shape and size using hydraulic/mechanical press to obtain "green compact." Compaction can be carried out at room temperature and therefore is known as cold pressing.

Cold pressing involves the pressure-mediated denser packing of loose powders, while being contained in dies/punches. Herein, the rearrangement of powder packaging followed by mechanical deformation results in compaction. Subsequently, the uniaxially compacted powder body is consolidated to dense compacts via high-temperature sintering.

A representative compaction cycle is shown in Figure 2.3. In this process step, ball milled powders are poured into a die with predefined shape or size and then compacted to obtain a green body by external application of uniaxial pressure to die-punch assembly. In order to obtain smooth outer surface of the cold compact, the die wall is often lubricated with stearic acid. Also, the compaction pressure can be tailored to obtain a compact with sufficient strength for further handling. Typically, a cold compact has an apparent density of 50–55% theoretical density.

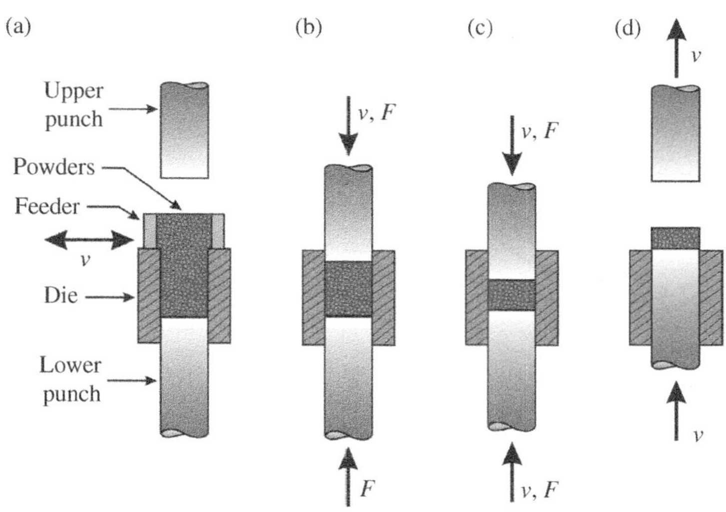

Figure 2.3 Conventional method of compaction. (a) Filling die cavity with powder by automatic feeding system, (b) initial, (c) final positions of upper/lower punches, and (d) ejection of part.

The fundamental issue with uniaxial compaction is the density variation along the thickness direction. To address this issue, a high pressure, typically ~300 MPa, is applied to the sealed container of powder/compact in water or oil, a process known as cold isostatic pressing (CIP). CIP is particularly useful to fabricate complex shaped parts without density gradient. The fundamental law of physics, i.e. Pascal's law, is followed by CIP and this allows the external pressure to be uniformly transmitted to all the parts of the ceramic compact in equal manner. This leads to compaction of similar extent.

The CIP process can be classified into two schemes: wet-bag and dry-bag processes. In the wet bag process, the ceramic powder is filled in a rubber mold, and then this bag is immersed in the fluid for compaction. The isostatic pressure around the powder compact is achieved as the pressure is transferred from pressurized fluid to the rubber mold. In case of the dry-bag process, the external pressure is radially applied between the flexible mold and the rigid shell, while the bag can rest on its top/bottom surface. Overall, the CIP process is useful to obtain compacts with minimum density gradient for complicated parts.

2.2.2.2 Pressureless Sintering

In pressureless sintering, a powder compact is heated to sintering temperature in a furnace and in the absence of external gas or mechanical pressure. Depending on whether the compact would be densified via solid-state or liquid-phase sintering (LPS), both sintering temperature and time are to be optimized. Sintering at high temperature is commonly referred to as "firing" in the industrial context.

Generally, heat treatment cycle includes a first stage of preheating to allow removal of volatile or organic components entrapped within a powder compact. This stage is followed by sintering, where the ceramic component is held at sintering temperature for a specified dwell time. It is important to note that the selection of temperature and time of sintering, and the sintering environment decide the evolution of microstructure, porosity elimination, and grain growth. Also, the cooling at the end of heating cycle induces thermal stress in the final component. High thermal stresses may also lead to cracking in the ceramic component and hence a proper control of cooling cycle becomes essential. Typically, furnace cooling is adopted to minimize the thermal stresses.

In industries, a continuous sintering furnace is used to densify a large number of samples in the batch process. This process involves feeding the ceramic compacts to be delivered to a batch sintering furnace via the conveyer belt to allow them to experience an identical heating cycle. For many ceramics, particularly non-oxide materials, pressureless sintering cannot be adopted to achieve full densification and it is necessary to adopt pressure-assisted sintering (e.g. HIP).

2.2.3 Advanced Processing of Ceramics

2.2.3.1 Hot Pressing

HP is a densification process, involving concurrent application of pressure and high temperatures to powder compact contained within a (graphite) die-punch assembly.

> High pressure and temperature synergistically allow rearrangement of particles, allow plastic flow, and facilitate faster mass transport, leading to enhanced neck growth in the HP process.

The representation of the HP process is shown in Figure 2.4. For HP, the typical pressure of ~10–50 MPa at temperatures ranging between 1000 and 2200 °C for minutes to few hours are used to obtain a dense ceramic, often with densities of more than ~95% of the theoretical density for difficult-to-sinter ceramics.

2.2.3.2 Microwave Sintering

Microwave (MW) sintering is regarded as an advanced sintering process, which involves rapid heating of a powder compact to full density.

> MW sintering involves uniform volumetric heating of powder compact using free electrons/ions oscillation in a MW cavity.

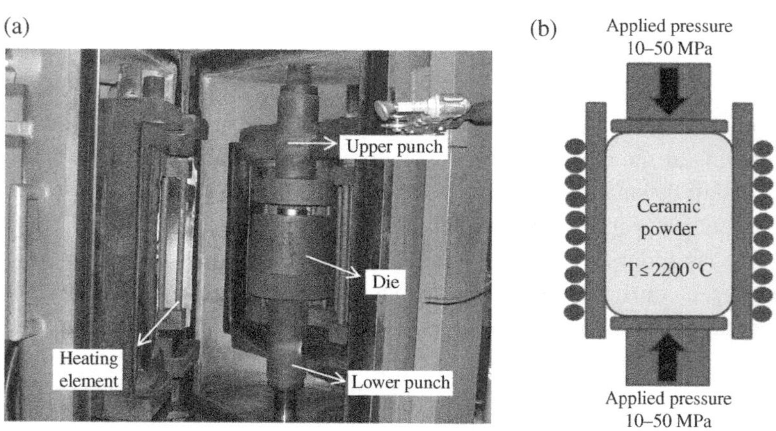

Figure 2.4 (a) Industrial hot-pressing setup and (b) schematic showing the heating and pressure application in HP.

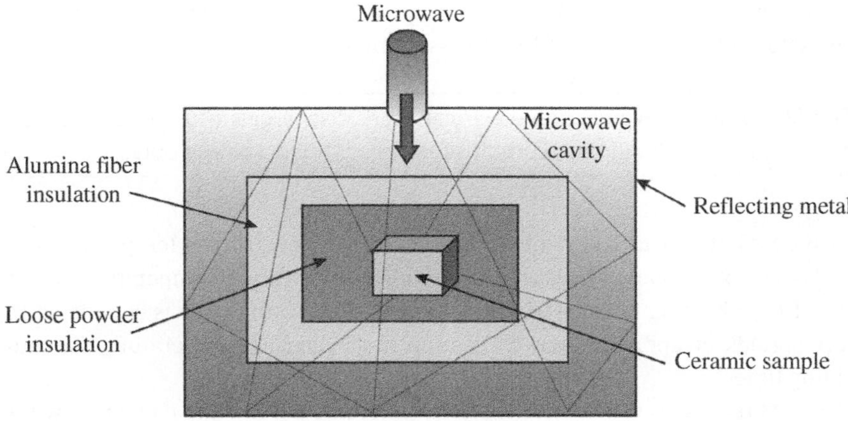

Figure 2.5 Schematic of microwave sintering setup.

In order to substantiate further the advantages of such technique with respect to conventional pressureless sintering, it is instructive to mention that the heating rate of 100 °C/min or higher can be achieved for MW/spark plasma sintering (SPS), while it is 2–5 °C/min for the pressureless sintering. This allows significant reduction in the overall process cycle time.

Volumetric heating is particularly more feasible in the MW sintering route even for complicated shapes. In this process, the ceramic powder compact is placed in a MW cavity and the sample achieves rapid yet uniform heating, as shown in Figure 2.5. Since high frequency allows skin effect, by preferential heating of the surface, a combination of low and high frequency during MW is required to achieve uniform heating. It is important to control the degree of MW interaction specific to the material's absorption property. The presence of a non-absorbing material around the compact to prevent heat loss is an interesting approach exploited in MW sintering. For insulating ceramics, like ZrO_2 or Al_2O_3, hybrid heating using SiC rods enclosing such powder compact in MW cavity can be adopted. SiC can absorb MW radiation very fast and subsequently transfers the heat to insulating ceramics.

Summarizing, the advantages associated with the MW sintering are short heating and cooling cycle, i.e. rapid volumetric heating, which restricts grain growth during sintering. Unlike conventional sintering, uniform heating of the material does not allow generation of steep surface to bulk thermal gradients in case of MW sintering.

2.2.3.3 Spark Plasma Sintering

SPS has been widely investigated as a superfast densification route for metals and ceramics. In this book, some case studies also report the use of SPS to process ceramics or ceramic composites for tribological applications. Therefore, this section exclusively discusses the SPS process of some model ceramics.

> The SPS process involves the electrical spark discharges in the interparticle neck regions involving high current together with low voltage pulse.

Using SPS, the materials can be consolidated to full density at temperatures of 200–500 °C lower than the respective conventional sintering temperature [1]. In view of high heating rates (100 °C/min), the entire heating cycle is completed in short periods of approximately 5–20 minutes, including temperature rise and holding times.

The SPS process is schematically shown in Figure 2.6 and the current flow for three-particle model system is also shown therein. The nature of the current flow and constriction resistance at particle–particle interface causes Joule's heating, leading to faster internal heating of the compact. In view of the fact that plasma formation (local high temperature state) during SPS heating remains to be conclusively proven/authenticated, other terms are often used in the literature for this technique, such as pulse electric current sintering (PECS) or field assisted sintering technique (FAST). The FAST process involves ON–OFF DC pulse to activate mass transport in the powder compact. During repeated application of an ON–OFF DC pulse, voltage to the powder compact leads to spark discharges at the particle–particle interfaces or at neck regions. The repeated spark discharges result in efficient sintering at low power consumption. All these restrict grain growth, often resulting in improved mechanical [2, 3], physical [4], or optical [5] properties, while eliminating the need of sintering aids.

2.3 Thermal Spray-Based Coating Deposition

The above-described manufacturing approaches are applicable to make dense ceramics in bulk format. As mentioned in Chapter 1, both bulk and ceramic coatings are used in tribological applications. Now, we shall describe various coating deposition techniques. Thermal spraying is widely used for depositing ceramic coatings on a material surface.

> In the thermal spray process, powders, wire, or rod shape of a material is heated up to a molten/semi-molten state. These heated particles are accelerated toward a target material surface, where they impact and interlock with the sample surface or with an already deposited splat.

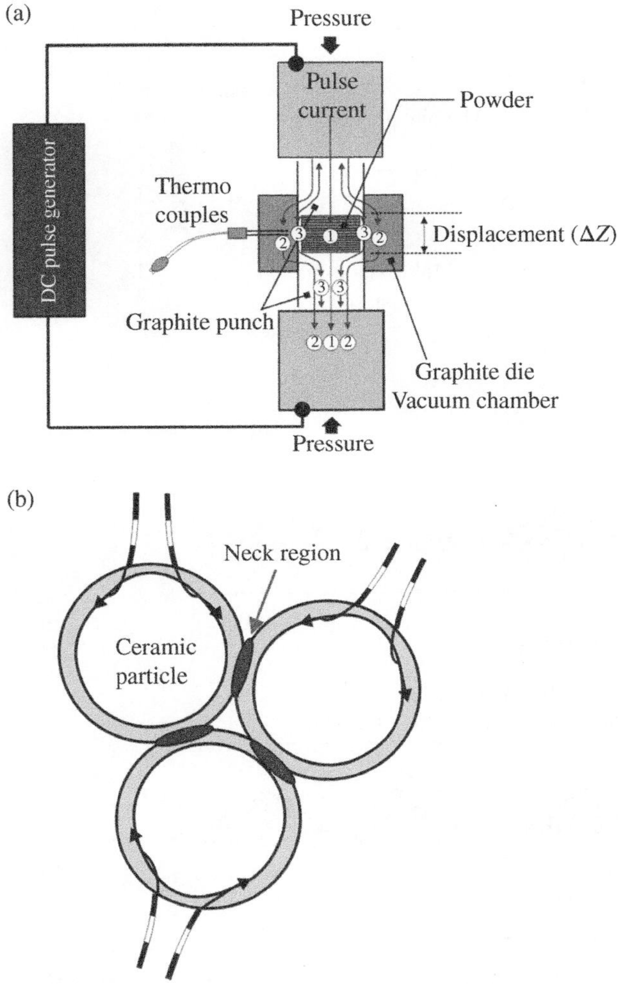

Figure 2.6 Spark plasma sintering (SPS): (a) schematic showing the DC current flow during the sintering process and (b) mechanism of temperature rise at the interface of particles due to Joule heating effect.

Schematic representations of the thermal spray process and the mechanism of coating structure formation are provided in Figure 2.7. A coating with uniform thickness and adherence can be obtained by carefully tailoring the relative movement between the spray gun and the substrate and other process parameters, as discussed later.

Several thermal spray techniques are now available to deposit coatings on material surfaces [6–10]. These techniques are distinguished by the temperature

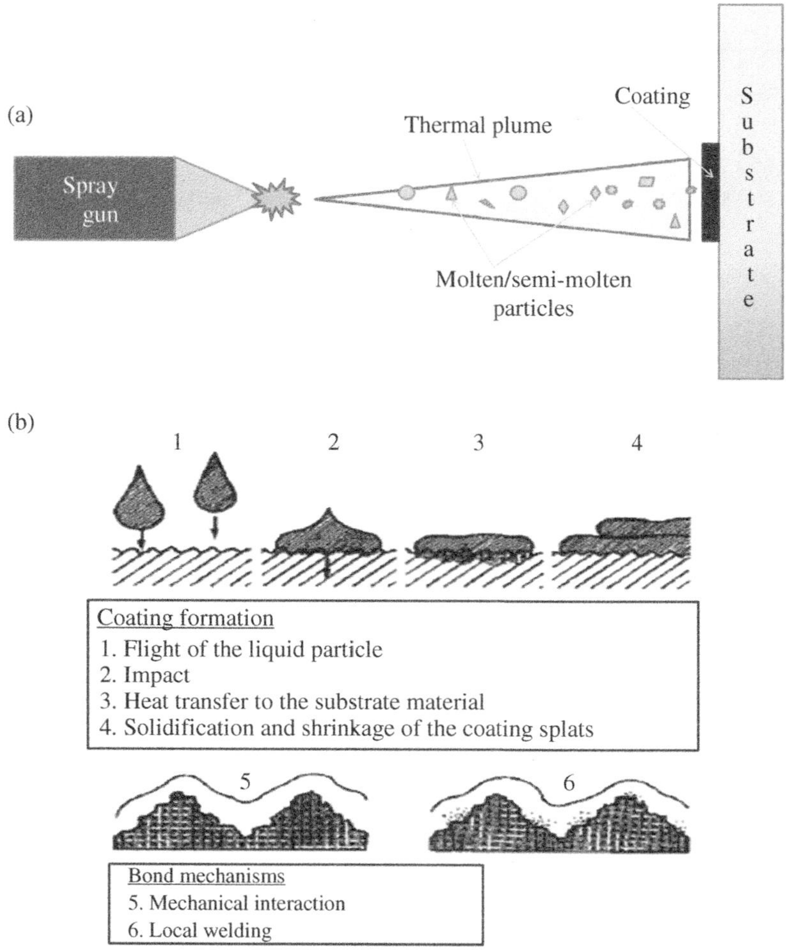

Figure 2.7 Generic description of the thermal spray process (a) and the mechanism of the coating structure formation (b).

that develops in the flame and acceleration that the particle achieves in the flame (see Figure 2.8).

While a number of coating deposition techniques are widely investigated to develop coatings in a number of ceramic systems, the detonation spraying will be discussed at length. This is primarily because of the fact that two chapters in this book are devoted to describe the abrasive and erosive wear of WC-Co coatings.

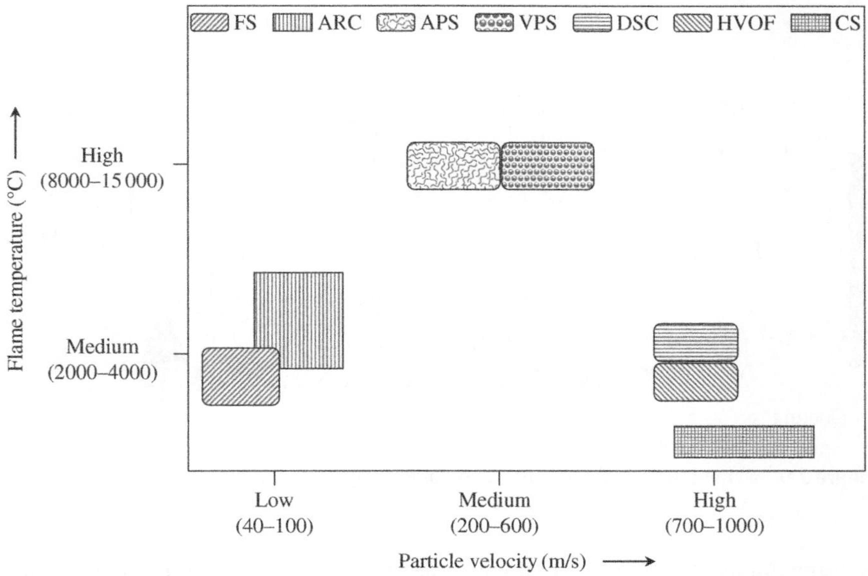

Figure 2.8 Schematic presentation of the combination of process temperature and velocity for various spray parameters. FS, flame spraying; ARC, arc spraying; APS/VPS, atmospheric/vacuum plasma spraying; DSC, detonation spray coating; and HVOF, high-velocity oxy-fuel spraying.

2.3.1 Basics of Thermal Spray Deposition

2.3.1.1 Plasma Spray Deposition

The plasma spray technique is very popular for depositing a wide spectrum of materials due to the high temperatures involved in the process [6–9, 11]. Plasma is a conductive gas that consists of positively charged ions, electrons, and neutral gas atoms (see Figure 2.9). Plasma can be generated either by interaction of an electric arc with a gas or by high radio frequency excitation. The plasma torch consists of water-cooled copper anode and water-cooled tungsten cathode. A direct current, maintained between the anode and cathode, ionizes the gas that flows around the concentric anode. This results in gas plasma with core temperature of around 15 000 °C. When the powder is delivered in the plasma, it gets heated and thereafter it is accelerated toward the substrate and forms the coating. The final temperature and heat transfer property of the plasma depend on the plasma power, anode-cathode design, and the type of the gas. An increase in the operating power level increases the plasma volume, but the temperature rise is not so significant. In general, argon is used most widely due to its low-energy plasma, i.e. ionization potential and thermal heat capacity, and N_2 to obtain the hottest plasma gases.

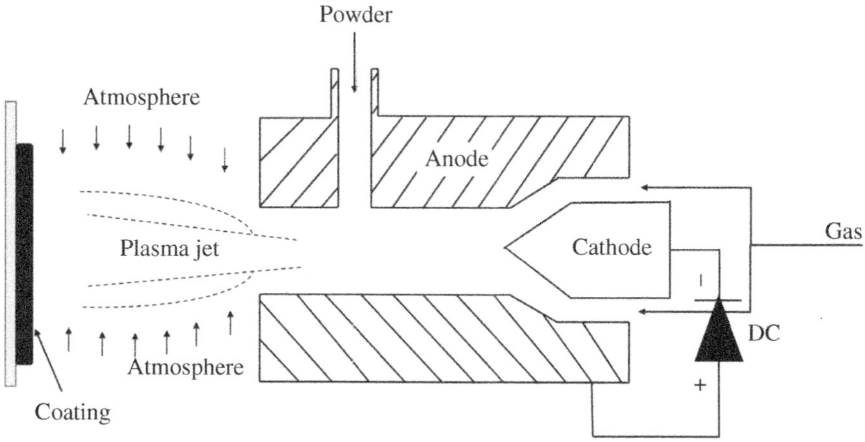

Figure 2.9 Schematic representation of the plasma spray torch.

Helium and hydrogen are generally used as secondary gases to enhance the heating capabilities of the plasma that will assist the particle melting.

2.3.1.2 Flame Spray Deposition

Flame spray is the oldest of thermal spray processes and is still widely used today due to its simple design and low operational costs compared with other techniques [6–8].

> In flame spraying, the material to be coated may be in the form of powder, wire, or rod and the high-temperature zone for melting the material is provided by the combustion of an oxygen–fuel gas mixture.

The most common fuel gas is acetylene, but propane or hydrogen can also be used. Flame spray is used for depositing a wide variety of materials ranging from polymers to ceramics and refractory metals.

Figure 2.10 shows a typical powder/wire flame spray gun. In powder flame spraying, the powder to be coated is normally introduced into the flame region by a compressed air or inert gas stream. In the case of wire flame spraying, the wire is continuously fed into the flame. A supply of compressed air atomizes the molten material into fine spherical particles and propels these particles at high velocity onto a specially prepared substrate. In the case of materials that cannot be drawn into wires, fine powder contained in plastic tubes has also been used in conjunction with the wire spray torch. Though the flame spray technique is very simple, it has its own limitations. The maximum temperature that can be obtained in the

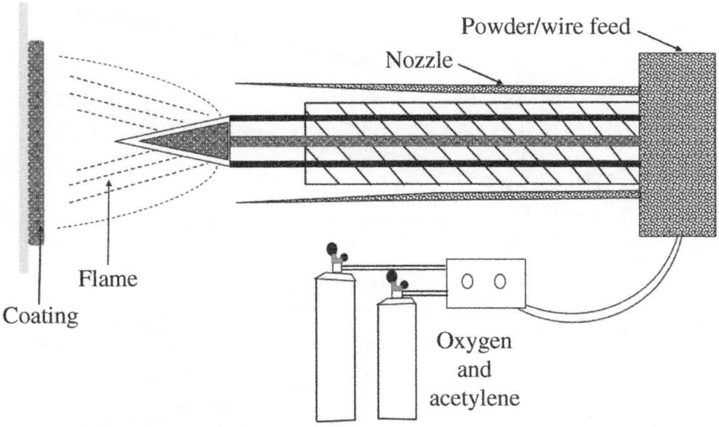

Figure 2.10 Schematic of the flame spray process.

flame spray is around 3000 °C and may not be sufficient to melt some ceramics, except for fine ceramics. Similarly, the direct exposure of the flame to the substrate results in dimensional distortion of the samples. This restricts the deposition of coatings on low melting temperature materials. This technique is not suitable for reactive materials due to the high oxidizing nature of the flame.

2.3.1.3 Wire Arc Spray Deposition

In wire arc spraying, a DC arc is struck between the two continuously fed consumable electrodes resulting in molten material at the tips of the electrodes. A high-velocity air jet, located behind the intersection of the two electrodes, separates the molten metal from the electrodes. This separated molten material is atomized and is broken into fine particles. Such fine particles create a fine distribution of molten droplets in the air stream, which are subsequently accelerated toward the substrate, where they impact and form coating. The schematic of the wire arc spraying process is shown in Figure 2.11.

> The direct melting involved in the wire arc spraying process results in high heat efficiency and the maximum attainable temperature is also higher than that which can be obtained in the flame spray technique.

This enables the wire arc technique to deposit a wide range of materials. The high deposition rate with respect to other thermal spray techniques makes the process well suited for depositing coatings on large components.

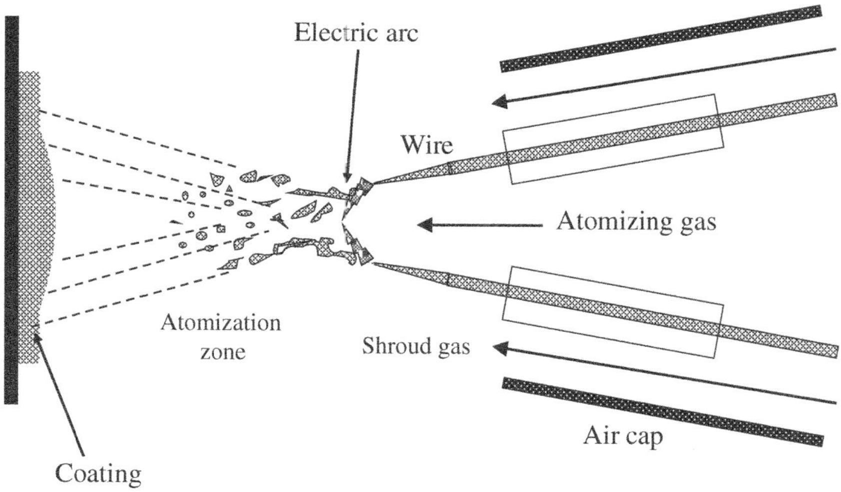

Figure 2.11 Schematic of the wire arc spraying process.

2.3.1.4 High-Velocity Oxy-Fuel Spray Deposition

The invention and commercialization of high-velocity oxy-fuel (HVOF) process (Jet Kote) in early 1980s successfully fulfilled the demands of the industry by providing high-quality dense coatings, which previously could be obtained only by proprietary processes.

> In the HVOF process, high-volume combustible gases are fed into a combustion chamber to obtain a high temperature (around 3000 °C). Also, a high-velocity gas stream (in the range 1500–2000 m/s) is essential to obtain dense coatings.

Hydrogen, propylene, propane, and acetylene are the most commonly used gases. The powder to be coated, fed from the pressurized nozzle in the high-velocity combustion flame, gets heated up and accelerates toward the substrate at high velocity to form an adherent and dense coating as in the case of detonation spraying. Owing to the high back pressure created by the combustion process, a pressurized powder feeder is required. However, the relatively low gas temperatures preclude the use of some refractory metals/ceramics in HVOF. This represents a significant shortcoming with respect to the detonation spray systems. Figure 2.12 schematically represents high-velocity oxy-fuel spray deposition.

2.3.1.5 Detonation Spray Coating

The detonation spray coating (DSC) technique is widely adopted to produce ceramic coatings [12–14]. A typical DSC process consists of the following stages

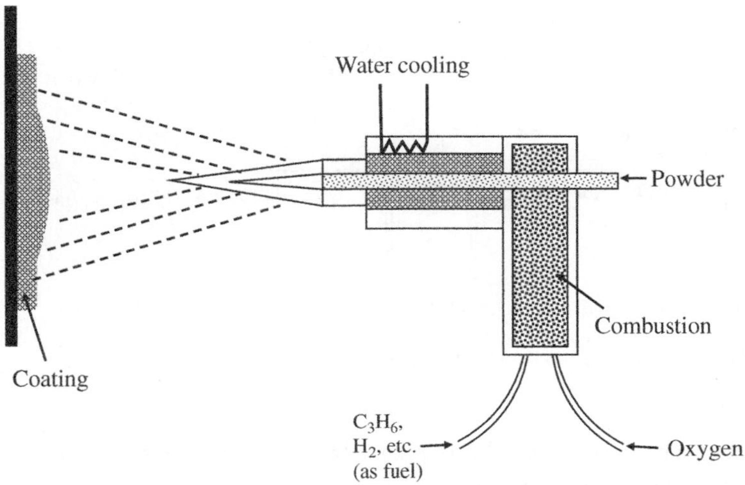

Figure 2.12 Schematic representation of high-velocity oxy-fuel spray deposition.

(see Figure 2.13). A metered quantity of a flammable gas mixture (oxygen and acetylene, for example) is fed through a tubular barrel, and concurrently, powders of the sprayed material are injected through a powder feeder. This allows the explosion to be triggered by a spark plug. The detonation wave attains a velocity of around 2600–3500 m/s and a temperature of around 3500 °C [15]. This wave accelerates the powder particles at velocities of 700–1000 m/s and impinges on the surface being coated. Consequently, the coatings are dense and exhibit high bond strength. Depending upon the type of material, detonation gas spraying units can repeat the process at a rate of up to 10 cycles per second. The most widely used design for gas mixtures to explosives is one-ended closed steel tube with approximately 25 mm internal diameter and 1350 mm length (Figure 2.13).

The kinetics of particle acceleration by a detonation wave is distinctly different from particle acceleration in other methods of spraying, such as plasma and HVOF.

The major contribution to acceleration of particle in the detonation spray techniques is the interaction between the particle and the products of detonation [13, 15]. For a pure oxy/acetylene mixture, the velocity of detonation increases with an increased amount of acetylene. It is important to note that the velocity of the detonation reduces sharply with an increase in content of

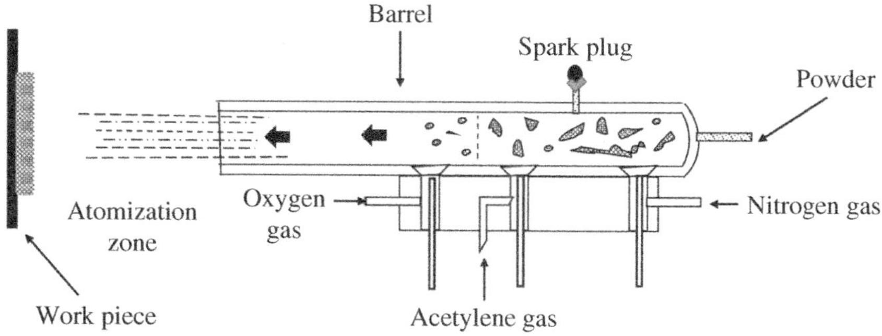

Figure 2.13 Schematic representation of the detonation spray process.

nitrogen present. The final velocity of the particle will depend on the gas mixture composition, powder loading distance, and total barrel length while keeping remaining operating parameters constant.

During the DSC process, the full combustion reaction between acetylene and oxygen is as follows [13, 17, 18]:

$$C_2H_2 + 2.5O_2 = 2CO_2 + H_2O + Q(\text{energy}) \tag{2.2}$$

When the temperature is above 800 °C, CO is more stable than CO_2 and the reactions will be as follows:

$$C_2H_2 + 1.5O_2 = 2CO + H_2O + Q(\text{energy}) \tag{2.3}$$

The other possible reactions from the reaction products are:

$$\text{Heat} + 2H_2O = 2H_2 + O_2 + \text{Explosive} \tag{2.4}$$

$$\text{Heat} + 2CO_2 = 2CO + O_2 \tag{2.5}$$

It can be inferred from Eqs. (2.4) and (2.5) that the water dissociation will enhance the detonation velocity and temperature. Also, CO formation will reduce the detonation velocity and temperature. Free carbon (C) will form when the oxygen–acetylene ratio is 1 : 1,

$$C_2H_2 + O_2 = CO + H_2O + C + Q(\text{energy}) \tag{2.6}$$

Hence, the final temperature and velocity of detonation products indirectly depend on many factors.

> The measurement of in-flight particle velocity and temperature as a function of processing conditions will help in a better understanding of the coating formation.

2.3.2 Bond Strength of Thermal Spray Coatings

One of the most important properties of the thermal spray coatings is bond strength, which is the adhesion strength of coating to the substrate surface [6, 7, 19–22]. According to ASTM, adhesion is defined as the state in which two surfaces may consist of valence forces or interlocking forces or both.

2.3.2.1 Bond Mechanism

The cross-sectional microstructure of a thermally sprayed material is shown in Figure 2.14, where unmelted particles are embedded within the layered structure and pores exist either totally within or between each layer.

> The basic bonding mechanisms of thermal spray coatings can be categorized into three major groups: mechanical interlocking, metal to metal bonding, and chemical bonding.

The adhesion strength of the deposit will vary with respect to the specific thermal spray process that influences the microstructure of the coatings [6, 7, 19–22]. For example, when we consider HVOF and detonation spray techniques, which produce a very dense microstructure, the porosity is typically less than 2% in comparison to around 5% porosity for a flame-sprayed or an atmospheric plasma-sprayed material. The size and distribution of pores in a coating directly affects coating characteristics, such as cohesive strength, permeability, receptivity, and machined surface finish.

Thus, the spray parameters, which influence the size and distribution of porosity, oxide content, residual stresses, and macro or microcracks, ultimately influence the performance of coatings. A coating with a heterogeneous layered-type structure or a composite mixture may include several materials like polymers, metals, and/or ceramics. A substrate surface preparation also influences an important role in coating integrity.

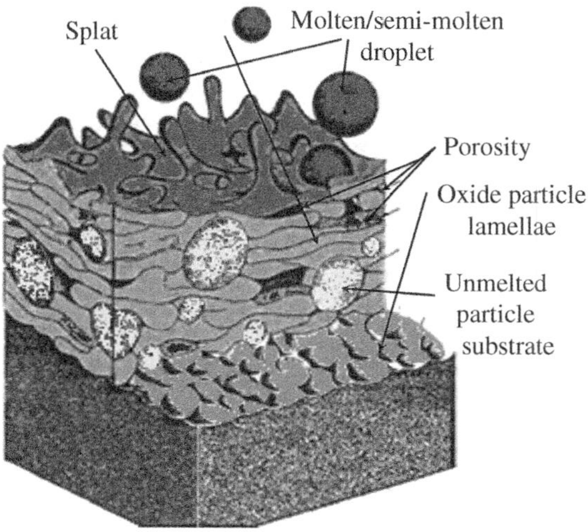

Figure 2.14 Schematic microstructure of a thermal spray coating [23].
Source: Reproduced with permission of Springer Nature.

The reliability or service life of the coatings is the most challenging problem being faced by the thermal spray scientists. The service life failure mode can be described as interfacial, cohesive, or mixed interfacial/cohesive. To assess the service life or the reliability of these thermal spray coatings, various testing methods are being used and measurement of the adhesion strength of the thermal spray coatings is one of them.

2.3.2.2 Test Methods

Several techniques have been used to measure the coating-substrate adhesion, like strain method, movement method, shears technique, and tensile techniques [20–22]. Some of these are clearly too simplistic, while the others are more sophisticated. Direct tensile pull-off test, which is widely used, will be discussed in this section.

Direct tensile pull-off test (ASTM C633-79) is widely used by thermal spray community to evaluate the bond strength of thermal spray coatings. Tests are done by applying coating to the end of a cylindrical test specimen of 25.4 mm diameter by 25.4 mm length (see Figure 2.15). The coating is bonded to another sample of the same geometry. The bond strength is calculated by dividing force by the one inch diameter area of cylinder assuming that the stress distribution is uniform over the area where the coatings are bonded. The ASTM C633-79 standard method was supported by FEM analysis. Based on the fracture criterion and FEM analysis, a relation between the bond strength of the elongated specimen and the bond strength of the standard specimen was developed. The FEM analysis resulted in better

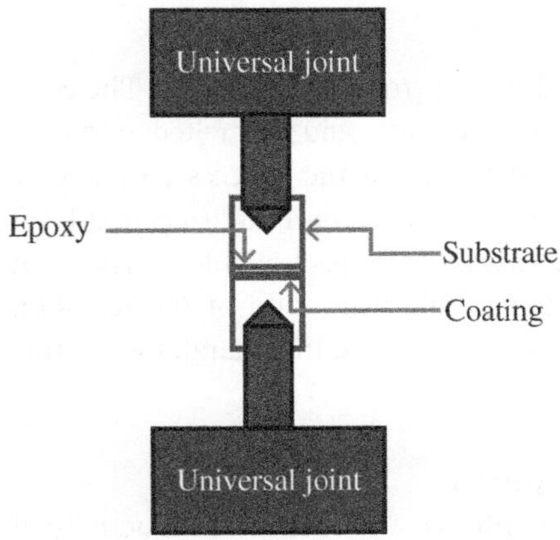

Figure 2.15 Direct tensile pull-off adherence test as per ASTM C633-79.

understanding of the effect of specimen size on stress distribution and thereby on the bond strength of the specimens. This was compared with the experimental interface stress analysis results, obtained by strain gauges. This resulted in standardization of the technique.

Though this technique is used widely, there are few limitations. They are as follows:

1) Unavailability of suitable tape or epoxy for measuring true bond strength values of DSC deposited coatings.
2) Limited shelf life of epoxy.
3) If the coating is porous and adhesive is sufficiently fluid, the adhesive may penetrate the coating, which will give erroneous results.
4) Residual stresses can be set up in the coating substrate system prior to testing because of the attachment of the pull rod to the exposed coating surface.
5) Another important drawback is whenever the high adhesion strength (>80 MPa) coatings are subjected to this test, the tape or epoxy only fails always at lower loads, while coating layers are still intact with the substrate.

Most of the times, the failure at the adhesive–coating interface during the bond strength measurements of detonation sprayed WC-Co coatings indicate that the bond strength of coatings is higher than the adhesive strength [25, 26]. As an interesting solution to overcome the limiting strength of an adhesive tape in direct tensile pull-off test, a new pin-type test (similar to modified Ollard method) was adopted to measure the adhesion strength of DSC in the present study. More details about the method are provided in the experimental section. At this point, it needs to be mentioned here that the coating microstructure and properties significantly influence the tribological properties.

2.3.3 Coating Structure

The structure of thermal spray coating is different from bulk material. The coating structure consists of a layer with pores, oxides, and unmelted particles. The formation of homogeneous structure depends on the process parameters. The overlapping splats lock onto one another to form a continuous coating layer [23]. The deposit or coating is built up by successive impact of solid particles or molten droplets as shown in Figure 2.14. The microstructure of the resultant splats will ultimately affect the properties of the coatings. In general, the coating structure is influenced by two key factors.

2.3.3.1 Particle and Substrate Material Properties

In particular, the melting point, ductility, hardness, and thermal conductivity of the particle and substrate material influence the coating microstructure. For example, low melting point materials are more likely to melt in the gun or flame during the acceleration of powder material. The relative hardness of the particle and substrate materials will affect the relative deformation of the particle and the substrate. The thermal conductivity of the particle and substrate materials will influence how the heat is dissipated after the splat has come to rest.

2.3.3.2 Particle Temperature and Velocity

The temperature and velocity of a particle stream during the thermal spray process relative to the melting point of the particle will determine if the particle is molten before impact or if it becomes molten during impact. The particle gets heated within the accelerating apparatus and heat is gained from conversion of kinetic energy during the impact process. In case of reactive material coatings, the availability of free oxygen will influence the final coating structure.

The final microstructure of coating will depend on the gas mixture composition, feedstock size, feedstock type, standoff distance, etc. In the present study, oxy-fuel ratio is varied while keeping all other parameters constant in order to study the influence of degrees of decarburization on the properties of detonation sprayed WC-Co coatings.

2.3.4 Case Study: WC-Co Coatings

It is important to reiterate here that two chapters in this book are focused on wear properties of WC-Co coatings. Therefore, a case study on microstructure and phase stability in WC-Co coatings is described in this section. A comparison will be made with WC-Co bulk alloys.

In general, WC-Co alloys are manufactured by powder metallurgy route, which consists of powder preparation, compaction, and sintering. Cemented carbide systems are ideal for LPS as they exhibit the necessary conditions to obtain full dense components in LPS, like (i) an appreciable amount of liquid, (ii) solubility of the solid in the liquid, and (iii) complete wetting. This process results in a final particulate composite that is essentially free of pores or voids, as the liquid binder phase assisted by capillary action is able to permeate the carbide grains.

Bulk WC-Co cermet is usually obtained by LPS. Time and temperature are important parameters for LPS. During sintering, Co binder undergoes melting and more importantly, the interaction between the molten Co phase and the solid WC cuboids is minimal. Thus, the final sintered WC-Co material essentially consists of WC and Co phases. The hardness (H) and the elastic modulus (E) of the sintered bulk WC-Co depends on the Co content and also on WC cuboid size. For example, as illustrated in Figure 2.16a, the Vickers hardness of sintered WC-Co lies in the range HV 800–HV 2300, while the elastic modulus spans the range 450–650 GPa. The indentation fracture toughness of sintered WC-Co on the other hand decreases with increasing hardness of WC-Co (primarily due to a decrease in Co content), as illustrated in Figure 2.16b. The toughness values of liquid-phase sintered WC-Co span the range 10–18 $MPa{\cdot}m^{1/2}$.

In contrast to LPS, the coating formation is less controlled in the case of thermal spray coatings. In all the thermal spray techniques, the feed stock is in the form of powders (e.g. WC-Co powder) and they are entrained in the high-velocity, high-temperature gas/air stream. As the particles travel with the gas/air stream, they get heated up and also undergo acceleration. As explained earlier, the nature of the thermal spray process and the powder particle size and density determine the final temperature and velocity of the powder particles prior to impacting the component surface. At one extreme, air plasma spray results in extremely high particle temperatures such that WC-Co particles become molten during flight, but the WC-Co particle velocities are low in the range 100–150 m/s. In contrast, in the case of detonation spray and HVOF coating techniques, the temperature of WC-Co particles lie in the range 1600–2300 °C (i.e. the Co binder phase alone becomes molten during flight), while the WC-Co particle velocities lie in the range 500–800 m/s.

> The semi-molten WC-Co particles impact the component/substrate surface, get deformed extensively to form disc-shaped pancakes, which are usually called as "splats."

Continuous impact of such particles result in the formation of thick layered coatings, composed of splats, superimposed over each other. A typical cross-sectional view of the detonation sprayed WC-Co coating is shown in Figure 2.17.

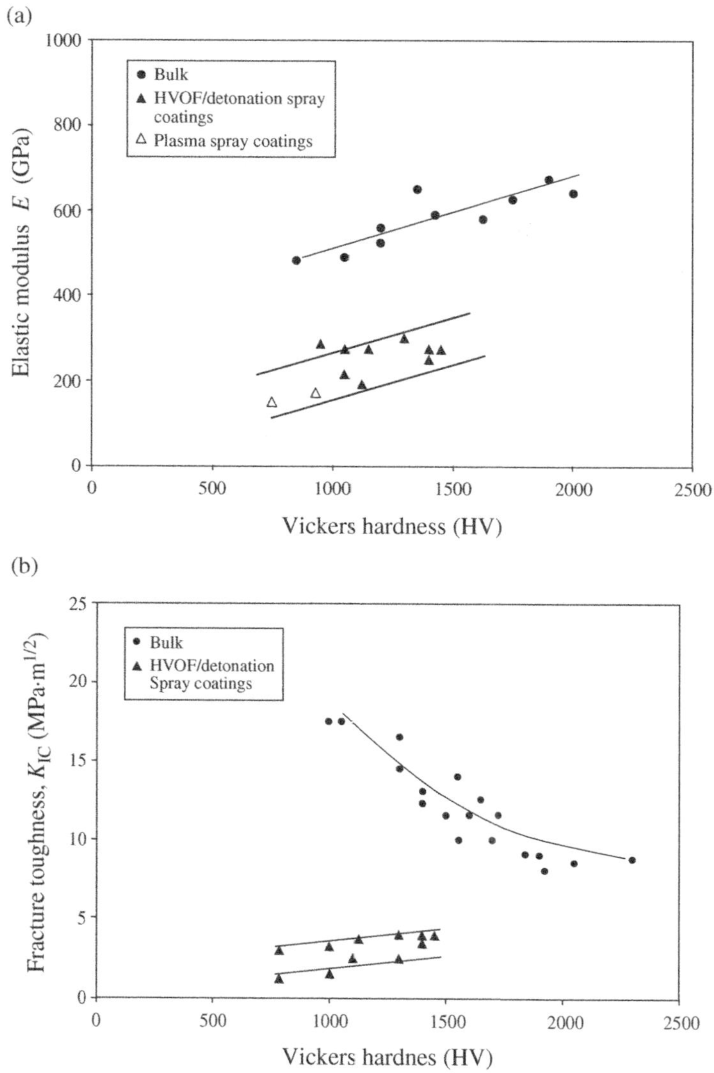

Figure 2.16 (a) Elastic modulus of bulk and thermally sprayed WC-Co coatings as a function of Vickers hardness [27] and (b) fracture toughness (K_{IC}) of bulk and thermally sprayed WC-Co coating as a function of Vickers hardness [27]. *Source:* Reproduced with permission of Springer Nature.

The presence of splats with each splat representing one original powder particle is clearly evident. It is also clear that the inter-splat boundaries are well-bonded in certain locations (see locations marked "A" in Figure 2.17) but not bonded at all in certain other areas (see locations marked "B" in Figure 2.17). In addition, the

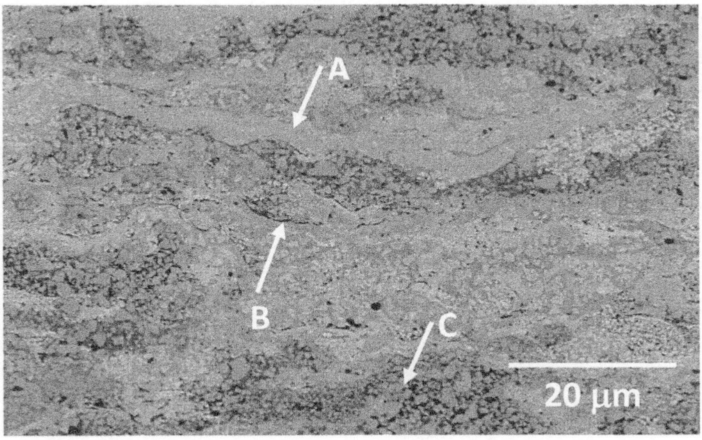

Figure 2.17 Cross-sectional SEM image of detonation sprayed WC-Co coating. A: Well-bonded inter-splat boundary. B: Loosely bonded inter-splat boundary. C: Porosity [27]. *Source:* Reproduced with permission of Springer Nature.

presence of pores can also be observed at various locations (see locations marked "C" in Figure 2.17). Thus, it can be concluded that the detonation sprayed WC-Co coatings have a layered structure with a mix of both weak and strong inter-splat boundaries. Such a structure has enormous influence on the mechanical properties of the WC-Co coating, since the weak inter-splat boundaries can be treated as cracks and the presence of numerous such cracks, parallel to the coating plane, can substantially reduce the elastic modulus and indentation toughness of WC-Co coatings. In fact, the modulus, hardness, and toughness of WC-Co coatings are presented as filled triangles, in Figure 2.16a and b. Compared with bulk WC-Co, represented as filled circles in Figure 2.16a and b, the elastic modulus of WC-Co coating is nearly half in value, while toughness is lower by a factor of 4 or so. In conclusion, it can be stated that the bulk WC-Co and detonation sprayed WC-Co coatings are quite different from each other in terms of microstructure and mechanical/tribological behavior.

The common microstructural features of thermally sprayed WC-Co coatings observed by several researchers [24, 27–30] are as follows:

1) Low WC cuboid volume fraction compared with the starting powder.
2) The formation of other phases like W_2C and elemental W; W_2C is often observed to encapsulate the WC cuboid.
3) Low carbon content in the coatings against starting powder.
4) The formation of an amorphous phase or nanocrystalline binder phase.

5) The formation of two distinct regions of binder phase, one being significantly richer in tungsten than other due to progressive decarburization and dissolution of WC cuboid from the particle surface to center.

Various mechanisms have been proposed for the formation of such microstructure characteristics. Vinayo et al. [30] described the reactions that take place during the thermal spraying of WC-Co:

$$WC \rightarrow W_2C + C \tag{2.7}$$

$$W_2C + O_2 \rightarrow W_2(C,O) \tag{2.8}$$

$$W_2(C,O) \rightarrow 2W + CO \tag{2.9}$$

Further modification to the reaction sequence was proposed by Fincke et al. [31], who believed that during thermal spraying, carbon diffuses from the WC into the metallic matrix and that Co diffuses from the matrix into the carbide as follows:

$$4Co + 4WC + O_2 \rightarrow 2Co_2W_4C + 2CO \tag{2.10}$$

$$3Co + 3WC + O_2 \rightarrow Co_3W_3C + 2CO \tag{2.11}$$

$$12Co + 12WC + 5O_2 \rightarrow 2Co_6W_6C + 10CO \tag{2.12}$$

Tu et al. [29] confirmed that the WC phase undergoes a degradation process to produce W_2C, which is then transferred to metallic W when high temperatures are attained in an oxygen-rich atmosphere. Similarly, Verdon et al. [32] and Stewart et al. [33] analyzed the WC-Co coatings in detail and proposed the coating formation models, which are discussed below. While the reaction one (Eq. 2.7) is in agreement with many experimental results described in the literature, no experimental evidence or description by other methods of the presence of $W_2(C,O)$ has appeared in the literature.

As mentioned above, based on the detailed analysis, several authors have proposed a comprehensive graphical description of the processes during spraying of WC-Co [9, 28, 30, 32, 33] and some of these are shown in Figures 2.18–2.21. During the spray process, the WC-Co powder particles are accelerated and heated as they travel through the flame. This results in melting of Co matrix. Subsequently, WC cuboid starts dissolving in liquid Co. At the same time, carbon rapidly diffuses through the Co liquid and gasifies at the particle surface by oxidation reaction. Finally, quenching of the particle on impact with the substrate will result in the formation of amorphous/nano binder phase. The formation of W_2C and W phases

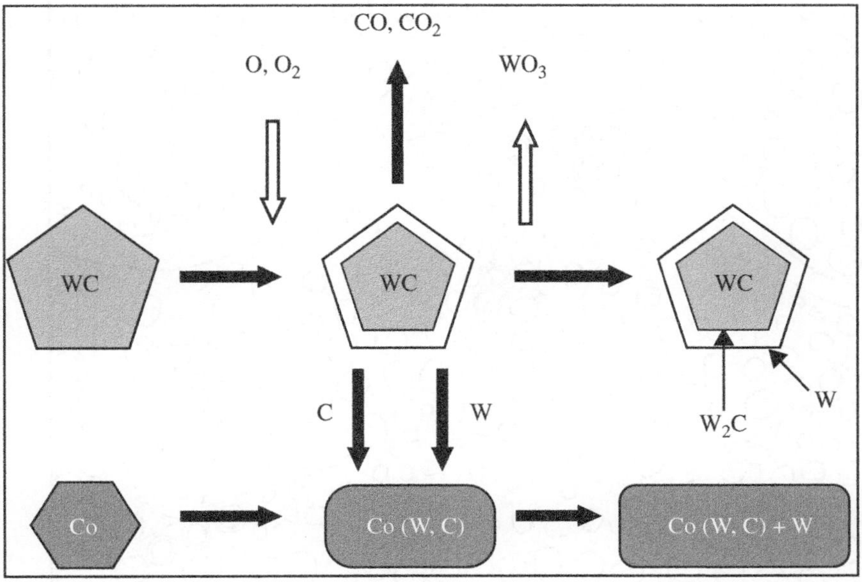

Figure 2.18 Chemical and metallurgical processes occurring during thermal spraying of WC-Co [9]. *Source:* Reproduced with permission of Taylor & Francis.

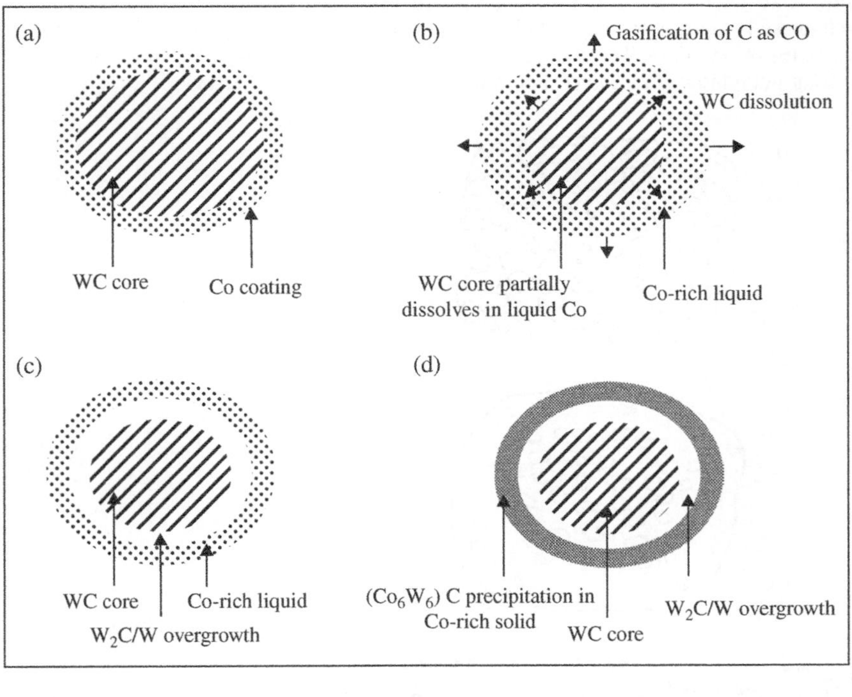

Figure 2.19 Schematic representation of the sequence of events leading to decarburization of HVOF sprayed nano-WC-Co powder, accompanied by the formation of W_2C and W phases. (a) Initial structure of WC grain with Co coating. (b) After superheat above WC-Co eutectic temperature (~1350 °C). (c) After cool-down to WC-Co eutectic temperature. (d) After cool-down to ambient temperature. *Source:* Adapted from Vinayo et al. [30] and Kear et al. [28].

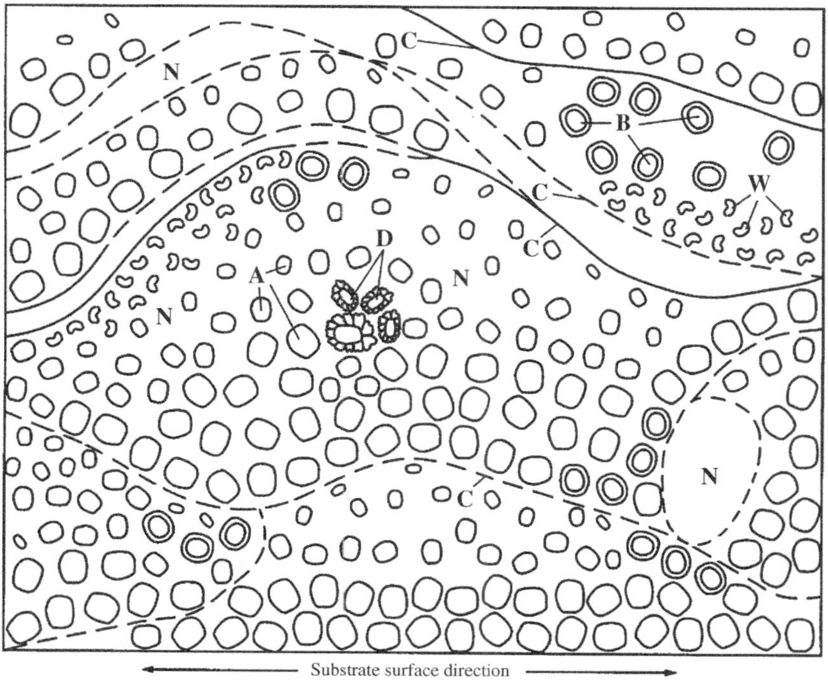

Figure 2.20 Schematic of the coating microstructure perpendicular to the substrate. Each phase is represented by a different alphabet. (A) WC carbide, (N) nano-crystalline matrix, (B) WC+W_2C, W-W crystallites, (D) Co dendrites surrounding WC grains, and (C) splat boundaries [32]. *Source:* Reproduced with permission of Elsevier.

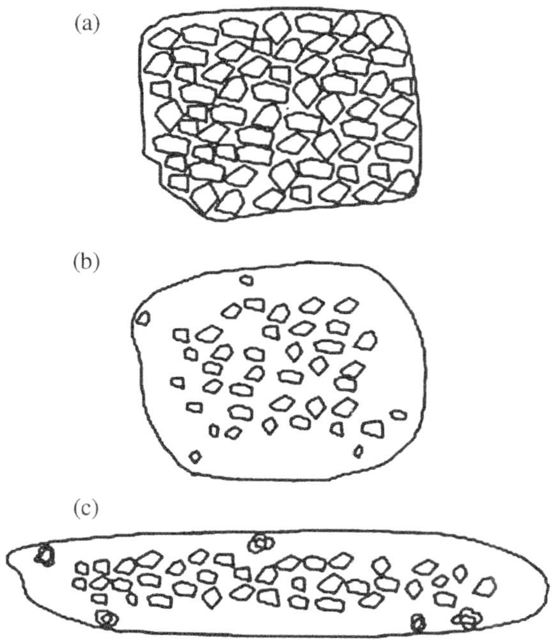

Figure 2.21 Schematic illustration of the processes involved in melting. WC dissolution (a) and decarburization (b) of a WC-Co powder particle. The formation of a lenticular splat on impact with the substrate is illustrated in (c) [33]. *Source:* Reproduced with permission of Elsevier.

depend on the extent of carbon loss (decarburization), experienced by the WC-Co particles. Some speculations exist in the formation of W_2C and elemental W. Some researchers [27, 28, 30, 34], proposed that the direct oxidation of solid WC can lead to formation of W_2C and W. However, Stewart et al. [33] suggested that W_2C, W, and the Co(W,C) nanocrystalline/amorphous phase form during rapid solidification of the spray.

Several researchers studied the influence of degree of decarburization on the coating tribological performance of coatings [28, 32, 33]. Direct comparison of all results is not possible as they are studied using different techniques and various feedstock powders. Hence, identifying the decarburization range that will influence the coating's properties is very essential. In addition to above, the complex nature of the coating microstructure and the understanding of structure–property relation are also very limited. Deposition technique and feedstock type will influence the degree of decarburization and coating's microstructure and its performance. The above studies are conducted on HVOF and plasma sprayed coatings.

2.4 Closure

This chapter summarizes various densification processes for ceramics, including pressureless and pressure-assisted sintering. In addition to presenting briefly the process physics, this chapter also summarizes various process variables to obtain bulk ceramics with near-full densification or ceramic coatings with desirable physical properties.

References

1 Tokita, M. (2004). Trends in advanced SPS spark plasma sintering systems and technology. *J. Soc. Powder Technol. Jpn.* 30 (11): 790–804.
2 Shen, Z., Zhao, Z., Peng, H., and Nygren, M. (2002). Formation of tough interlocking microstructures in silicon nitride ceramics by dynamic ripening. *Nature* 417: 266–268.
3 Kumar, R., Prakash, K.H., Cheang, P., and Khor, K.A. (2005). Microstructure and mechanical properties of spark plasma sintered zirconia-hydroxyapatite nano-composite powders. *Acta Mater.* 53: 2327–2335.
4 Orru, R., Licheri, R., Locci, A.M. et al. (2009). Consolidation/synthesis of materials by electric current activated assisted sintering. *Mater. Sci. Eng. R* 63: 127–287.
5 Aman, Y., Garnier, V., and Djurado, E. (2009). Influence of green state processes on the sintering behavior and the subsequent optical properties of spark plasma sintered alumina. *J. Eur. Ceram. Soc.* 29 (16): 3363–3370.

6 Pawlowski, L. (1995). *The Science and Engineering of Thermal Spray Coatings*. Chichester: Wiley.

7 Davis, J.R. (ed.) (2004). *Handbook of Thermal Spray Technology*. Materials Park, OH: ASM International.

8 Tucker, R.C. Jr. (1994). Thermal spray coatings. In: *ASM Handbook* (eds. C.M. Cotell, J.A. Sprague and F.A. Smidt), 497–509. Materials Park, OH: ASM International.

9 Berger, L.M. (2007). Technology overview hardmetals as thermal spray coatings. *Powder Metall.* 50: 205–214.

10 Chagnon, P. and Fauchais, P. (1984). Thermal spraying of ceramics. *Ceram. Int.* 10 (4): 119–131.

11 Heimann, R.B. (1994). *Plasma – Spray Coating*. Weinheim: Wiley-VCH.

12 Guest, C.J.S. (1986). Plasma and detonation gun coatings. *Trans. IMF* 64: 33–38.

13 Kadyrov, E. and Kadyrov, V. (1995). Gas dynamical parameters of detonation powder spraying. *J. Therm. Spray Technol.* 4 (3): 280–286.

14 Tucker, R.C. Jr. (1974). Structure property relationships in deposits produced by plasma spray and detonation gun techniques. *J. Vac. Sci. Technol.* 11 (4): 725–734.

15 Kadyrov, E. (1996). Gas-particle interaction in detonation spraying systems. *J. Therm. Spray Technol.* 5 (2): 185–195.

16 Bertenev, S.S., Fedhko, Y.P., and Grigorov, A.I. (1982). *Detonation Spray Coatings in Machine Construction* (in Russian) (ed. V.N. Goldfain). Leningrad, USSR: Machine Construction.

17 Shesternenkov, V.I. (1966). Detonation coatings. Institute of Material Science, Academy of Science of the UKRSSR, 30–36, Translated from *Metasllurgiya*, No. 1, 1968, 37–47, Original article submitted October 1966.

18 Rama Krishna, L., Sen, D., Rao, Y.S. et al. (2002). Thermal spray coating of aluminum nitride utilizing the detonation spray technique. *J. Mater. Res.* 17 (10): 2514–2523.

19 Berndt, C.C. and McPherson, R. (1979). The adhesion of flame and plasma sprayed coatings – a literature review – AWRA report p11-1-78. *Aust. Weld. Res.* 6: 75–85.

20 Rickerby, D.S. (1988). A review of the methods for the measurement of coating-substrate adhesion. *Surf. Coat. Technol.* 36: 541–557.

21 Chalker, P.R., Bull, S.J., and Rickerby, D.S. (1991). A review of the methods for the evaluation of coating-substrate adhesion. *Mater. Sci. Eng. A* 140: 583–592.

22 Berndt, C.C. and Lin, C.K. (1993). Measurement of adhesion for thermally sprayed materials. *J. Adhes. Sci. Technol.* 7: 1235–1264.

23 Herman, H. and Sampath S. (1996). *Thermal spray coatings*. (ed. Stern. K. H.). Metallurgical and Ceramic Protective Coatings. Springer, Dordrecht.

24 ASTM C633-79 (1979). *Standard Test Method for Adhesion or Cohesive Strength of Flame Sprayed Coatings*. Reprint from *Annual Book of ASTM Standard*. Philadelphia, PA: ASTM International.

25 Kharlamov, Y.A. (1978). Bonding of detonation-sprayed coatings. *Thin Solid Films* 54: 271–278.

26 Du, H., Hua, W., Lu, J. et al. (2005). Influence of process variables on the qualities of detonation gun sprayed WC-Co coatings. *Mater. Sci. Eng. A* 408: 202–210.

27 Sundararajan, G. and Suresh Babu, P. (2009). Detonation sprayed WC-Co coatings: unique aspects of their structure and mechanical behavior. *Trans. IIM* 62 (2): 95–103.

28 Kear, B.H., Skandan, G., and Sadangi, R.K. (2001). Factors controlling decarburization in HVOF sprayed nano-WC/Co hardcoatings. *Scr. Mater.* 44: 1703–1707.

29 Tu, D., Chang, S., Chao, C., and Lin, C. (1985). Tungsten carbide phase transformation during the plasma spray process. *J. Vac. Sci. Technol. A* 3: 2479–2482.

30 Vinayo, M.E., Kassabji, F., Guyonnet, J., and Fauchais, P. (1985). Plasma sprayed WC-Co coatings: influence of spray conditions (atmospheric and low pressure plasma spraying) on the crystal structure, porosity, and hardness. *J. Vac. Sci. Technol. A* 3: 2483–2489.

31 Fincke, J.R., Swank, W.D., and Haggard, D.C. (1994). *Proceedings of 7th NTSC, 1994*, 325–330. Boston, MA: ASM International.

32 Verdon, C., Karimi, A., and Martin, J.L. (1998). A study of high velocity oxy-fuel thermally sprayed tungsten carbide based coatings. Part 1: microstructures. *Mater. Sci. Eng. A* 246: 11–24.

33 Stewart, D.A., Shipway, P.H., and McCartney, D.G. (2000). Microstructural evaluation in thermally sprayed WC-Co coatings: comparison between nanocomposite and conventional starting powders. *Acta Mater.* 48: 1593–1604.

34 Babu, P.S., Basu, B., and Sundararajan, G. (2008). Processing–structure–property correlation and decarburization phenomenon in detonation sprayed WC–12Co coatings. *Acta Mater.* 56: 5012–5026.

3

Conventional and Advanced Machining Processes

In the previous chapter, the processing of ceramics and ceramic coatings is described. Since the consolidation or thermal deposition of coatings and machining constitute the entire manufacturing spectrum, this chapter describes the machining of ceramics. As ceramics-based cutting tool inserts represent a major application domain, the physics of the conventional machining process is briefly discussed in the initial part of the chapter with a brief process description of a few processes. This is followed by a detailed discussion on advanced machining processes with a major emphasis on electro-discharge machining (EDM).

3.1 Introduction

Machining is widely being utilized as a major manufacturing process in a large number of engineering industries [1, 2]. Machining also involves the materials' removal from the contacting surfaces. The use of ceramics as tool inserts in machining of steel is widely appreciated [3].

> Among different classes of engineering materials, one of the reasons for a better acceptability of metals and alloys is the easier adaptability to various machining techniques.

To substantiate the above statement, making a hole on a metallic plate without fracture or significant shape distortion is possible with conventional machining techniques, which can however not be adopted for brittle ceramics. In order to obtain desired shapes, both conventional and nonconventional machining processes, like EDM, can be used for metals.

Friction and Wear of Ceramics: Principles and Case Studies, First Edition.
Bikramjit Basu, Mitjan Kalin, and B. Venkata Manoj Kumar.

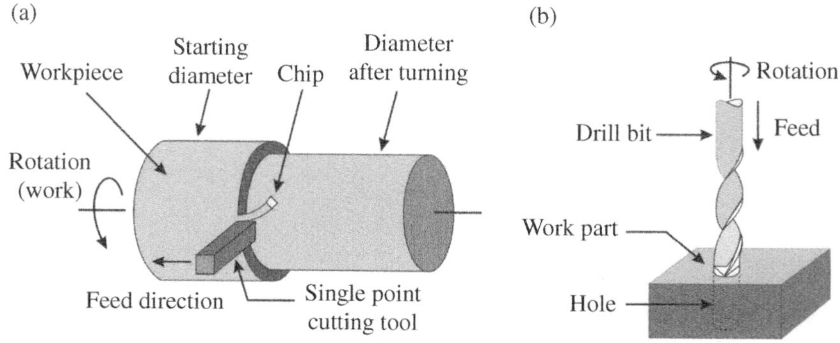

Figure 3.1 Conventional machining process for metals: (a) turning and (b) drilling.

In order to shape different engineering materials, the machining processes, play an important role in removing excess material to obtain desired geometry. Two generic types of machining processes, i.e. conventional and nonconventional machining processes, are used [4].

The conventional machining processes include tuning, drilling, milling, and grinding. Some widely used conventional machining processes are shown in Figure 3.1. In all conventional processes, excess material is removed in the form of a chip. In the case of turning a cylindrical workpiece, the diameter is constantly reduced during the progression of material removal (see Figure 3.1a). For making a hole in a metallic plate, the drilling process can be adopted, as shown in Figure 3.1b. In the conventional machining processes, highly complex stress conditions with high levels of shear lead to controlled fracture of the material.

3.2 Conventional Machining

The physics of the conventional machining process can be better described using Figure 3.2. As shown in Figure 3.2, the conical-shaped tool with a sharp edge removes the material from a flat workpiece and the tool edge is known as the flank.

The flank area of the ceramic tool, therefore, experiences maximum wear.

For the progression of smooth machining, the tool wear needs to be minimized with the use of ceramic tool inserts with high wear resistance. The material removal from the workpiece takes place by shear-induced chip formation, as shown in Figure 3.2. Various angles are mentioned in Figure 3.2 to describe the shear processes and evolution of chip. Three types of chips are normally reported:

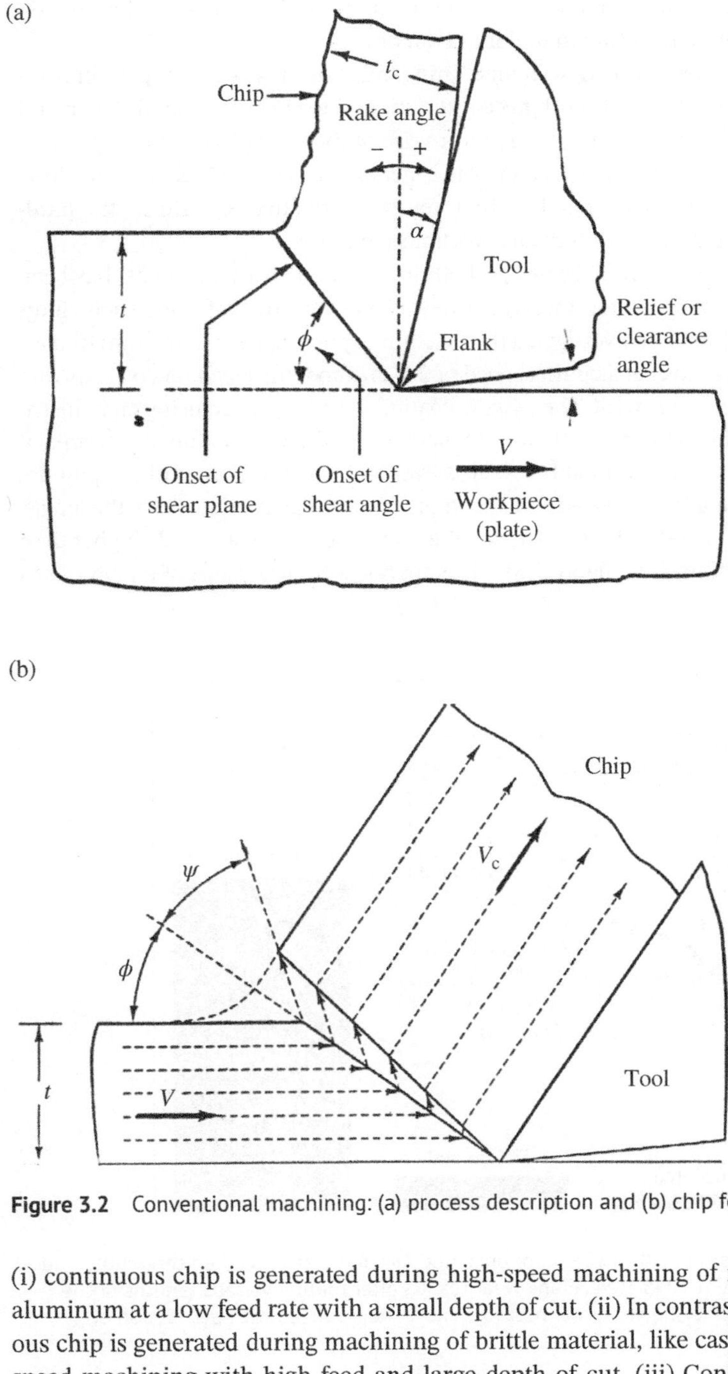

Figure 3.2 Conventional machining: (a) process description and (b) chip formation.

(i) continuous chip is generated during high-speed machining of mild steels or aluminum at a low feed rate with a small depth of cut. (ii) In contrast, discontinuous chip is generated during machining of brittle material, like cast iron, at low-speed machining with high feed and large depth of cut. (iii) Continuous chips with built-up edge are generated due to another cut, which is deeper than the tip of the cutting tool, resulting in tool life reduction.

As far as the energy involved in machining, most of the energy is spent in overcoming friction at the tool/workpiece interface and to shear the metal. The result is the chip generation and the temperature rise of the workpiece/tool/chip.

At the shear zone, 80–85% heat energy is generated, while 15–20% at the chip–tool interface. It is perceivable that high temperature adversely affects the hardness and wear resistance of tool and workpiece material.

The machining of several ceramics, both in bulk and coating formats, has been studied. Those studies often involved a new class of coatings. In one such study, Chhowalla and Unalan investigated the machining potential of c-Zr_3N_4 thin film for milling AlSl 1045 steel at a cutting speed of 200 m/min with a feed rate of 10 m/min and depth of 1.5–2.0 mm [5]. The c-Zr_3N_4 coatings were deposited using an industrially viable modified filtered cathodic arc method. As shown in Figure 3.3, results of milling with standard lubricants revealed less flank wear (an order of magnitude) of c-Zr_3N_4 coating (hardness ~36 GPa) compared with TiN coating. Also, the length of cut on the milled steel for c-Zr_3N_4 coating is two orders of magnitude higher than that of TiN coatings (see Figure 3.3). The superior wear resistance of c-Zr_3N_4 films

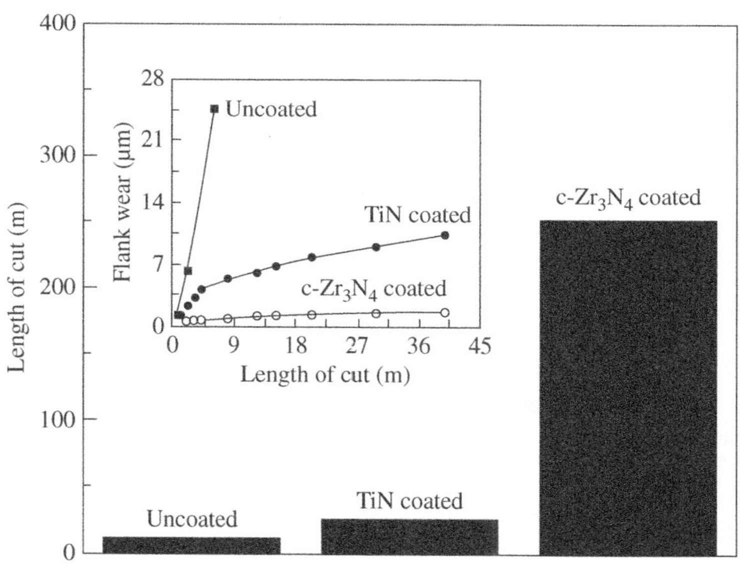

Figure 3.3 Length of cut results from uncoated TiN and c-Zr_3N_4-coated three flute end mills in milling AISI 1045 steel. Flank wear results after milling various lengths of the low carbon steel are shown in the inset [5]. *Source:* Reproduced with permission of Springer Nature.

deposited by the filtered cathode arc method implicates their potential for use as automotive parts and tools components, which remain to be explored.

3.3 Advanced Machining Processes

In the context of tribological applications, e.g. cutting tool inserts, the wear-resistant ceramic components are to be machined to different sizes and complex shapes, which is time-consuming and often difficult for conventional machining processes. The nonconventional machining approaches are now described in the following sections, in particular, referring to process physics and materials' aspects.

3.3.1 Electro-Discharge Machining

EDM is a potential manufacturing route to fabricate dies and moulds of conductive materials [6–8]. The erosive effect of electrical discharge was first discovered by Priestley in 1770. Since the resistance–capacitance system was used to store and define discharge energy, a long pause between discharges was in practice during the early years of development. Later on, controllable machining was achieved using rapid and repetitive spark discharges between the electrode tool and workpiece to produce accurate and high-quality products [9]. Earlier investigations of spark machining to erode the material using dielectric liquids proved the improved efficiency with high material removal rates (MRR) [10].

EDM is an important example of a nonconventional machining process to shape complex components of conductive materials [11–13]. Phenomenologically, EDM can be described as a thermoelectric process involving repetitive electrical discharge (sparks) to carefully remove electrically conductive workpiece and electrode materials. One of the greatest advantages of EDM is its use to machine hard, brittle, and difficult-to-machine electroconductive materials, with desired shape and closer dimensional tolerance in a shorter time.

Two competing EDM processes are die sinking and wire EDM. In the case of the former, the die cavity is directly impressed on the workpiece while accomplishing large material removal (see Figure 3.4) [14]. Typical electrode materials in the die sinking process are copper, copper-tungsten, copper-chromium alloy, graphite, tungsten, brass, steel, etc. Materials with good electrical and thermal conductivity are preferred to reduce electrode wear. During die sinking, the voltage, current, and duty cycle are varied to obtain the machined materials' good surface roughness properties. The efficacy of the EDM for a given material is also assessed in quantifying the MRR and tool wear rate. For wire EDM, highly conductive oxygen-free high conductivity copper (OFHC) is widely used as the wire to cut intricate or complex-shaped product from a workpiece (see Figure 3.4b).

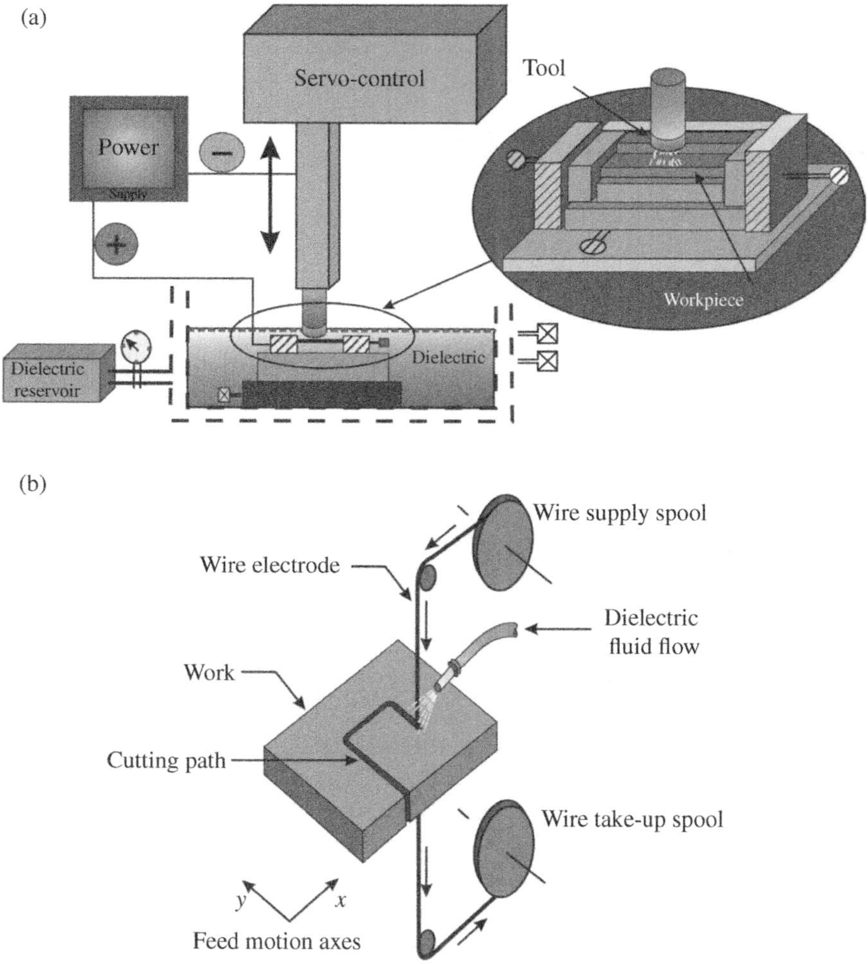

Figure 3.4 Schematic illustration of the EDM processes: (a) die sinking [14] and (b) wire EDM for machining electroconductive materials. *Source:* Reproduced with permission of Elsevier.

EDM has good applications to cover a wide range of materials including all metals, ceramics, and composites. Since EDM has the advantage of having no-contact between the electrode and workpiece, mechanical stresses corresponding to chatter and vibrational problems are eliminated [15]. In addition, machining can be done independent of material hardness as long as the electrical conductivity is maintained. A maximum resistivity of 100 Ω·cm is considered to be a limit for the material to be used as an electrode in EDM [16]. Further, the process is able to drill holes at very incident angles on curved surfaces without drill wander [14]. EDM has replaced traditional machining processes such as milling of heat-treated tool steels with full hardness before machining [17]. EDM is capable of making not only small dies, holes, pins, shafts, and nozzles as small as 5 μm in diameter but also complex three-dimensional microcavities [18–21]. Therefore, the applications of EDM lie in different fields like automotive, aerospace, mould, tool, and die making industries. Further, EDM is useful in medical, optical, and jewellery industries as well [8]. Also, micro-EDM as an alternative method for producing photo-masks in the integrated circuit industry has been reported [22]. Based on the application, several variations of EDM processes have been developed, such as wire EDM [23], ultrasonic machining [24, 25], electrical discharge texturing [26], electrical discharge grinding [27], hybrid machining process [28], etc.

Although several EDM systems are in the research stage, two primary processes are usually applied: die-sink EDM and wire EDM. The die-sink EDM system utilizes a pre-shaped tool electrode to make a cavity into the workpiece.

> The wire EDM process utilizes a constantly moving strand of wire as shaped electrode that passes through the workpiece in the vertical axis and cuts a contour in the horizontal plane. In both processes, multiple cuts are used when close tolerances and fine finishes are needed.

3.3.1.1 Working Principle

> In EDM, the tool and workpiece are connected through capacitor and charged from pulsed direct current (DC) source.

If the tool is made as cathode and workpiece as anode, then circuit is referred to as normal polar circuit, otherwise negative polar. When the potential across the electrode exceeds the breakdown voltage (V_b), sparking occurs at a point of least electrical resistance through the narrowest possible gap. Consequently, after discharge, capacitor recharges and spark appear at the next narrowest gap. The heat energy generated from series of such sparks is utilized in removing material from surfaces of workpiece and/or tool. The dielectric is useful in localizing thermal energy into small areas, cooling electrodes, and flushing the debris away.

The mechanism of material removal in EDM can be explained with sequential phases of electrical discharges (Figure 3.5). The tool is moved toward the workpiece until the gap is small enough (typically: 0.01–0.50 mm), so that the pulsed DC creates an intensive electric field at the surface irregularities of electrode

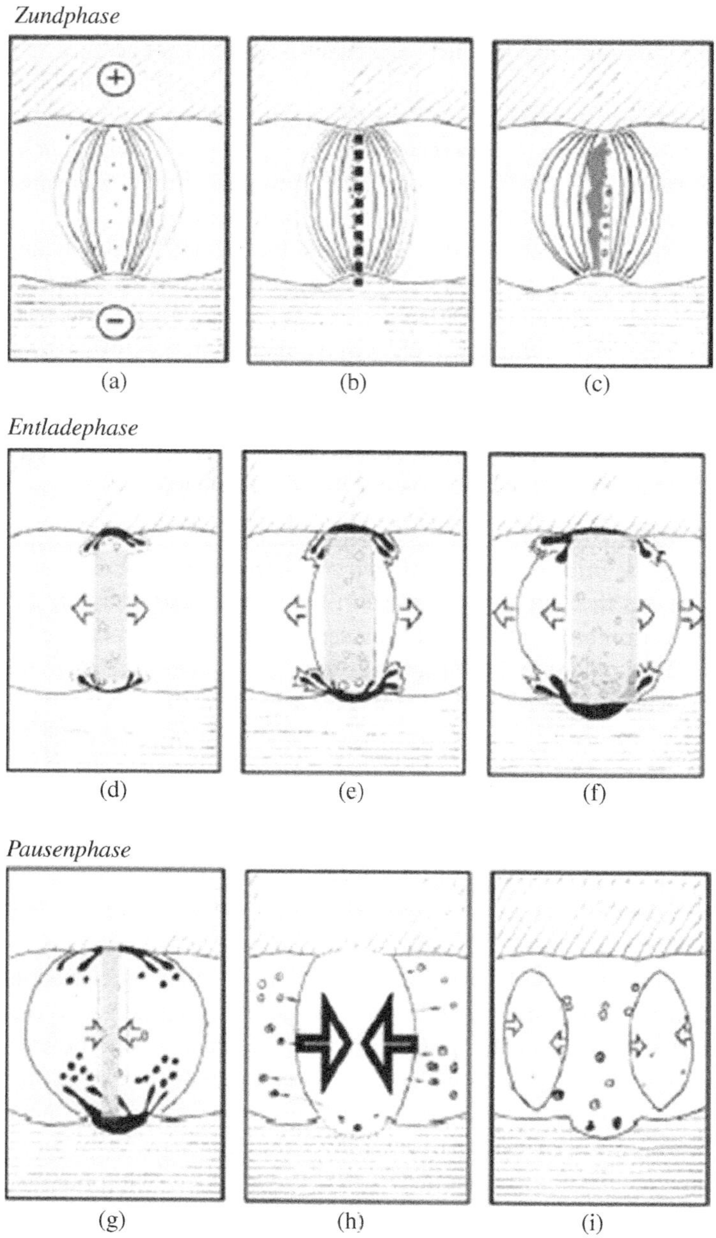

Figure 3.5 Sequential phases occurring during one pulse of EDM cycle [29]. *Source:* Reproduced with permission of Springer Nature.

materials. A series of voltage pulses of 40–400 V and of frequency about 5 kHz is usually applied. The naturally occurring microscopic containments suspended in the dielectric fluid migrate and concentrate at the strongest point and form a high-conductivity bridge with negatively charged particles (electrons) from the workpiece, across the gap (Figure 3.5a). When the voltage increases, temperature will be high enough to ionize a small portion of dielectric fluid and electrons, resulting in spark discharge channels between the two surfaces.

At approximately the midpoint of the electrical pulse, the voltage decreased but current increased. As a result, extremely high temperature is generated, which melts and vaporizes the workpiece at the points of spark contact. The high voltage induces a plasma channel between two conductive electrodes and generates a temperature of ~20 000 °C [29]. Further, the gaseous bubble of by-products of vaporization expands outward from the spark channel (Figure 3.5f).

The spark channel and vapor bubble will collapse when the pulse current is terminated. Further, the relatively cool and flowing dielectric cause explosive expulsion of molten metal from both electrode surfaces and as a result, small craters are formed on both surfaces (see Figure 3.5i).

Series of such discharges at very fine intervals occur throughout the surface, resulting in uniform erosion of material from both electrode tool and workpiece. The gap between the tool and workpiece is maintained with the help of a servo-system and eventually a cavity is generated as exactly reverse image of the tool electrode in the workpiece [30].

> This process of melting and evaporation from the workpiece makes EDM different from the conventional machining process, as chips are not mechanically generated.

Therefore, it can be said that the migration of elements from tool and workpiece under pulsed electric field and subsequent alloying at the contacting surface along with disintegrated products of dielectric fluid are responsible for material removal [31–33].

3.3.1.2 EDM Process Variables

The performance of EDM material is assessed often by three variables: material removal rate (MRR), tool wear ratio (TWR), and surface quality (SQ). The diameter of eroded hole and depth of cut are used to determine the volume of the removed workpiece material.

> The MRR is defined as the workpiece volume removed per unit time.

The MRR depends on physical properties such as melting point and latent heat and geometrical properties such as shape and size of workpiece and tool materials. Van Dyck correlated thermodynamics and heat transfer with erosion of materials during EDM.

> The TWR is defined as the ratio (expressed in percentage) of volume of electrode wear to volume of the workpiece.

The TWR is enhanced by the failure of the carbon to precipitate at certain difficult-to-reach regions of the electrode [34, 35]. A wear inhibitor black carbon layer deposition on the electrode surface using specific process parameters was suggested to reduce the electrode wear [36]. The orbiting of electrode relative to the workpiece was reported to reduce the TWR [37].

The SQ characteristics in terms of roughness of tool and workpiece are expressed through various parameters such as arithmetic roughness average (R_a), mean roughness depth (R_z), etc. Surface roughness decreases with rise in pulse frequency and with reduction in current. Roughness values of EDM surface lie in a wide range from 0.05 to 25 μm. Experimental results indicate the variation of centre line average (CLA) of the EDM surface with process parameters with the following relations:

$$\mathrm{CLA} = K'V_\mathrm{b}^{0.5}C^m \tag{3.1}$$

where K' is a constant and the value of exponent m varies between 0.31 and 0.36.

3.3.1.3 EDM Parameters

In order to enhance the performance of the EDM process, proper selection of operating parameters like polarity, electrode materials, dielectric, etc., is essential. The selection of EDM parameters depends on the specific application. Increase in current or voltage increases the MRR and the surface roughness of the workpiece.

The closer gap between the electrode and workpiece results in improved surface finish and lower MRR. The available energy can be shared by number of sparks with increasing spark frequency, resulting in reduction in surface roughness. As the pulse duration increases, the MRR and surface roughness decrease the electrode wear [38].

A decrease in MRR with an increase in pulse duration is attributed to unsteady machining conditions, usually observed for pulse greater than 20 μs. It has been reported that major part of the tool erosion occurs during initial few micro seconds of the pulse and therefore, use of pulse with low initial current can reduce tool wear [39]. The present research trends in EDM include ultrasonic vibration, dry EDM machining, EDM with powder additives, EDM in water and modeling technique, etc. [8, 40].

> The better EDM ability of a tool is determined by higher MRR and lower TWR. Also, lower the R_a or R_z values of EDM surface, better is the SQ of material.

A good combination of these attributes, using proper selection of different parameters, is always preferred in selecting a tool or workpiece material. The analysis of variance (ANOVA) techniques are used to measure the amount of deviation from the desired combination of performance measures. Ho and Newman [6] discussed the EDM research by classifying into four major groups: (i) optimizing the process variables, including electrical and nonelectrical parameters, and electrode design and manufacture; (ii) monitoring and control over the process using proper pulse/time domain, frequency, etc.; (iii) improving the performance measures like MRR, TWR, and SQ; and (iv) EDM applications and developments.

3.3.1.4 Surface Analysis

The microstructural details usually reveal three different layers [41–43] on the electrically discharged component surface.

> During the EDM process, molten material solidifies at a faster rate due to cooling effect of dielectric and adheres in the form of a thin layer to the machined surface, known as recast layer.

The recast layer is observed to be extremely hard (>HRC 65) and brittle with porous structure.

> In order to produce components with high fatigue resistance, recast layers containing poor physical properties have to be removed either mechanically or electrochemically.

The continuous and simultaneous heating, cooling, and diffusion processes result in a thin heat affected zone (HAZ), surrounding the recast layer. The presence of grain boundary cracks and thermal residual stresses are usually found in HAZ. The third layer, namely converted zone, lies below the HAZ, containing change in grain structure from the original structure.

Cusanelli et al. reported microstructural features of recast white layer and subsurface layers of HAZ of steel EDM surface layer, using SEM and TEM analyses [44].

The presence of columnar and dendritic structure in white recast layer was attributed to different cooling rates [43, 45]. The HAZ is composed of tempered martensite and is identified as the origin of microcracks [43, 46–48].

Proper selection and control over process parameters such as pulse duration and type of electrode can result in improving the SQ of the EDM surface [49]. The usage of different techniques such as surface alloying to produce composite electrode [50–53], powder additives in dielectric to reduce breakdown strength of dielectric and initiate the ignition [54–56], and ball burnishing to harden and modify the microstructure of EDM surface [57, 58] were reported.

3.3.1.5 EDM of Ceramic-Based Composites

In recent times, conductive ceramics and ceramic-based composites have been attractive tools due to their high temperature properties.

> Materials with appreciable electrical and thermal conductivity, together with a high melting point, are preferred to minimize electrode wear in EDM. The maximum resistivity of 100 Ω·cm is considered to be a limit for the material to be used as an electrode in EDM.

Ceramic materials have become significant for various industrial applications [16]. However, research and development of ceramic materials, especially suited for EDM, is still limited [48, 59, 60]. Sanchez et al. discussed about the feasibility of EDM of boron carbide (B_4C) and silicon infiltrated silicon carbide (SiSiC) [13]. Lee et al. studied ultrasonic-assisted EDM of engineering ceramics [61]. In order to use alumina as EDM tool, silicon nitride or zirconia ceramics as electrodes, an electrically conductive reinforcement such as TiB_2, TiC, or NbC is reported to be essential [62]. Mohri et al. used assisted electrode method for sparking in insulating ceramics [63]. In this direction, certain electroconductive ceramics, such as TiB_2, TiC, SiC, TiN [64, 65], and ceramic composites, such as SiC-TiB_2, Si_3N_4-TiN, Si_3N_4-TiCN, Al_2O_3-SiC_w-TiC, Sialon-TiN, etc., have been investigated in EDM [66–70]. Also, the influence of EDM on strength reliability of Si_3N_4-TiN is studied [67]. Li et al. found that increasing the densification would result in TiC-Cu electrode with improved EDM performance [64]. Mamalis et al. reported the feasibility of machining ceramic-metal composite steel plate coated with WC-Co using plasma spraying [71]. Mahdavinejad and Mahdavinejad investigated the instability of die-sinking EDM of WC-Co with copper electrode [72]. The instability of the process increases with increasing pulse duration because of enhanced melting and resolidification. Gadalla and Tsai reported the material removal mechanism of WC-Co composite as melting and evaporation of disintegrated Co, followed by the release of WC gains [73]. While investigating the EDM performance of several ceramic-based composite electrodes, Lauwers et al. also identified melting and evaporation as well as oxidation and dissolution as dominant material removal mechanisms [12] (Figures 3.6 and 3.7).

Chen et al. found that the MRR of $Ti_{50}Ni_{49.5}Cr_{0.5}$ and $Ti_{35.5}Ni_{49.5}Zr_{15}$ alloys in the EDM process can display a reverse relationship to the product of the alloy's melting temperature and thermal conductivity [74].

Another nonconventional machining process is laser beam machining, wherein a high-power laser source is focused on a flat material and the metal locally melts and vaporizes (see Figure 3.8). Depending on the traversing path of the laser beam, the specific area of the metal will be machined. Here, the laser power can be varied depending on the material. The laser absorption ability, i.e. the ability of

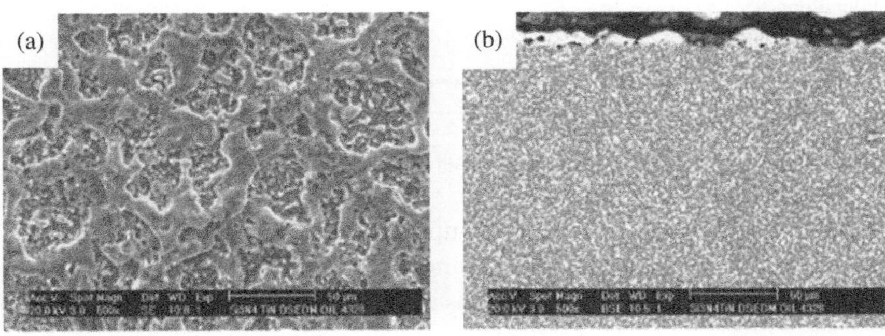

Figure 3.6 (a) Topography and (b) cross-section of Si_3N_4-TiN by die-sinking EDM in oil [73]. *Source:* Reproduced with permission of Taylor & Francis.

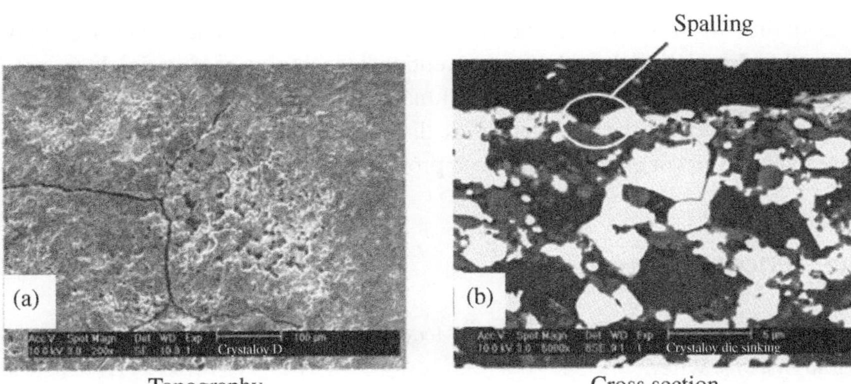

Figure 3.7 (a) Topography and (b) cross-section of Al_2O_3-SiC_w-TiC by die-sinking EDM in oil [73]. *Source:* Reproduced with permission of Taylor & Francis.

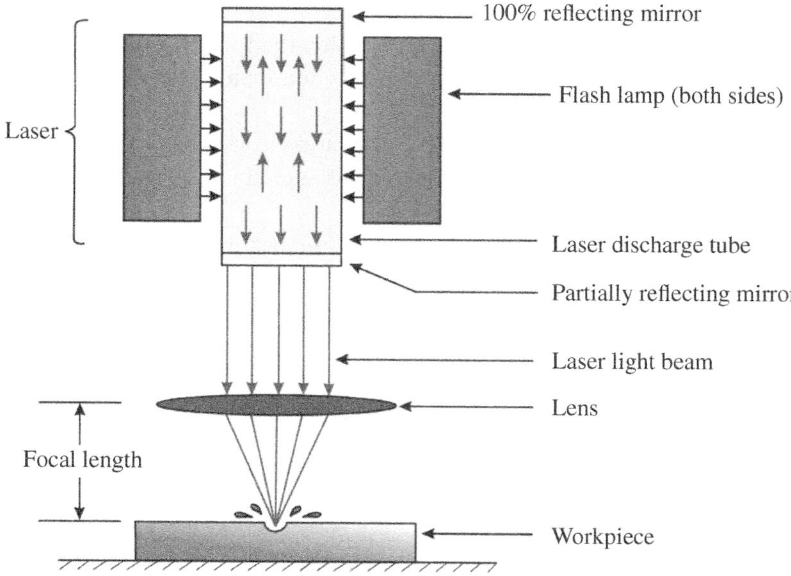

Figure 3.8 Schematic illustration of the laser beam machining process.

the metal to get spontaneously heated upon laser beam interaction, is the only property requirement in case of laser beam machining. For metals, the laser beam machining is to be conducted in inert gas atmosphere to prevent oxidation.

3.4 Closure

Considering the importance of the machining process to obtain tribological components, this chapter discusses the nonconventional machining of ceramics. In order to discuss the differences, the conventional machining approach is also described. In addition to describing the process physics, the materials' aspects are also presented in this chapter. A major emphasis has been provided to discuss the EDM of ceramics.

References

1 Groover, M.P. (2006). *Fundamental of Modern Manufacturing: Materials, Processes and Systems*. Wiley.
2 Ghosh, A. and Mallik, A.K. (2010). *Manufacturing Science*. East-West Press.
3 Narutaki, N., Yamane, Y., Tashima, S., and Kuroki, H. (1997). A new advanced ceramic for dry machining. *CIRP Ann.* 46: 43–48.

4 Jain, V.K. (2010). *Advanced Machining Processes*. Allied Publishers.
5 Chhowalla, M. and Unalan, H.E. (2005). Thin films of hard cubic Zr_3N_4 stabilised by stress. *Nat. Mater.* 4: 317–322.
6 Ho, K.H. and Newman, S.T. (2003). State of the art electrical discharge machining (EDM). *Int. J. Mach. Tools Manuf.* 43: 1287–1300.
7 Abbas, N.M., Solomon, D.G., and Bahari, M.F. (2007). A review on current research trends in electrical discharge machining (EDM). *Int. J. Mach. Tools Manuf.* 47: 1214–1228.
8 Schumacher, M.B. (2004). After 60 years of EDM the discharge process remains still disputed. *J. Mater. Process. Technol.* 149: 376–381.
9 Drabkina, S.J. (1951). Theory on the development of the discharge channel in electroerosion (in Russian). *J. Technol. Phys.* 4: 25.
10 Crowe, R.W. (1956). Formative timelags in the electric breakdown of liquid hydrocarbons. *J. Appl. Phys.* 27 (2): 156–160.
11 Petrofes, N.F. and Gadalla, A.M. (1988). Processing aspects of shaping advanced materials by electrical discharge machining. *Adv. Mater. Manuf. Process.* 3: 127–153.
12 Lauwers, B., Kruth, J.P., Liu, W. et al. (2004). Investigation of material removal mechanisms in EDM of ceramic composite materials. *J. Mater. Process. Technol.* 149: 347–352.
13 Sanchez, J.A., Cabanes, I., Lopez de Lacalle, L.N., and Lamikiz, A. (2001). Development of optimum electro discharge machining technology for advanced ceramics. *Int. J. Adv. Manuf. Technol.* 18 (12): 897–905.
14 Manoj Kumar, B.V., Ram Kumar, J., and Basu, B. (2007). Electro-discharge machining performance of TiCN based cermets. *Int. J. Refract. Met. Hard Mater.* 25: 293–299.
15 Kalpajian, S. and Schmid, S.R. (2003). Material removal processes: abrasive, chemical, electrical and high-energy beam. In: *Manufacturing Processes for Engineering Materials*, 4e, 541. Upper Saddle River, NJ: Prentice Hall.
16 Konig, W., Dauw, D.F., Levy, G., and Panten, U. (1988). EDM – future steps towards the machining of ceramics. *CIRP Ann.* 37: 623–631.
17 Arthur, A., Dickens, P.M., and Cobb, R.C. (1996). Using rapid prototyping to produce electrical discharge machining electrodes. *Rapid Prototyp. J.* 2 (1): 4–12.
18 Rajurkar, K.P. and Yu, Z.Y. (2000). 3D micro-EDM using CAD/CAM. *CIRP Ann.* 49 (1): 127–130.
19 Masuzawa, T., Fujino, M., and Kobayashi, K. (1985). Wire electro-discharge grinding for micro-machining. *CIRP Ann.* 34 (1): 431–434.
20 Masuzawa, T., Kuo, C.L., and Fujino, M. (1994). A combined electrical machining process for micro nozzle fabrication. *CIRP Ann.* 43 (1): 189–192.
21 Yu, Z.Y., Masuzawa, T., and Fujino, M. (1998). Micro-EDM for three dimensional cavities – development of uniform wear method. *CIRP Ann.* 47 (1): 169–172.

22 Yeo, S.H. and Yap, G.G. (2001). A feasibility study on the micro electro-discharge machining process for photomask fabrication. *Int. J. Adv. Manuf. Technol.* 18 (1): 7–11.

23 Puri, A.B. and Bhattacharyya, B. (2003). An analysis and optimization of the geometrical inaccuracy due to wire lag phenomenon in WEDM. *Int. J. Mach. Tools Manuf.* 43 (2): 151–159.

24 Kremer, D., Lhiaubet, C., and Moisan, A. (1991). A study of the effect of synchronizing ultrasonic vibrations with pulses in EDM. *CIRP Ann.* 40 (1): 211–214.

25 Jia, Z.X., Zhang, J.H., and Ai, X. (1997). Study on a new kind of combined machining technology of ultrasonic machining and electrical discharge machining. *Int. J. Mach. Tools Manuf.* 37 (2): 193–197.

26 Aspinwall, D.K., Wise, M.L.H., Stout, K.J. et al. (1992). Electrical discharge texturing. *Int. J. Mach. Tools Manuf.* 32 (1–2): 183–193.

27 Kozak, J., Rajurkar, K.P., and Wang, S.Z. (1994). Material removal in EDWM of PCD blanks. *J. Eng. Ind. (Trans. ASME)* 116 (3): 363–369.

28 Aspinwall, D.K., Dewes, R.C., Burrows, J.M., and Paul, M.A. (2001). Hybrid high speed machining (HSM): system design and experimental results for grinding/HSM and EDM/HSM. *CIRP Ann.* 50 (1): 145–148.

29 Norasetthekul, S., Eubank, P.T., Bradley, W.L. et al. (1999). Use of zirconium diboride–copper as an electrode in plasma applications. *J. Mater. Sci.* 34: 1261–1270.

30 Benedict, G.F. (1987). Electrical discharge machining (EDM). In: *Nontraditional Manufacturing Processes*, 211–213. New York and Basel: Marcel Dekker, Inc.

31 Gangadhar, A., Sunmugam, M.S., and Philip, P.K. (1992). Pulse train studies in EDM with controlled pulse relaxation. *Int. J. Mach. Tools Manuf.* 32 (5): 651–657.

32 Erden, A. (1983). Effect of materials on the mechanism of electric discharge machining (EDM). *J. Eng. Mater. Technol.* 105: 132–138.

33 Soni, J.S. and Chakraverti, G. (1996). Experimental investigation on migration of material during EDM of die steel (T215 Cr12). *J. Mater. Process. Technol.* 56 (1–4): 439–451.

34 Mohri, N., Suzuki, M., Furuya, M., and Saito, N. (1995). Electrode wear process in electrical discharge machining. *CIRP Ann.* 44 (1): 165–168.

35 Marafona, J. and Wykes, C. (2000). A new method of optimising material removal rate using EDM with copper–tungsten electrodes. *Int. J. Mach. Tools Manuf.* 40 (2): 153–164.

36 Marafona, J. (2007). Black layer characterisation and electrode wear ratio in electrical discharge machining (EDM). *J. Mater. Process. Technol.* 184 (1–3): 27–31.

37 Snoeys, R., Staelens, F., and Dekeyser, W. (1986). Current trends in non-conventional machining techniques. *CIRP Ann.* 35 (2): 467–480.

38 Jilani, S.T. and Pandey, P.C. (1984). Experimetnal investigations into the performance of water as dielectric in EDM. *Int. J. Mach. Tool Des. Res.* 24 (1): 31–43.

39 De Bruijn, H.E., Delft, T.H., and Pekelaring, A.J. (1978). Effect of a magnetic field on the gap cleaning in EDM. *CIRP Ann.* 27 (1): 93–95.

40 Jilani, S.T. and Pandey, P.C. (1982). Analysis and modelling of EDM parameters. *Precis. Eng.* 4 (4): 215–221.

41 Lee, L.C., Lim, L.C., Narayanan, V., and Venkatesh, V.C. (1988). Quantification of surface damage of tool steels after EDM. *Int. J. Mach. Tools Manuf.* 28 (4): 359–372.

42 Lee, L.C., Lim, L.C., Wong, Y.S., and Lu, H.H. (1990). Towards a better understanding of the surface features of electro-discharge machined tool steels. *J. Mater. Process. Technol.* 24: 513–523.

43 Lim, L.C., Lee, L.C., Wong, Y.S., and Lu, H.H. (1991). Solidification microstructure of electrodischarge machined tool steels. *Mater. Sci. Technol.* 7: 239–248.

44 Cusanelli, G., Wyser, A.H., Bobard, F. et al. (2004). Microstructure at submicron scale of the white layer produced by EDM technique. *J. Mater. Process. Technol.* 149: 289–295.

45 Ghanem, F., Braham, C., and Sidhom, H. (2003). Influence of steel type on electrical discharge machine surface integrity. *J. Mater. Process. Technol.* 142: 163–173.

46 Rebelo, J.C., Dias, A.M., Kremer, D., and Lebrun, J.L. (1998). Influence of EM pulse energy on the surface integrity of martensitic steel. *J. Mater. Process. Technol.* 84: 90–96.

47 Cabanillas, E.D., Desimoni, J., Punte, G., and Mercader, R.C. (2000). Formation of carbides by electro-discharge machining of alpha iron. *Mater. Sci. Eng. A* 276: 133–140.

48 Yadav, V., Jain, V.K., and Dixit, P.M. (2002). Thermal stresses due to electrical discharge machining. *Int. J. Mach. Tools Manuf.* 42 (8): 877–888.

49 Thomson, P.F. (1989). Surface damage in electrodischarge machining. *Mater. Sci. Technol.* 5: 1153–1157.

50 Tsai, H.C., Yan, B.H., and Huang, F.Y. (2003). EDM performance of Cr/Cu-based composite electrodes. *Int. J. Mach. Tools Manuf.* 43: 245–252.

51 Tsuekawa, Y., Okumiya, N., Mohri, N., and Takahashi, I. (1994). Surface modification of aluminium by electrical discharge alloying. *Mater. Sci. Eng. A.* 174: 193–198.

52 Gangadhar, A., Sunmugam, M.S., and Philip, P.K. (1991). Surface modification in electrodischarge processing with a powder compact tool electrode. *Wear* 143: 45–55.

53 Simao, J., Lee, H.G., Aspinwall, D.K. et al. (2003). Workpiece surface modification using electrical discharge machining. *Int. J. Mach. Tools Manuf.* 43 (2): 121–128.

54 Jeswani, M.L. (1981). Effect of the addition of graphite powder to kerosene used as the dielectric fluid in electrical discharge machining. *Wear* 70: 133–139.

55 Quan, Y.M. and Liu, Y.H. (1995). Powder-suspension dielectric fluid for EDM. *J. Mater. Process. Technol.* 52 (1): 44–54.

56 Schumacher, B.M. (1990). About the role of debris in the gap during electrical discharge machining. *CIRP Ann.* 39 (1): 197–199.

57 Yan, B.H., Wang, C.C., Chow, H.M., and Lin, Y.C. (2000). Feasibility study of rotary electrical discharge machining with ball burnishing for Al_2O_3/6061Al composite. *Int. J. Mach. Tools Manuf.* 40 (10): 1403–1421.

58 Hassan, A.M. (1997). The effect of ball- and roller-burnishing on the surface roughness and hardness of some non-ferrous metals. *J. Mater. Process. Technol.* 72 (3): 385–391.

59 Martin, C., Cales, B., Vivier, P., and Mathieu, P. (1989). Electrical discharge machinable ceramic composites. *Mater. Sci. Eng. A* 109: 351–356.

60 Tokura, H., Kondoh, I., and Yoshikswa, M. (1989). Ceramic material processing by electrical discharge in electrolyte. *J. Mater. Sci.* 24: 991–998.

61 Lee, T.C., Zhang, J.H., and Lau, W.S. (1998). Machining of engineering ceramics by ultrasonic vibration assisted EDM method. *J. Mater. Manuf. Process.* 13 (1): 133–146.

62 Put, S., Vleugels, J., Van der Biest, O. et al. (2001). Die sink electrodischarge machining of zirconia based composites. *Br. Ceram. Trans.* 100 (5): 207–213.

63 Mohri, N., Fukuzawa, Y., Tani, T. et al. (1996). Assisting electrode method for machining insulating ceramics. *CIRP Ann.* 45 (1): 201–204.

64 Li, L., Wong, Y.S., Fuh, J.Y.H., and Lu, L. (2001). EDM performance of TiC/copper-based sintered electrodes. *Mater. Des.* 22: 669–678.

65 Luis, C.J., Puertas, I., and Villa, G. (2005). Material removal rate and electrode wear study on the EDM of silicon carbide. *J. Mater. Process. Technol.* 164–165: 889–896.

66 Ramulu, M. (1988). EDM sinker cutting of a ceramic particulate composite, SiC–TiB_2. *Adv. Ceram. Mater.* 3 (4): 324–327.

67 Ramulu, M. and Taya, M. (1989). EDM machinability of SiC_w/Al composites. *J. Mater. Sci.* 24: 1103–1108.

68 Liu, C.-C. and Huang, J.-L. (2003). Effect of the electrical discharge machining on strength and reliability of TiN/Si_3N_4 composite. *Ceram. Int.* 29: 679–687.

69 Jiang, D., Vleugels, J., Van der Biest, O. et al. (2005). Electrically conductive and wear resistant Si_3N_4-based composites with $TiC_{0.5}N_{0.5}$ particles for electrical discharge machining. *Mater. Sci. Forum* 492–493: 27–32.

70 Salehi, S., Van der Biest, O., and Vleugels, J. (2006). Electrically conductive ZrO_2–TiN composites. *J. Eur. Ceram. Soc.* 26 (15): 3173–3179.

71 Mamalis, G., Vaxevanidis, N.M., and Pantelis, D.I. (1992). On the electrodischarge machining of ceramic plasma-sprayed steel plates. *CIRP Ann.* 41 (1): 235–238.

72 Mahdavinejad, R.A. and Mahdavinejad, A. (2005). ED machining of WC–Co. *J. Mater. Process. Technol.* 162–163: 637–643.

73 Gadalla, A.M. and Tsai, W. (1989). Machining of WC–Co composites. *Mater. Manuf. Process.* 4 (3): 411–423.

74 Chen, S.L., Hsieh, S.F., Lin, H.C., and Huang, J.S. (2007). EDM of TiNiCr and TiNiZr ternary SMA. *Mater. Sci. Eng. A* 445–446: 486–492.

4

Mechanical Properties of Ceramics

The wear resistance of materials largely depends on the mechanical properties or their differences among the mating solids. To this end, the discussion in this chapter starts with the fundamental definition of stress and strain. After sufficient description of the tensorial components of stress and strain, a brief discussion is made on the response of materials under three different modes of loading. The difference in mechanical behavior, in terms of the stress–strain response, of materials under different modes of loading is also explained. This is then followed by a brief discussion on the relevant theories to explain the brittle behavior of ceramics. The concept of the criticality in fracture conditions, leading to the definition of the fracture toughness, has been emphasized. Several experimental methodologies together with guidelines to conduct mechanical property evaluation are also summarized. Overall, this chapter establishes a theoretical framework to understand and experimentally determine the mechanical properties of the ceramics for tribological applications. Some contemporary discussion can be found elsewhere [1–3].

4.1 Defining Stress and Strain

To start with, it is important to define the concept of stress and strain. When a solid experiences or is subjected to mechanical load, the stress is developed. For a solid with complex shape, like the one shown in Figure 4.1a, one may be interested to know the stress at a plane and stress at a point *O*.

> The stress on a plane can be described as the ratio of the net force acting on the cross-sectional area of that plane. In simplistic terms, stress at a point can be defined as the limiting ratio of the resultant force acting on the hypothetical plane containing the specific point to an infinitesimal area around a specific point in a solid.

Friction and Wear of Ceramics: Principles and Case Studies, First Edition.
Bikramjit Basu, Mitjan Kalin, and B. Venkata Manoj Kumar.

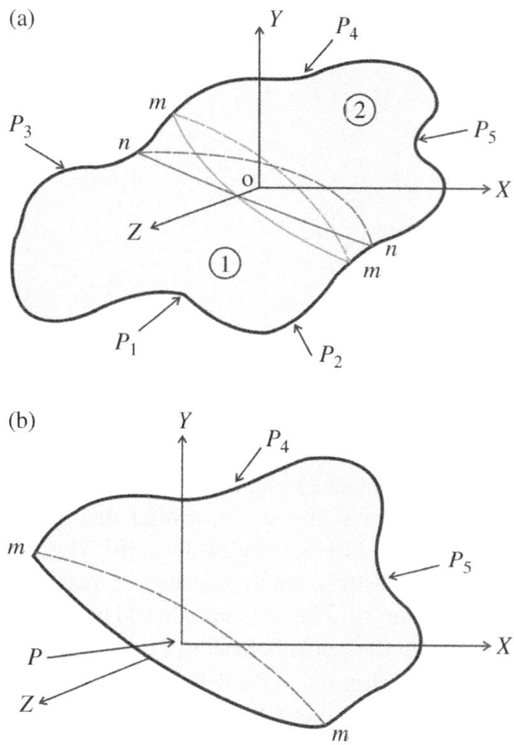

Figure 4.1 Conceptual definition of stress at a plane in a solid under equilibrium of external forces: (a) solid experiencing forces from various direction and (b) free-body diagram of a section with plane *mm*. *Source:* Adapted from Dieter [4].

The stress at a point depends on the orientation and cross-sectional area of the hypothetical plane containing the point of interest, with respect to the net force acting on the plane. This needs to be determined while constructing the free-body diagram of the part of the solid containing that specific point. For example, one can consider two hypothetical planes "mm and nn" around the point *O* in Figure 4.1a. In defining stress, one has to first determine net force from the equilibrium of the free-body diagram of the solid under the action of the external forces. In reference to Figure 4.1a, the area of "mm" or "nn" plane and its orientation is different with respect to the net force.

One such example of defining stress at point *O* with hypothetical plane (mm) is shown in Figure 4.1b. First, one has to consider different forces acting on the free-body diagram and then to determine the net force, under equilibrium of the external forces acting on the solid.

Mathematically, it can be expressed as

$$\sigma = \lim_{A \to 0} \frac{P}{A} \tag{4.1}$$

where ΔP is the resultant of all load components (P_1, P_2, P_3) that the solid experiences under equilibrium (see Figure 4.1). The stress at a point O can therefore be defined as a measure of net force on a hypothetical plane around the point of interest, when the area of the plane is assumed to be negligible or zero.

Next, we define three different types of stresses (see Figure 4.2). Two generic variants of stress are normal and shear stress. Normal stresses can be either tensile or compressive in nature. The tensile force can be defined as the force of equal magnitude being applied in opposite direction and under such loading, a cylinder will extend in length with a decrease in diameter (Figure 4.2a). The reverse happens in case of compressive force, which is applied with equal magnitude in the same direction (see Figure 4.2b). Under compression, a cylinder will experience an increase in diameter, but will shorten in length. The shear loading is however different with equal forces being applied at the two opposite faces of a cube. While both tensile and compressive stresses will cause volume change, the shear stress leads to a shape change or angular distortion, measured by angle θ (Figure 4.2c). Following the definition of stress, both compressive and tensile stress can be defined as the ratio of force (F) to the initial cross-sectional area (A_0) of the cylinder (F/A_0). Similarly, shear stress can be quantified as F/A_0, where A_0 is the area of one of the parallel faces of a cube along which the shear force is being applied.

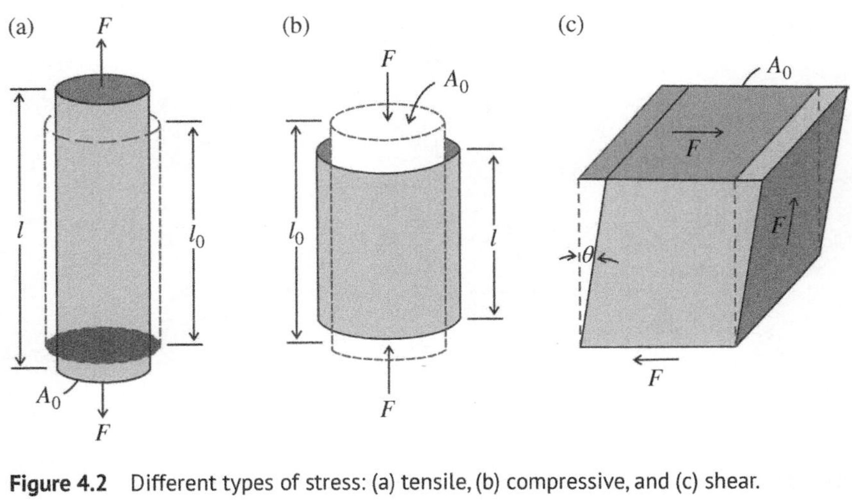

Figure 4.2 Different types of stress: (a) tensile, (b) compressive, and (c) shear.

Note that tensile/compressive force is also applied perpendicular to the marked plane of a cylinder with area, A_0 (see Figure 4.2).

> It should be clear that the stress is to be represented not only by magnitude and direction but also by its orientation with respect to a plane.

This brings us to define stress as a tensorial quantity. A scalar needs only magnitude, while a vector is to be represented by both magnitude and direction. In contrast, a tensor requires magnitude, direction, and a reference plane. While defining stress, one therefore needs to know the magnitude, direction, and the plane across which it acts. Since stress is a tensor of rank 2, it needs a total of $3^2 = 9$ terms. With reference to Figure 4.3, the stress state at a point (O) can further be described by constructing a cube around point O and then one has to consider various stress components along each face of the cube.

> Stress is not a vector, but a tensor of rank 2.

In defining various stress components of stress tensor (σ_{ij}), one has to remember that the first subscript refers to the reference plane and the second subscript denotes the direction. The normal stress component has double identical subscript, indicating that force/stress is acting on the planes along the plane normal.

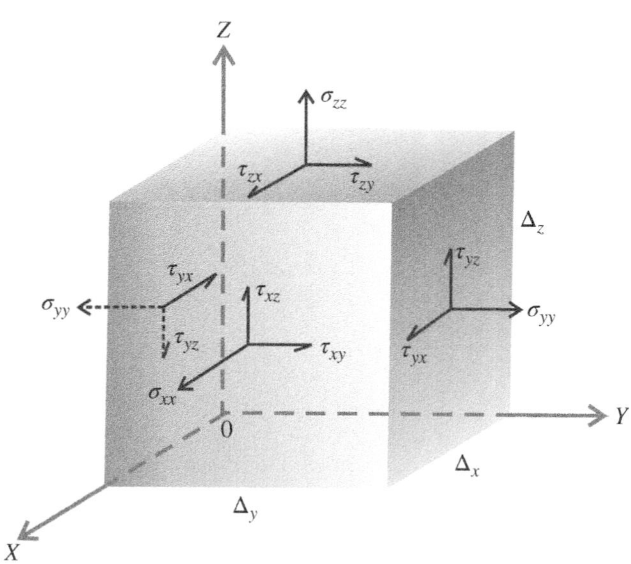

Figure 4.3 Stress acting on an elemental unit cube.

In contrast, the shear stress has different subscripts. In reference to the volume elements shown in Figure 4.3, the three stress components acting on the X-plane (the plane perpendicular to X-axis) are σ_{xx}, τ_{xy}, and τ_{xz}. Similarly, specific stress components for Y-plane or Z-plane are also shown in Figure 4.3. If one considers the areas of the cube faces small enough (change in stress over faces is therefore negligible), then the moments created by the shear forces would be equal. Thus, one would be able to realize that $\tau_{xy} = \tau_{yx}$; $\tau_{yz} = \tau_{zy}$; $\tau_{xz} = \tau_{zx}$. Therefore, the stress tensor, σ_{ij}, will have six independent stress components. The stress tensor, σ_{ij}, can therefore be defined as

$$\sigma_{ij} = \begin{pmatrix} \sigma_{xx} & \tau_{xy} & \tau_{xz} \\ \tau_{yx} & \sigma_{yy} & \tau_{yz} \\ \tau_{zx} & \tau_{zy} & \sigma_{zz} \end{pmatrix} \tag{4.2}$$

Similar to stress, the strain at a point in a solid continuum will now be explained. In a simplistic sense, strain denotes the ratio of change in the dimension to the original dimension, when a solid experiences a stress state. With reference to Figure 4.4, the strain can be described in terms of displacement of a point P in a solid continuum. While P has a coordinate (x, y, z), the displaced point P' has a different coordinate $(x+u, y+v, z+w)$. Therefore, the displacement component

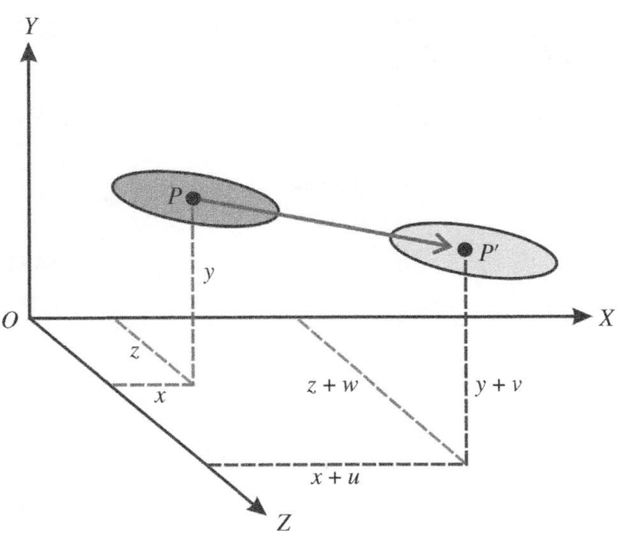

Figure 4.4 Defining strain due to the displacement of a point P to P' in a solid continuum.

along X, Y, and Z direction is u, v, and w, respectively. In accordance with the convention, the normal strain components are defined as

$$e_{xx} = \frac{\partial u}{\partial x},\ e_{yy} = \frac{\partial v}{\partial y},\ \text{and}\ e_{zz} = \frac{\partial w}{\partial z} \tag{4.3}$$

Similarly, the shear strain components are defined as the rate of change of displacement component along X-axis to the y-component position and likewise as follows:

$$e_{xy} = \frac{\partial u}{\partial y},\ e_{yx} = \frac{\partial v}{\partial x} \tag{4.4}$$

$$e_{yz} = \frac{\partial v}{\partial z},\ e_{zy} = \frac{\partial w}{\partial y} \tag{4.5}$$

$$e_{xz} = \frac{\partial u}{\partial z},\ e_{zx} = \frac{\partial w}{\partial x} \tag{4.6}$$

Based on the above definition of strain components and similar to stress tensor, the entire displacement tensor can be defined as follows:

$$e_{ij} = \begin{pmatrix} e_{xx} & e_{xy} & e_{xz} \\ e_{yx} & e_{yy} & e_{yz} \\ e_{zx} & e_{zy} & e_{zz} \end{pmatrix} = \begin{pmatrix} \frac{\partial u}{\partial x} & \frac{\partial u}{\partial y} & \frac{\partial u}{\partial z} \\ \frac{\partial v}{\partial x} & \frac{\partial v}{\partial y} & \frac{\partial v}{\partial z} \\ \frac{\partial w}{\partial x} & \frac{\partial w}{\partial y} & \frac{\partial w}{\partial z} \end{pmatrix} \tag{4.7}$$

The displacement tensor can be further described as

$$e_{ij} = \frac{1}{2}\left(e_{ij} + e_{ji}\right) + \frac{1}{2}\left(e_{ij} - e_{ji}\right) \tag{4.8}$$

or

$$e_{ij} = \varepsilon_{ij} + \omega_{ij}$$

In the above, strain tensor is

$$\varepsilon_{ij} = \frac{1}{2}\left(\frac{\partial u_i}{dx_j} + \frac{\partial u_j}{\partial x_i}\right) \tag{4.9}$$

and rotation tensor is

$$\omega_{ij} = \frac{1}{2}\left(\frac{\partial u_i}{\partial x_j} - \frac{\partial u_j}{\partial x_i}\right) \tag{4.10}$$

In terms of the complete description, the strain (ε_{ij}) and rotation (ω_{ij}) tensor can be further expressed as

$$\varepsilon_{ij} = \begin{pmatrix} \varepsilon_{xx} & \varepsilon_{xy} & \varepsilon_{xz} \\ \varepsilon_{yx} & \varepsilon_{yy} & \varepsilon_{yz} \\ \varepsilon_{zx} & \varepsilon_{zy} & \varepsilon_{zz} \end{pmatrix} = \begin{pmatrix} \frac{\partial u}{\partial x} & \frac{1}{2}\left(\frac{\partial u}{\partial y} + \frac{\partial v}{\partial x}\right) & \frac{1}{2}\left(\frac{\partial u}{\partial z} + \frac{\partial w}{\partial x}\right) \\ \frac{1}{2}\left(\frac{\partial u}{\partial y} + \frac{\partial v}{\partial x}\right) & \frac{\partial v}{\partial y} & \frac{1}{2}\left(\frac{\partial v}{\partial z} + \frac{\partial w}{\partial y}\right) \\ \frac{1}{2}\left(\frac{\partial u}{\partial z} + \frac{\partial w}{\partial x}\right) & \frac{1}{2}\left(\frac{\partial v}{\partial z} + \frac{\partial w}{\partial y}\right) & \frac{\partial w}{\partial z} \end{pmatrix} \tag{4.11}$$

$$\omega_{ij} = \begin{pmatrix} \omega_{xx} & \omega_{xy} & \omega_{xz} \\ \omega_{yx} & \omega_{yy} & \omega_{yz} \\ \omega_{zx} & \omega_{zy} & \omega_{zz} \end{pmatrix} = \begin{pmatrix} 0 & \frac{1}{2}\left(\frac{\partial u}{\partial y} - \frac{\partial v}{\partial x}\right) & \frac{1}{2}\left(\frac{\partial u}{\partial z} - \frac{\partial w}{\partial x}\right) \\ \frac{1}{2}\left(\frac{\partial v}{\partial x} - \frac{\partial u}{\partial y}\right) & 0 & \frac{1}{2}\left(\frac{\partial v}{\partial z} - \frac{\partial w}{\partial y}\right) \\ \frac{1}{2}\left(\frac{\partial w}{\partial x} - \frac{\partial u}{\partial z}\right) & \frac{1}{2}\left(\frac{\partial w}{\partial y} - \frac{\partial v}{\partial z}\right) & 0 \end{pmatrix} \tag{4.12}$$

Following the above analysis, the shear strain, which determines the shape change involved in the plastic deformation of metals, is defined as

$$\begin{aligned} \gamma_{xy} &= \frac{\partial u}{\partial y} + \frac{\partial v}{\partial x} \\ \gamma_{xz} &= \frac{\partial w}{\partial x} + \frac{\partial u}{\partial z} \\ \gamma_{yz} &= \frac{\partial w}{\partial y} + \frac{\partial v}{\partial z} \end{aligned} \tag{4.13}$$

Similar to the definition of stress components, any strain component with identical subscripts defines normal strain, whereas shear strain is represented by unequal subscripts. Like stress tensor, the shear components have the following relationship:

$$\gamma_{xy} = \gamma_{yx}, \gamma_{yz} = \gamma_{zy}, \ \gamma_{xz} = \gamma_{zx} \tag{4.14}$$

Therefore, the final description of strain tensor is

$$\varepsilon_{ij} = \begin{pmatrix} \varepsilon_{xx} & \frac{1}{2}\gamma_{xy} & \frac{1}{2}\gamma_{xz} \\ \frac{1}{2}\gamma_{xy} & \varepsilon_{yy} & \frac{1}{2}\gamma_{yz} \\ \frac{1}{2}\gamma_{xz} & \frac{1}{2}\gamma_{yz} & \varepsilon_{zz} \end{pmatrix} \tag{4.15}$$

The above analysis indicates that strain tensor has six dependent components. Further, total volume change is zero during plastic or permanent deformation of a metal and this can be expressed as

$$V = \varepsilon_{xx} + \varepsilon_{yy} + \varepsilon_{zz} = 0 \tag{4.16}$$

Summarizing, the strain tensor has five independent components with two normal and three shear components.

4.2 Comparison of Tensile Behavior

In this section, the difference in the mechanical response of metals/ceramics/polymers under tensile loading will be discussed. Prior to this, it is instructive to recall the nature of bonding in these materials. The metals are characterized by metallic bonds with atoms floating in a cloud of electrons. While such bonding explains good electrical and thermal transport properties, this specific bonding is also isotropic in nature, implying the mechanical properties to be isotropic as well. The same is not true for ceramics, which predominantly either have ionic (oxides) or covalent (non-oxides) bonding. The ionic bonding requires electroneutrality conditions to be maintained during deformation. The covalent bonding, being directional in nature, would impart anisotropic mechanical properties. Polymers, on the other hand, have mixed bonding. Along the backbone chains in a polymer, the "mer" units are covalently bonded, while two neighboring chains are weakly bonded by either *H*-bonds or van der Waals bonds.

In view of the differences in bonding characteristics, the three different classes of materials, viz. metals, ceramics, and polymers, respond in a distinguishable way, when mechanically loaded.

The differences in the tensile response, in terms of engineering stress versus strain plot, are shown in Figure 4.5. The fundamental difference is that while the

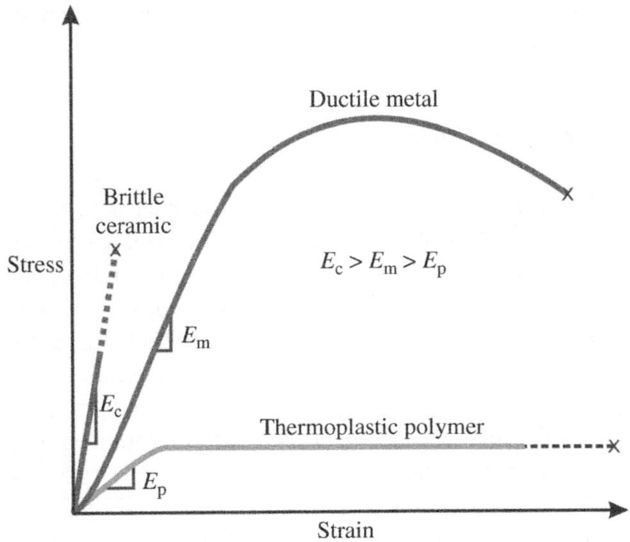

Figure 4.5 Schematic illustration comparing the stress–strain response of three primary classes of material under uniaxial tension.

linear elastic response is exhibited by ceramics, the metals experience linear elastic to nonlinear plastic behavior, under tension. The property of major relevance to wear resistance property is the elastic modulus, which is determined by the slope of the stress–strain response in the linear region. Another property that is also important is the strength, particularly under compression, for metals. The strength property limits the load that can be applied at a tribological contact. While a ceramic material mostly exhibits linear response up to the point of fracture, a metal experiences initial linear response followed by nonlinear deformation behavior up to the point of fracture. The slope of the linear part of such response determines the Young's modulus or elastic modulus of the materials. In contrast to metals and ceramics, a thermoplastic polymer experiences a constant deformation at a lower stress level, which is preceded by linear stress strain response. The slope of initial linear response, a measure of elastic modulus, is however markedly different in metals/ceramics/polymers.

> Although tensile strength is low for ceramics, the compressive strength is typically eight times larger than the tensile strength. This has the greatest significance for wider acceptability of ceramics in many demanding tribological applications.

Also, the elastic modulus is the highest for ceramics and this also has the great relevance toward Hertzian contact damage resistance.

4.3 Brittle Fracture of Ceramics

It has been mentioned earlier in this chapter that ceramics, under tension, show linear elastic response up to fracture. The absence of nonlinear deformation is the signature of brittleness. In a simplistic sense, the ceramic microstructure, unless otherwise designed with toughening elements, cannot resist the crack propagations, i.e. the pre-existing cracks or cracks developed during loading will not find any resistance in their path. This leads to catastrophic or ungraceful failure of ceramics.

> The brittleness of ceramics is a major concern for ceramics in engineering applications including the field of tribology.

Let us now revisit the atomistic description of the fracture process. As shown in Figure 4.6, the interatomic bonds can be described as "spring-like" structures. The elastic response of the stretching of such structures can be described as $F = -Kx$, where F is external load, K is the spring constant, and x is the spring displacement. As shown in Figure 4.6b, the bonds stretch from the equilibrium spacing of a_0 to a ($a > a_0$) and as explained below, the stress concentration at the crack tip will

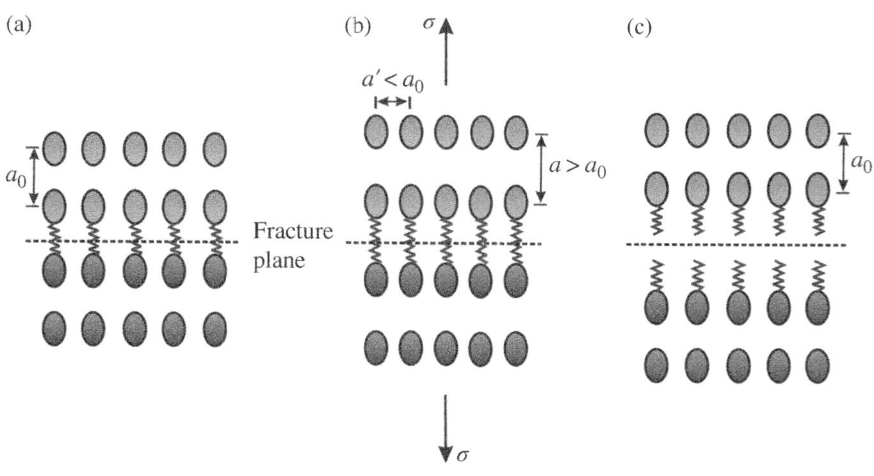

Figure 4.6 Atomistic description of fracture of a solid: (a) undeformed state, (b) deformed under tensile load, and (c) fracture due to the breakage of interatomic bonds. The interatomic bonding accessing the fracture plane is represented by spring-like behavior. *Source:* Adapted from Lawn [5].

subsequently rupture the spring (bond) (Figure 4.6c), leading to continued crack propagation. This entire scenario of physical events takes place ahead of any advancing crack in a ceramic microstructure under external tensile loading.

The microscopic view of the crack size distribution, their orientation w.r.t. loading axis, and their influence on fracture progression/failure under tensile and compression loading is shown in Figure 4.7. These cracks often vary in size and their orientation w.r.t. tensile axis. The largest crack with most favorable orientation (close to 90° w.r.t. tensile axis) would propagate in a direction perpendicular to tensile axis (Figure 4.7a and b). Once such crack grows to criticality, it leads to failure. The scenario is strikingly different in case of compression loading. Under compression, the cracks, oriented closer to the axis of compression, are more likely to grow (see Figure 4.7c and d).

If the pre-existing flaws are oriented at an angle to compression axis, then it will first grow in a manner to orient itself to as close to that of compression axis before

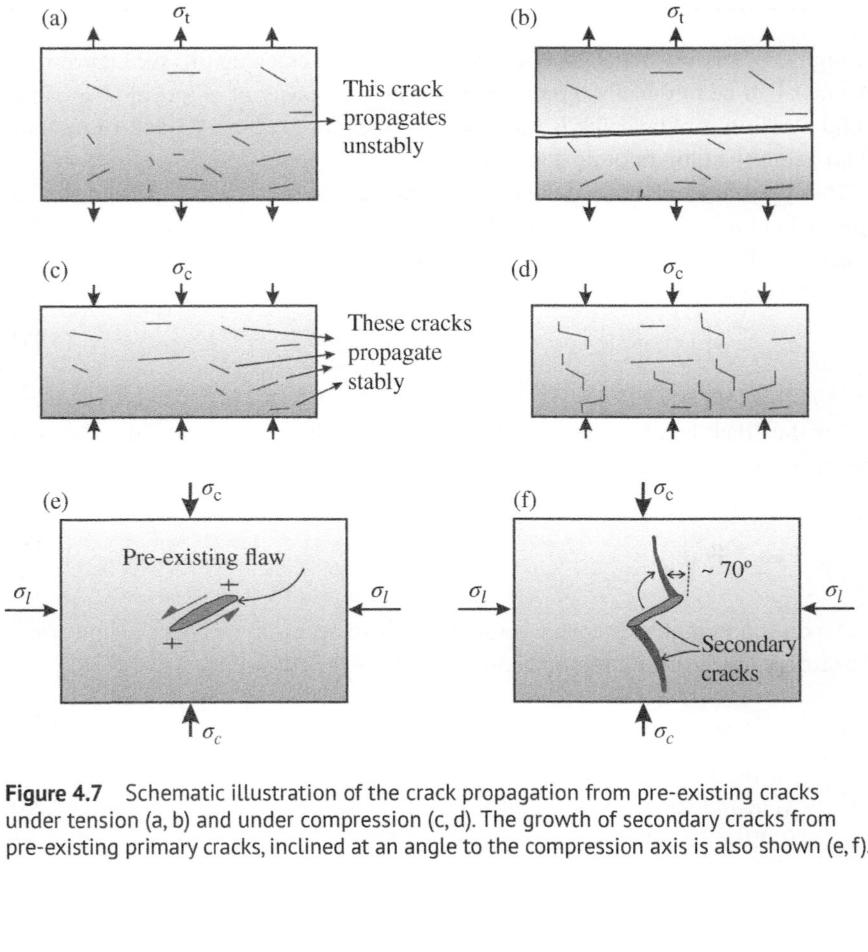

Figure 4.7 Schematic illustration of the crack propagation from pre-existing cracks under tension (a, b) and under compression (c, d). The growth of secondary cracks from pre-existing primary cracks, inclined at an angle to the compression axis is also shown (e, f).

further growth. This is shown in the formation of secondary cracks (see Figure 4.7f). The difficulty in crack propagation under compression also explains why ceramics have exceptional compressive strength, i.e. eight times that under tension. It may be worthwhile to mention that for metals, the tensile strength and compressive strength are nearly equal. For example, the tensile strength and compressive strength of a stainless steel is 500–550 MPa. In contrast, the tensile strength of Al_2O_3 ceramic can be 150–200 MPa, but the compressive strength of Al_2O_3 can be as high as 1200–1600 MPa. Therefore, ceramics can outperform metals in terms of compressive strength and are better appreciated in the field of tribology.

In the following texts, the brittle fracture theories of ceramics/glass materials will be briefly reviewed, in particular reference to quantitatively explain the criticality of fracture.

> Brittle fracture can be best described by interatomic bond breakage/rupture under external stress application.

Inglis's theory was based on stress concentration at crack tip [6] (see Figure 4.8). A crack can be physically represented as a hole or cavity of elliptical shape in a solid continuum. The crack is therefore assumed as a region of bond rupture or discontinuity in interatomic bonding.

The stress concentration at the crack tip can be expressed by the following equation, where σ is the externally applied stress, and c, ρ are, respectively, the edge crack length and crack tip radius of curvature:

$$\sigma_{\max} = \sigma\left(1 + 2\sqrt{\frac{c}{\rho}}\right) \tag{4.17}$$

For the crack length of greater than crack tip radius (i.e. $c >>> \rho$), the maximum stress at crack tip edge can be described by

$$\sigma_{\max} = 2\sigma\sqrt{\frac{c}{\rho}} \tag{4.18}$$

According to Inglis's theory, fracture is likely to occur in a scenario when crack tip stress can trigger interatomic bond rupture at the crack tip.

$$\sigma_{\max} = \sigma_{th} \tag{4.19}$$

$$2\sigma\sqrt{\frac{c}{\rho}} = \left(\frac{E\gamma}{a_0}\right)^{1/2} \tag{4.20}$$

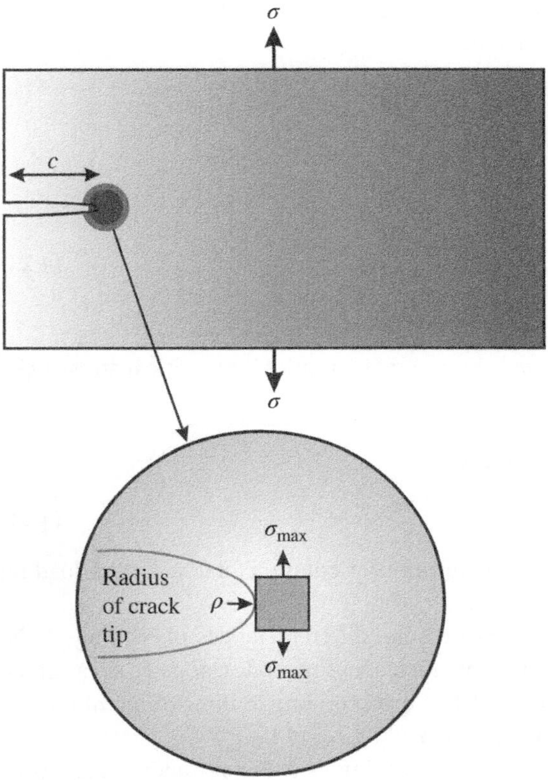

Figure 4.8 Schematic illustration showing the stress concentration at the crack tip edge.

$$\sigma_c = \left(\frac{E\gamma\rho}{4a_o c}\right)^{1/2} \tag{4.21}$$

Orowan proposed that the crack tip curvature radius can be approximated as interatomic distance (a_o) [7], i.e. $\rho = a_o$. This allows one to arrive at the expression for the critical fracture stress,

$$\sigma_f = \left(\frac{E\gamma}{4c}\right)^{1/2} \tag{4.22}$$

It should be clear now that the larger the crack length (c) or sharper the crack with reduced radius of curvature (ρ), the lower will be the fracture strength.

A different expression for fracture strength of ceramics can be obtained following Griffith's theory, based on the premise that the total energy of a ceramic plate with central through-thickness crack of length $2c$ is minimized at equilibrium and this leads to the following expression:

$$\sigma_c = \sqrt{\frac{2\gamma E}{\pi c^*}}$$

or

$$c^* = \frac{2\gamma E}{\pi \sigma_c^2} \tag{4.23}$$

> The fracture of ceramics depends on a combination of both, critical stress (σ_c) and critical crack length (c^*).

The stress intensity factor is defined as

$$K = Y\sigma\sqrt{\pi c} \tag{4.24}$$

where Y is a factor dependent on location/orientation of crack and loading condition.

In the classical fracture mechanics theory, different modes of loading of the crack faces are recognized as tensile or crack opening mode (mode I), shear mode (mode II), and tearing mode (mode III). Corresponding to three different modes, the stress intensity factor can be defined as K_I, K_{II}, and K_{III}.

At the critical condition, the crack tip stress intensity factor under mode I loading is defined as

$$K_{IC} = Y\sigma_c\sqrt{\pi c^*} \tag{4.25}$$

where c^* is the half of the through-thickness critical crack length

The value of the geometric factor Y varies around 1–1.1. The K_{IC} defined in Eq. 4.25 is widely known as a parameter, which describes fracture toughness. From the above discussion, it should be clear that K_{IC} is a material property, i.e. each ceramic that has a unique K_{IC} value. In various tribological applications, the external force-induced K_I value should not exceed K_{IC} of the biomaterial and therefore, K_{IC} is one of the important parameters to assess the structural reliability.

4.4 Cracking in Brittle Materials

In view of the paramount importance of cracking on wear of ceramics, it is important to have some knowledge on the types of cracking and their characteristics. This is best studied by using two different types of indenters – one is a sharp

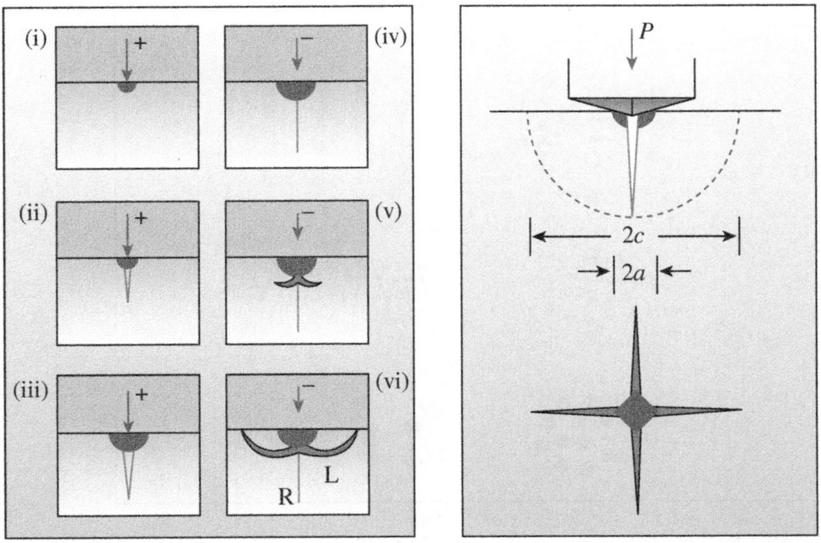

Figure 4.9 Schematic showing the development of radial-median (marked as "R") and lateral crack (marked as "L") during loading and unloading of a sharp indenter against a flat surface of a brittle material, respectively (left), and the geometrical parameter illustrating the dimension of highly deformed zone and crack length (right).

indenter, which leads to concentrated loading, and the other is a spherical indenter, which leads to distributed loading. Also, early studies on indentation cracking were conducted using soda-lime glass as a model system as its transparent nature allows one to capture the initiation and growth of cracking under sharp or spherical indenters. As shown in Figure 4.9, the development of radial-medium cracks can be clearly observed. During initial loading at a tribological contact, the radial-medium cracks grow with time during loading. When unloaded, a lateral crack pattern develops along the transverse direction. During abrasion by a harder solid sliding on a relatively softer surface, the mechanical interaction between two solids can be considered as equivalent to multiple overlapping indentation process under concentrated loading. Therefore, multiple lateral cracks generated from overlapping indented regions will intersect each other as before meeting free surfaces. Such a physical phenomenon can cause material removal, that is, wear. One of the ways to enhance resistance against fracture or damage is to increase fracture toughness. Various mechanisms of fracture resistance are shown in Figure 4.10 and a specific case of fiber toughening is illustrated in Figure 4.11.

Based on fundamental fracture mechanics theory [5], Lawn observed the following relationship:

$$P \propto c_1^{3/2} \quad \text{or} \quad P \propto c_c^{3/2} \tag{4.26}$$

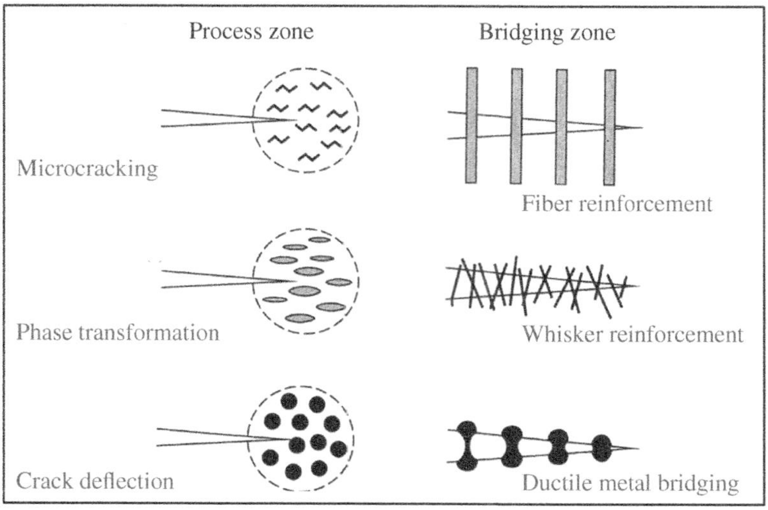

Figure 4.10 Summary of various toughening mechanisms in ceramic-based materials.

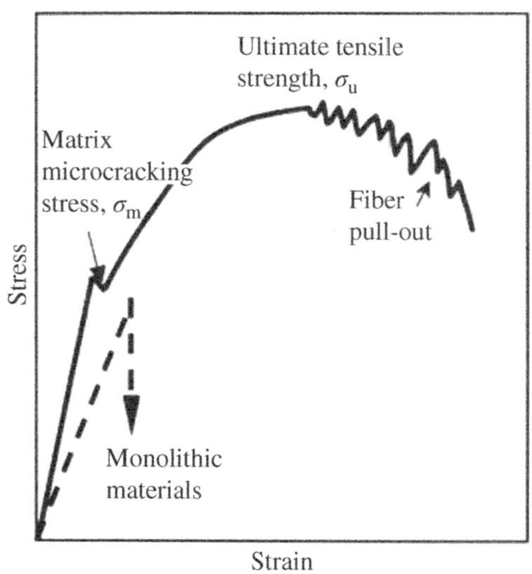

Figure 4.11 Schematic illustration of tensile stress–strain response of ceramic fiber reinforced ceramic composites vis-à-vis response of monolithic ceramic.

where P is the indentation load and c_l or c_c is the lateral or cone crack length, respectively. It clearly indicates that the severity of mechanical wear damage will increase with increase in indent load.

4.5 Experimental Assessment of Mechanical Properties

In this section, a brief discussion on the principles of measurement of mechanical properties of ceramics is presented. In view of inherent brittleness, the preparation of standard tensile samples from ceramics is extremely difficult. Therefore, different test methods are adopted to measure the basic mechanical properties of ceramics, as discussed below.

4.5.1 Hardness

Conventionally, the hardness is defined as the resistance to permanent (plastic) deformation.

This property is measured by taking indents on flat polished surfaces (Figure 4.12). Mostly for ceramic materials, hardness is measured using Vickers indentation technique. The following expression is used to assign the hardness value of a material after the indentation,

$$H_v = 1.854\left(\frac{P}{d^2}\right) \tag{4.27}$$

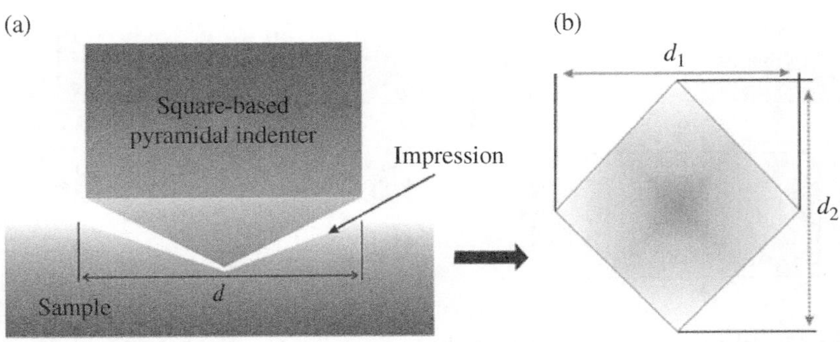

Figure 4.12 Schematic showing the (a) Vickers indentation and (b) measurement of impression diagonals for the hardness measurement.

where H_V is Vickers hardness (also called Vickers Pyramid Number), P is the applied load, and d is the average length of two diagonals of the indentation (e.g. average of d_1 and d_2 in Figure 4.12). It is recommended to measure the indent sizes (length scale of the order "μm") using SEM in order to avoid considerable errors in hardness measurement.

4.5.2 Compressive Strength

It is widely recognized that ceramics have much better compression property than metals and this is an advantage in tribological applications. In Figure 4.13a, the test configuration followed to evaluate compressive deformation properties is

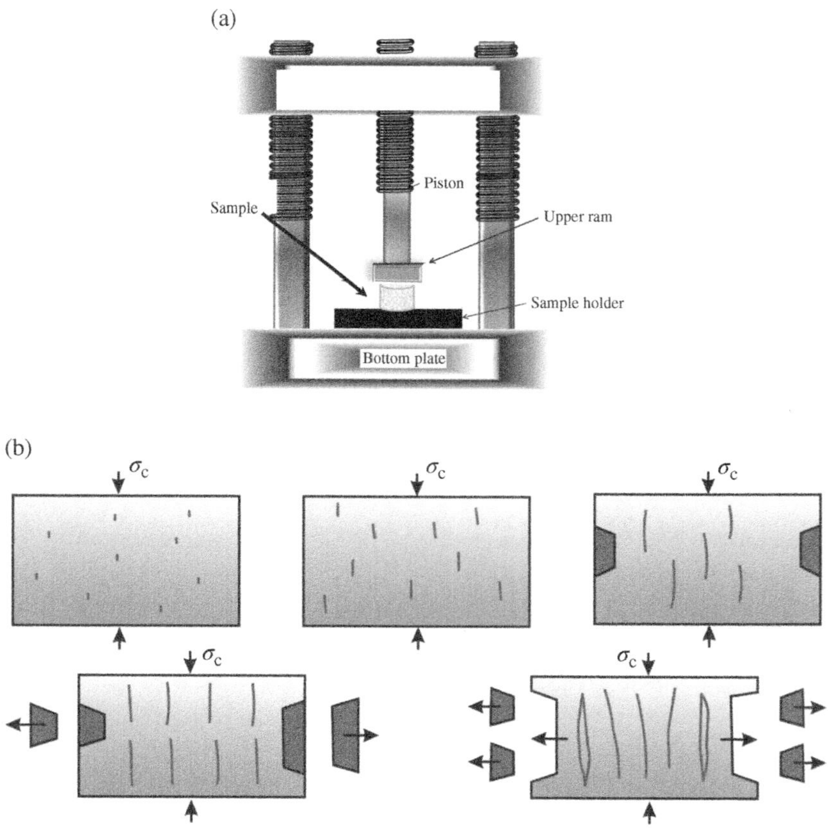

Figure 4.13 (a) Typical test setup showing the compressive strength measurement. The compression failure mechanisms are also shown schematically in terms of vertical crack propagation in (b).

schematically illustrated. Typically, cylindrical samples with an aspect ratio of 1.0 or above are used.

Compressive strength can be measured using a universal testing machine (UTM). The samples are placed in between two parallel plates of the machine and force is applied on the appropriately aligned samples with a constant cross-head speed (typically around 0.05 mm/s). During the entire compression test, the load–displacement response can be recorded using a computer attached to UTM. In reference to Figure 4.13b, the compressive strength of the samples can be calculated from the fracture load and dimension of the samples using the simple formula,

$$\sigma_{cs} = \frac{P}{A} \tag{4.28}$$

where P is the maximum load (fracture load) and A is the cross-sectional area.

4.5.3 Flexural Strength

Considering the fact that tensile sample (dog-bone geometry) preparation for ceramics is a challenging task, the strength of the ceramics is measured under flexure mode either by three-point or by four-point bend configuration. Either a concentrated load is applied in three-point configuration or a distributed load is applied at two different places in four-point configuration (see Figure 4.14). The flexural strength thereafter is calculated on the basis of measured fracture load and dimension of the test sample. For three-point loading (Figure 4.14b), the fracture strength can be computed using the following expression:

$$\sigma_{\mathrm{f}} = \frac{3PL}{2bd^2} \tag{4.29}$$

where σ_{f} is the flexural strength of the material, P is the fracture load, L is the span length, b is the width of the sample, and d is the thickness of the specimen. In a similar line, the flexural strength for four-point bend configuration (Figure 4.14c) can be obtained from the expression:

$$\sigma_{\mathrm{f}} = \frac{3PL}{4bd^2} \tag{4.30}$$

In case of four-point loading, maximum tensile stress is experienced over a larger sample area as opposed to three-point flexural mode. Therefore, a lower and a better estimate of the strength is measured using four-point flexural tests.

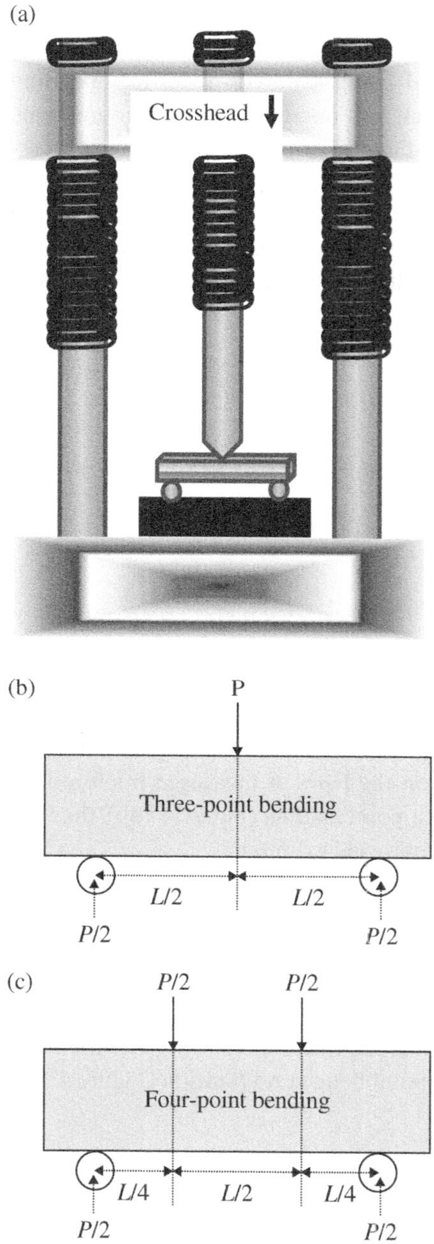

Figure 4.14 Schematic illustration of the experimental measurement of flexural strength (a). Also, the loading configuration for the three-point (b) and four-point (c) flexural mode is also shown.

4.5.4 Tensile Strength

For the metallic materials, the tensile strength can be commonly determined from uniaxial tensile testing.

In this method, a cylindrical disc with much larger diameter than thickness is loaded under uniaxial compression along its diameter (see Figure 4.15). The significance of this simple yet elegant method is that one actually applies compressive forces, but in reality measures the tensile strength of brittle ceramics [8],

$$\sigma = \frac{2P}{\pi Dt} \tag{4.31}$$

where P is the break load, D is the diameter of the disc, and t is the specimen thickness.

4.5.5 Elastic Modulus

The elastic modulus can be determined using dynamic elastic property analysis instrument. This instrument works on the basis of impulse excitation technique and the measured frequency of vibration is utilized to estimate Young's modulus. A steel ball impulser is used to excite the sample at rest. The sample undergoes transient vibrations, similar to spring mass system, in response to the excitation. The frequency of vibration is strongly dependent on the shape, dimension, mass, and stiffness of the sample. The vibration response of the system is recorded using an accelerometer (contact sensor) or a microphone (noncontact

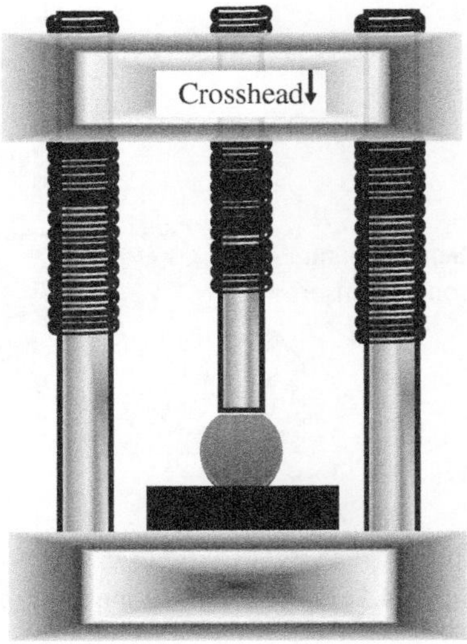

Figure 4.15 Schematic of tensile strength measurements of ceramics under diametral compression mode.

sensor). Typically, the sample locations, which experience maximum displacement (antinodal points), are used for excitation and sensing purpose, while locations with zero displacement (nodal points) are used to support the specimen in order to avoid rapid damping. The impact location, sensor location, and support locations are selected such that the desired mode of transient vibration is experienced by the sample. Impulse excitation technique is used in determining elastic modulus and damping properties of various metals [9], glasses [10], ceramics [11–13], etc.

Figure 4.16 shows the nodal and antinodal locations for bar/rod samples under flexural mode of vibration, wherein the displacement is in the plane normal to the length dimension. Out-of-plane flexure and in-plane flexure refer to flexural mode of vibration in which displacement is perpendicular and parallel to the major plane of the specimen, respectively. Figure 4.17 depicts the nodal and antinodal locations along with impact and sensing positions pertaining to disc-shaped specimens. Mode 1 corresponds to anti-flexural mode of vibration, wherein displacement in cross-sectional plane is symmetrical about two orthogonal diameters in the plane of the disc. In contrast, mode 2 corresponds to axisymmetric flexural mode of vibration, wherein displacement in cross-sectional plane is uniform for specified radial distance originating from the center. The above discussion is summarized in Figure 4.18.

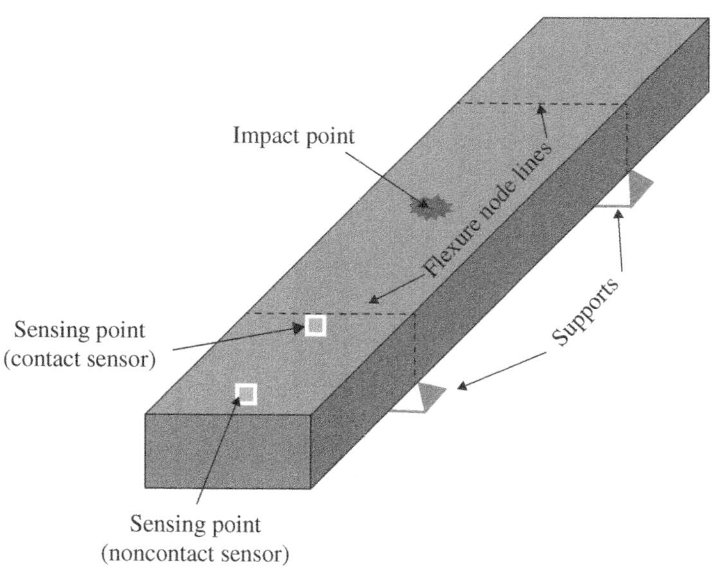

Figure 4.16 Nodal and antinodal locations in bar/rod for out-of-plane flexural vibration.

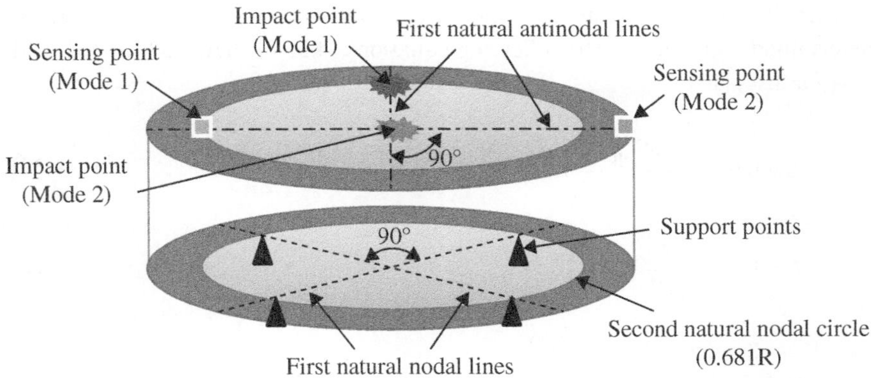

Figure 4.17 Nodal and antinodal locations in discs, while measuring elastic modulus.

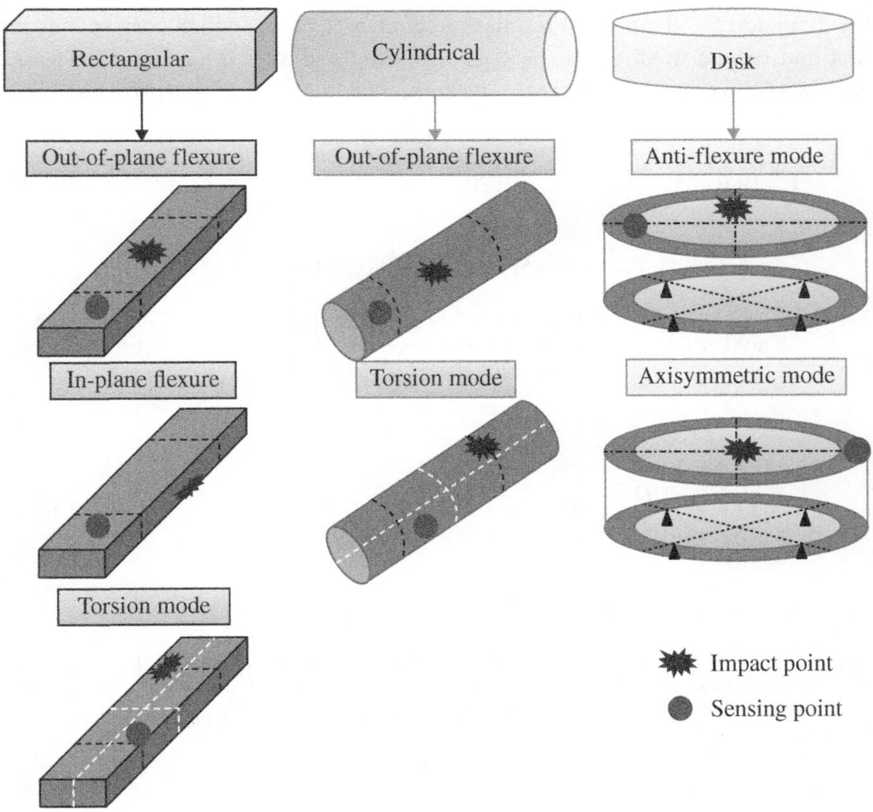

Figure 4.18 Impact and sensing locations for various specimen configurations used in elastic modulus measurement.

Once the vibration frequency is measured, Young's modulus (E) values can be determined using any of the following equations (4.32 for bar, 4.33 for rod, and 4.34 for disc) [14]:

$$E = 0.9465\left(\frac{mf_{\mathrm{f}}^2}{b}\right)\left(\frac{L^3}{t^3}\right)T \tag{4.32}$$

$$E = 1.6067\left(\frac{L^3}{D^4}\right)\left(mf_{\mathrm{f}}^2\right)T_1 \tag{4.33}$$

$$E = \frac{37.6991 f_i^2 D^2 m\left(1-v^2\right)}{\left(K_i^2 t^3\right)} \tag{4.34}$$

where m is mass, f_{f} is flexural frequency, b is width, L is length, t is thickness, D is the diameter, f_i is the natural resonant frequency (i = 1, 2, which corresponds to first and second mode), ν is the Poisson's ratio, and K_i is the geometric factor, obtained from empirical relations. The correction factor T and T_1 is given as,

$$T = 1 + 6.585\left(\frac{t}{L}\right)^2, \quad \text{if } L/t \geq 20 \tag{4.35}$$

$$T = 1 + 6.585\left(1 + 0.0752v + 0.8109v^2\right)\left(t/L\right)^2 - 0.868\left(t/L\right)^4 - \left[\frac{8.340\left(1 + 0.2023v + 2.173v^2\right)\left(t/L\right)^4}{1 + 6.338\left(1 + 0.1408v + 1.536v^2\right)\left(t/L\right)^2}\right], \quad \text{if } L/t < 20 \tag{4.36}$$

$$T_1 = 1 + 4.939\left(\frac{D}{L}\right)^2, \quad \text{if } D/t \geq 20 \tag{4.37}$$

$$T_1 = 1 + 4.939\left(1 + 0.0752v + 0.8109v^2\right)\left(D/L\right)^2 - 0.4883\left(D/L\right)^4 - \left[\frac{4.691\left(1 + 0.2023v + 2.173v^2\right)\left(D/L\right)^4}{1 + 4.754\left(1 + 0.1408v + 1.536v^2\right)\left(t/L\right)^2}\right], \quad \text{if } D/t < 20 \tag{4.38}$$

Alternatively, elastic modulus (E) and Poisson's ratio (ν) can be evaluated by ultrasonic method using lithium niobate crystals for reusing transmitting and receiving signals. The following relationships can be further used to determine the elastic modulus (E, GPa) and Poisson's ratio (ν),

$$E = \frac{(1+\nu)(1-2\nu)}{(1-\nu)}\left(\rho C_L^2\right) \tag{4.39}$$

$$\nu = \frac{1/2(C_L/C_S)^2 - 1}{(C_L/C_S)^2 - 1} \tag{4.40}$$

where ρ is the density of specimen (g/cm^3), C_L is the velocity of longitudinal wave (m/s), and C_S is the velocity of shear (transverse) wave (m/s). It is important to mention that the internal defects can reflect ultrasonic waves, leading to shorter travel time for the given sample thickness, ultimately resulting in overestimation of elastic modulus.

It is worthwhile to note that Young's modulus of ceramics can also be determined using static methods, like instrumented indentation, three-point bending, diametral compression, etc.

> While the initial slope of the unloading response (P-h plot) in case of instrumented hardness measurement provides an estimate of elastic modulus, the slope of the initial linear loading response under uniaxial compression/flexure/diametral compression can be used to determine the elastic modulus under respective loading condition.

Generally, all the static methods result in permanent deformation of the sample, whereas dynamic methods are nondestructive in nature. Though, both the methods can be used to experimentally estimate elastic modulus, the values obtained from both the methods differ, typically about 10%, due to inelastic effects.

4.5.6 Fracture Toughness

> The wear resistance of ceramic is particularly dependent on the crack–microstructure interaction, which can be quantitatively captured using many properties that include fracture toughness.

The higher the fracture toughness, one would expect a better wear resistance property of ceramics.

> The fracture toughness of brittle materials cannot be reliably measured using the indentation technique. One should use SENB, SEVNB, Chevron Notch Beam, etc., to measure reliable values of toughness for ceramics.

However, in order to compare the toughness property of the newly developed composites, the indentation technique can be used. Additionally, it is now widely recognized that a careful use can provide reproducible results for indentation toughness measurements. It should be mentioned here that the indentation method is now extensively used for convenience to determine the fracture toughness of small and relatively brittle specimens, which are otherwise hard to machine into standard test samples (for example, SENB, SEVNB, etc.). Broadly, there are two established ways to calculate the fracture toughness of brittle materials. One is to use toughness evaluation with large cracks and the other one is to evaluate by short crack methods.

4.5.6.1 Notched Beam Test

The success of this technique, as can be realized from the discussion in this section, is largely dependent on how one introduces sharp cracks of specified geometry at the notch tip in a ceramic sample.

1) Single edge notched beam (SENB) method: The testing geometry is shown in Figure 4.19. The mode I critical stress intensity factor (K_{IC}) can be determined using

$$K_{IC} = Y\left(3PL / h^2 d\right)c^{1/2} \tag{4.41}$$

In the above expression, $Y = 1.99 - 2.47(c/h) + 12.97(c/h)^2 - 23.17(c/h)^3 + 24.8(c/h)^4$, where P is the fracture load, h is the specimen thickness, d is the specimen width, c is the crack length, and L is the span length.

2) Single edge V-notched beam (SEVNB) method: A sharp V notch is introduced with the notch radius of smaller than 10 μm. Figure 4.19 presents the test configuration for SEVNB testing, which employs four-point bending configuration. The fracture toughness can be computed using the following equation [15, 16]:

$$K_{IC} = \frac{P_f\left(L_o - L_i\right)}{BW^{3/2}} \frac{3\alpha^{1/2}}{2\left(1-\alpha\right)^{1/2}} f\left(\alpha\right) \tag{4.42}$$

where P_f, L_o, L_i, B, and W are fracture load, outer span, inner span, specimen width, and specimen depth, respectively; $\alpha = a/W$, where a is the precrack size; and $f(\alpha)$ is

$$f\left(\alpha\right) = 1.9887 - 1.326\alpha - \frac{\alpha\left(1-\alpha\right)\left(3.49 - 0.68\alpha + 1.35\alpha^2\right)}{\left(1+\alpha\right)^2} \tag{4.43}$$

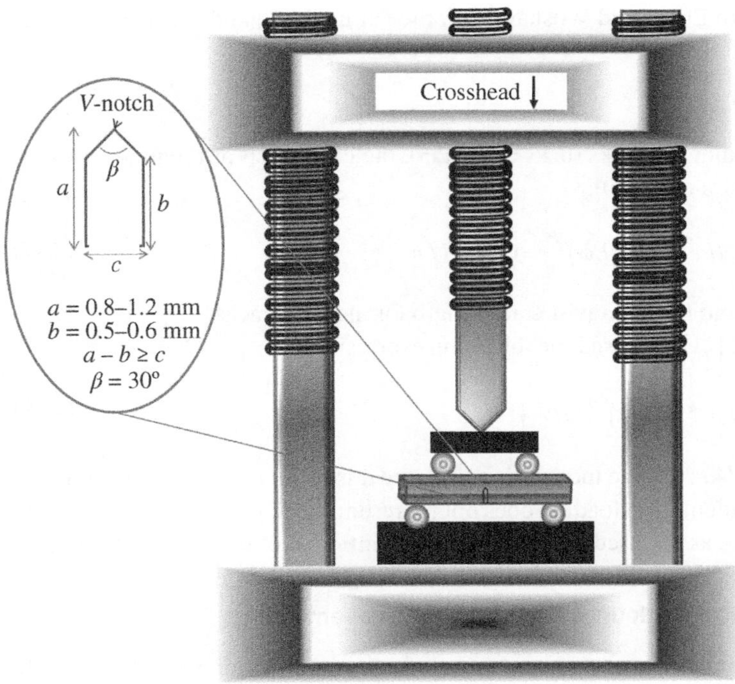

Figure 4.19 Typical geometry and loading configuration involved in SEVNB testing to evaluate fracture toughness of ceramic samples. Typical test sample dimensions are also mentioned (left).

4.5.6.2 Indentation Microfracture Method

As discussed later in this book, the wear mechanisms of ceramics involve cracking and subsequent physical events that occur as a result of interaction between cracks and elements of microstructure. This section briefly discusses how one can estimate the measure of toughness by indentation cracking method, when the cracks are artificially produced by indenting a smoothly polished ceramic surface with an indenter.

The fracture toughness can be calculated from the measurement of radial cracks around the Vickers indentations, according to the formula proposed by Anstis et al. [17]:

$$K_{IC} = 0.016(E/H)^{1/2} P/c^{3/2} \tag{4.44}$$

Kaliszewski and coworkers [18] modified the above equation to take into account the effect of the compressive stresses,

$$K_{IC} = 0.019(E/H)^{1/2} P/c^{3/2} \tag{4.45}$$

According to Evans and Wilshaw [19], the Palmqvist (l) and median cracks (c) are related by

$$l / a = c / a \quad 1 \tag{4.46}$$

For the Palmqvist cracks ($0.25 < l/a < 2.5$), the expression of toughness (K_{IC}) is given by Niihara et al. [20]:

$$\left(K_{IC}\varphi / Ha^{1/2}\right)\left(H / E\varphi\right)^{2/5} = 0.035\left(l / a\right)^{-1/2} \tag{4.47}$$

where $\varphi = 3$ and l = Palmqvist crack length for median cracks

Shetty et al. [21] proposed the following expression:

$$K_{IC} = 0.025\left(E / H\right)^{0.4}\left(HW\right)^{1/2} \tag{4.48}$$

where $W = P/4a$ (P is the indentation load and a is mean radial crack length).

Since the indentation loading does not represent mode I loading, it is advisable to mention K_C, as opposed to K_{IC}, although mentioned often in the literature.

4.5.7 Practical Guidelines for Reliable Measurements

The conventional manner of measuring elastic modulus of metals by tensile testing is generally not followed in the case of ceramics.

The following guidelines can be used to determine the elastic modulus in nondestructive manner:

1) Prior to carrying out impulse excitation experiment on the desired material, the validation experiments are to be performed on standard samples of similar sample configuration to ensure reliability.
2) The desired sample dimensions for impulse excitation technique are $b/t > 5$ for rectangular samples, $L/D > 10$ for cylindrical samples, and $D/t > 4$ for disk-type samples.
3) All the faces of the specimens need to be flat and parallel for samples with rectangular cross section, while constant diameter and roundness need to be ensured for samples with circular cross section.
4) Measurement of length, width, thickness, diameter, and mass of the samples are to be done with an accuracy of 0.1%. Care must be taken while recording the diameter and thickness, as these influence the elastic modulus estimation by factor D^4 and t^3, respectively.
5) In case of accelerometer-based contact sensor for acquisition of vibration signal, the placement of sensor is to be made such that it does not significantly

mass load the sample. To avoid this, the sensor should be placed only as far from the nodal location as necessary to record the desired output.

6) For disc-shaped specimens, striking for first mode (anti-flexure mode) is to be made such that it will minimize the excitation of second mode (axisymmetric mode).
7) The frequency must be determined to an accuracy of 10% for rectangular and cylindrical samples and 1% for disc-shaped samples.
8) If desired repeatability in frequency measurements is not achieved, iterative changes in sensor location, excitation location, sample support locations, etc., are to be made. Generally, repeatability is affected when two vibration modes exist with similar frequencies or due to inhomogeneity in the sample.
9) In case of rectangular (or cylindrical) samples, if L/t (or D/t)$\geq$20, elastic modulus can be calculated directly based on the measurement of flexural frequency. Else, torsional frequency needs to be measured and shear modulus is to be calculated, based on which elastic modulus will be estimated through iteration of Poisson's ratio.

   For disc-shaped specimens, the ratio of mode 2 and mode 1 vibration frequency is used to estimate Poisson's ratio, followed by calculation of elastic modulus for both the modes as represented in Figure 4.17. The dynamic shear modulus can also be calculated by knowing the elastic modulus and Poisson's ratio.

4.6 Closure

- **Hardness:** Resistance against permanent deformation
- **Strength:** Load bearing capability
- **Fracture toughness:** Resistance against crack growth
- **Elastic modulus:** Ratio of stress to strain in the elastic regime of the stress–strain curve.

This chapter provides a conceptual understanding of stress and strain as tensorial quantities. The principles of mechanical properties of three major material classes, metals, ceramics, and polymers, in relation to the stress–strain response are also reviewed. In case of ceramics, the larger compressive strength, elastic modulus, and good hardness are advantageous for tribological applications. However, poor fracture toughness remains a major concern.

References

1 Basu, B. (2017). *Biomaterials for Musculoskeletal Regeneration: Concepts.* Germany: Springer Nature.

2 Basu, B. and Balani, K. (2011). *Advanced Structural Ceramics*. Wiley and American Ceramic Society.

3 Basu, B. *Biomaterials Science and Tissue Engineering: Principles and Methods.* Cambridge University Press.

4 Dieter, G.E. (1986). *Mechanical Metallurgy*, Mcgraw-Hill Series in Mechanical Engineering, 3e. McGraw-Hill Education.

5 Lawn, B. (1993). *Fracture of Brittle Solids*, Cambridge Solid State Science Series, 2e. Cambridge University Press.

6 Inglis, C.E. (1913). Stresses in a plate due to the presence of cracks and sharp corners. *Trans. Inst. Nav. Archit.* 55: 219–241.

7 Orowan, E. (1952). Fatigue and fracture of metals. In: *Symposium at Massachusetts Institute of Technology, USA*. New York: Wiley.

8 Ovri, J. and Davies, T. (1987). Diametral compression of silicon nitride. *Mater. Sci. Eng. A* 96: 109–116.

9 Swarnakar, A.K., Van Der Biest, O., and Baufeld, B. (2011). Young's modulus and damping in dependence on temperature of Ti–6Al–4V components fabricated by shaped metal deposition. *J. Mater. Sci.* 46: 3802.

10 Swarnakar, A.K., Stamboulis, A., Holland, D., and Van Der Biest, O. (2013). Improved prediction of Young's modulus of fluorine-containing glasses using MAS-NMR structural data. *J. Am. Ceram. Soc.* 96 (4): 1271–1277.

11 Roebben, G., Basu, B., Vleugels, J. et al. (2000). *J. Alloys Compd.* 310 (1–2): 284–287.

12 Guicciardi, S., Swarnakar, A.K., Van Der Biest, O., and Sciti, D. (2010). The effect of the electrical properties on the pulsed electric current sintering behavior of ZrO_2 based ceramic composites. *Scr. Mater.* 62: 831–834.

13 Swarnakar, A.K., Donzel, L., Vleugels, J., and Van Der Biest, O. (2009). High temperature properties of ZnO ceramics studied by the impulse excitation technique. *J. Eur. Ceram. Soc.* 29: 2991–2998.

14 ASTM C1259-95 (1995). *Standard Test Method for Dynamic Young's Modulus, Shear Modulus and Poisson's Ratio for Advanced Ceramics by Impulse Excitation of Vibration*, 375. West Conshohocken, PA: ASTM International.

15 Choi, S.R. and Bansal, N.P. (2005). Mechanical behavior of zirconia/alumina composites. *Ceram. Int.* 31: 39–46.

16 Mizuno, M. and Okuda, H. (1995). VAMAS round robin on fracture toughness of silicon nitride. *J. Am. Ceram. Soc.* 78: 1793–1801.

17 Anstis, G.R., Chantikul, P., Lawn, B.R., and Marshall, D.B. (1981). A critical evaluation of indentation techniques for measuring fracture toughness: I, direct crack measurements. *J. Am. Ceram. Soc.* 64: 533–538.

18 Kaliszewski, M.S., Behrens, G., Heuer, A.H. et al. (1994). Indentation studies on Y_2O_2-stabilized ZrO_2: I, development of indentation-induced cracks. *J. Am. Ceram. Soc.* 77 (5): 1185–1193.

19 Evans, A.G. and Wilshaw, T.R. (1976). Quasi-static solid particle damage in brittle solids – I. Observations analysis and implications. *Acta Metall.* 24 (10): 939–956.

20 Niihara, K., Morena, R., and Hasselman, D.P.H. (1982). Evaluation of K_{Ic} of brittle solids by the indentation method with low crack-to-indent ratios. *J. Mater. Sci. Lett.* 1: 13–16.

21 Shetty, D.K., Wright, I.G., Mincer, P.N., and Heuer, A.H. (1985). Indentation fracture of WC-Co cermets. *J. Mater. Sci.* 20: 1873–1882.

Section II

Fundamentals of Tribology

5

Contact Surface Characteristics

In the preceding chapter, it has been emphasized that friction and wear properties are largely influenced by contact surface characteristics. In this context, the characterization of physical properties of tribological surfaces is provided in this chapter. An emphasis is given to understand the surface topographical features in terms of roughness parameters and the related contact-/noncontact-based measurement methods. The difference between nominal and real areas of contact is also highlighted. The Hertzian contact stresses in elastic and elastic–plastic contacts are also explained.

5.1 Nature and Roughness of Contact Surfaces

It is important to study the nature of contact surface–surface interactions that influences the real area of contact, friction, wear, and/or lubrication. The high levels of stresses at the real contact area of asperity junctions lead to work-hardening or deformation of metals and alloys; while the instances of deformation are also not uncommon in ceramics and polymeric surfaces. The friction and wear of ceramics are also affected by the presence of surface films, like lubricating oil or grease at the contact.

Any polished surface always appears as nominally flat. But they are not at the microscopic length scale. It is important to emphasize that real tribological surfaces always contain irregularities and deviations from their ideal counterparts (see Figure 5.1) [1]. While surface irregularities at the macroscopic scale refer to shape-related deviations, micro/nanoscale deviations refer to roughness [2].

A typical 2D profile of an engineering surface is provided in Figure 5.2. Such a profile can be distinguished from a similar, but not identical profile in terms of 2D surface roughness parameters (R_a, R_q, etc.).

Friction and Wear of Ceramics: Principles and Case Studies, First Edition.
Bikramjit Basu, Mitjan Kalin, and B. Venkata Manoj Kumar.

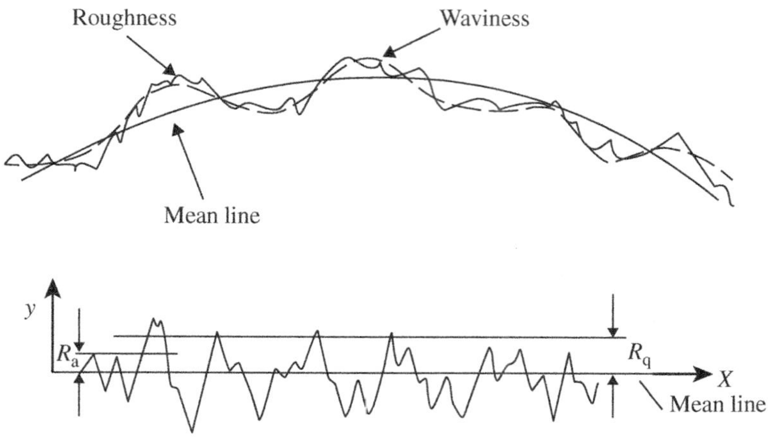

Figure 5.1 Schematic illustration of topography of nominal engineering surface (bold line) and real surface to show the microscale roughness/waviness on the surface (dashed line).

> For a known sample length, the surface profile mean height from peaks to inverted valleys is defined as average surface roughness or arithmetical mean deviation (R_a).

The average surface roughness or arithmetical mean deviation (R_a) for a profile length (l) is mathematically expressed as

$$R_a = \frac{1}{n}\sum_{i=1}^{n}\left|y(x_i)\right| \tag{5.1}$$

where $y(x_i)$ is the height of the surface profile from the mean line, at a given distance, x on the sample length.

The other important roughness parameter R_q is called the root mean square value of roughness profile, which can be mathematically expressed as

$$R_q = \sqrt{\frac{1}{l}\int_0^l y^2(x)\,dx} \tag{5.2}$$

As both these R_a and R_q parameters are statistical/mathematical function of measuring device readings, errors are commonly observed. Based on the application, suitable surface roughness parameter shall be used. It is worth noting that surface profiles with almost same R_a value may belong to two completely different

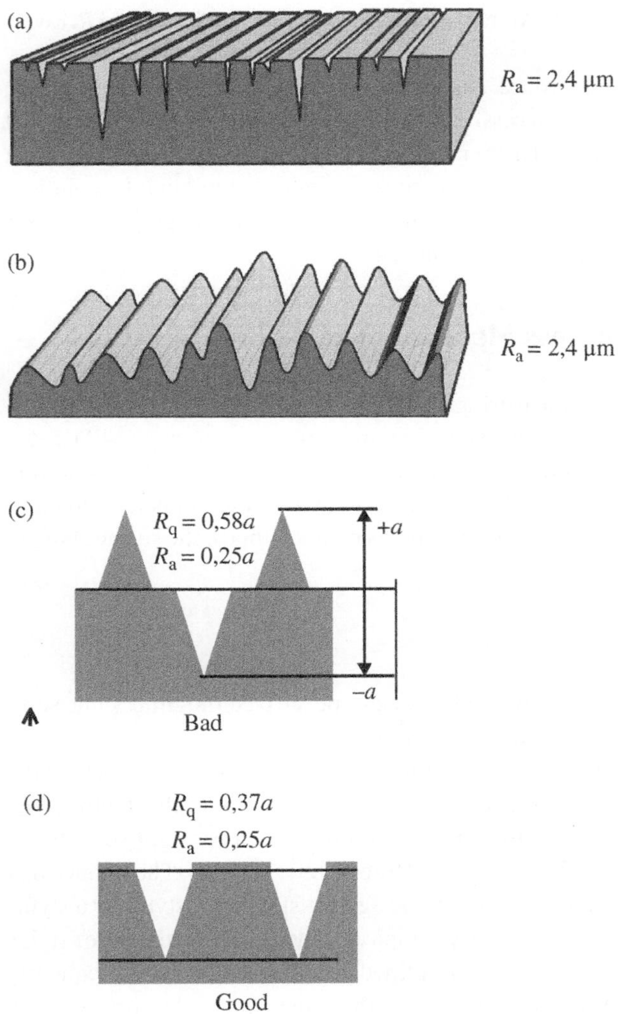

Figure 5.2 Two surfaces with identical R_a values can have very different surface topographical features (heights/slopes) as shown in (a, b). Also, the representative roughness profiles of good and bad surfaces in the context of tribological application (c, d).

types of surfaces. Thus, additional roughness parameter R_q is used to distinguish such surfaces.

In addition, skewness (Sk) and kurtosis (K) are other two rarely used statistical height descriptors. Other measures of surface roughness are extreme-value height descriptors such as R_p (maximum peak height), R_v (maximum valley depth), R_{pm}

(average peak-to-mean height), R_t (maximum peak-to-valley height), and R_z (average peak-to-valley height).

Three-dimensional (3D) surface characteristics can be referred to as surface topography. The collection of successive 2D surface profiles provides topographical information for given surfaces. The reliability of surface topographical information is better for closer 2D surface profiles. As there is limitation in lateral resolution, analytical models can be used to visualize the actual surface topography. 3D laser surface profiles are useful for surface visualization.

5.2 Surface Roughness Measurement

Several techniques used for measuring surface roughness parameters can be categorized broadly into contact type: stylus, atomic force microscopy (AFM), etc.; and noncontact type: optical interferometry, laser surface profilometry, scanning electron microscopy (SEM), etc. While contact-type measurements tend to damage the surface, noncontact-type measurements do not damage the surface but are relatively expensive.

5.2.1 Stylus Method

Stylus method is the most widely used contact-type surface roughness measurement method. A hard and stiff (diamond) stylus tip moves in response to asperities of contacting surfaces in measuring the roughness parameters. A schematic representation of a stylus profilometer is shown in Figure 5.3. In this method, the vertical movements of a stylus displaced by the surface at a constant velocity are recorded using a linear variable differential transformer (LVDT). The surface displacement is achieved by moving the sample against stationary stylus or a stylus cantilever is moved against a stationary sample. The vertical movement of stylus produces an analog signal, which is amplified, conditioned, and subsequently digitized to generate a number that represents the roughness of the surface [3, 4].

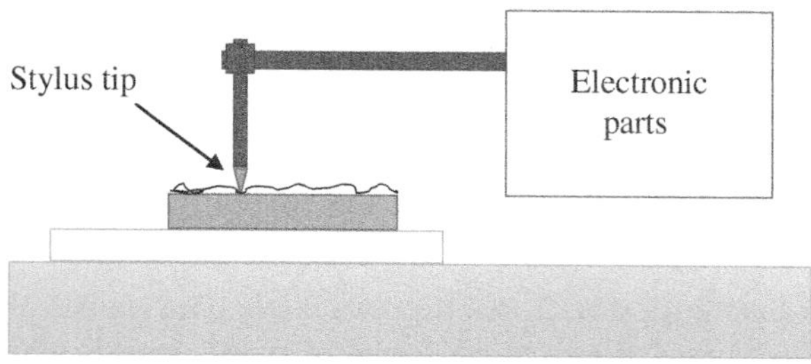

Figure 5.3 A schematic representation of a stylus profilometer.

5.2.2 Atomic Force Microscopy

> Atomic force microscopy analysis generates the surface profile of a solid surface by measuring the forces between a sharp probe tip (radius of curvature < 10 nm) and the material surface at a small distance of 0.2–10 nm.

Atomic force microscope (AFM) is widely used in tribology research for the surface features characterization, particularly the surface roughness, with nanometer resolution. With the AFM, one can also obtain force–displacement plots to estimate the surface stiffness. In the tribological studies of the engineering materials, surface profiles of the mating material before and after testing are analyzed by using the AFM. A lower scanning speed is better to obtain a good resolution.

The AFM probes are typically made of Si_3N_4 or Si. By rastering the tip on the sample surface and recording the change in force as a function of position, the AFM provides information on surface topography/roughness and on electrical and magnetic properties (depending on the probe tip) (Figure 5.4). The image resolution obtained in the AFM depends on the probe–sample interaction volume and is typically 1 nm in the x–y plane and 0.1 nm in the z-direction. The low scanning rate and the requirement of extreme smooth surface are considered as the notable drawbacks of the AFM. When the probe tip is in close proximity to the sample surface, the deflection of the cantilever occurs due to a force (F) between the sample and the probe tip. This force (F) is governed by Hooke's law.

$$F = -k \cdot x \tag{5.3}$$

where x is the cantilever deflection and k is the spring constant (~0.1–1 N/m) of the cantilever.

The detector measures the bending of the cantilever as the tip is rastered over the sample surface. The measured cantilever deflections are used to generate the surface topography.

Three primary imaging modes in the AFM are possible, as shown in Figure 5.4.

1) Contact mode: <0.5 nm separation between the probe tip and surface; in the strongly repulsive force regime; constant force or distance.
2) Noncontact mode: 0.1–10 nm probe tip to surface separation; weak attractive force regime, vibrating probe.
3) Tapping mode: 0.5–2 nm probe tip to surface separation; strongly repulsive; vibrating probe. Tapping mode should be used to obtain topography.

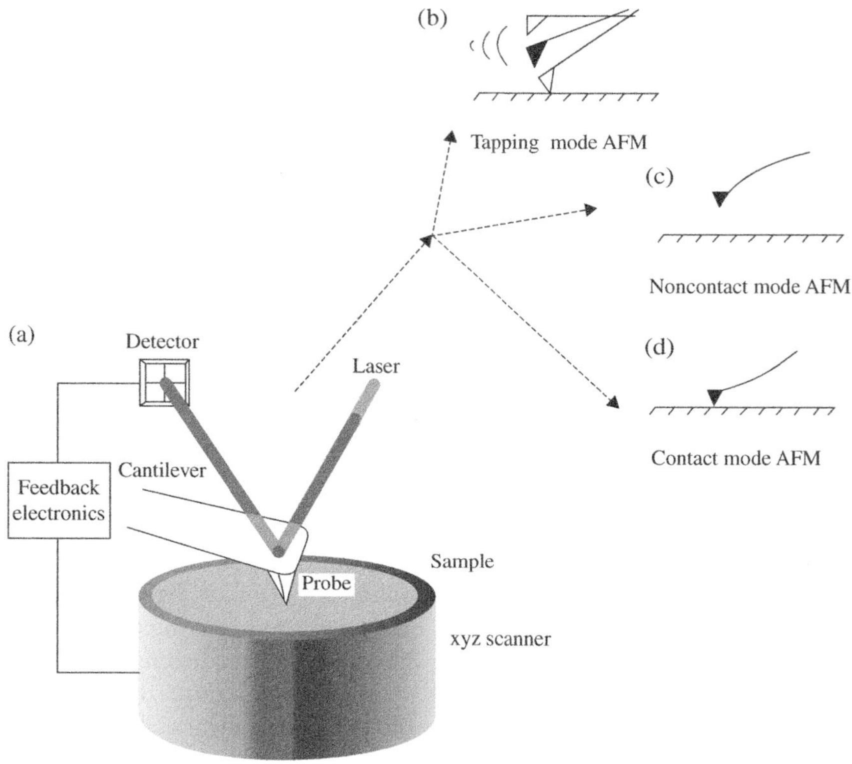

Figure 5.4 Basic working of an AFM showing how cantilever deflection due to probe–sample interaction is monitored by a feedback mechanism with the help of a diode laser (a). Also, the functioning of AFM is shown (b–d).

5.2.3 Optical Interferometry

When light wave (electromagnetic radiation) is incident on a surface, it is possible that the light wave is diffusely reflected and/or specularly reflected. As surface roughness increases, the intensity of diffracted radiation increases and light wave tends to diffuse. A smooth surface for light of a specific wavelength is possibly the rough surface for light of another wavelength.

Among the numerous optical methods, optical interferometry is a widely used noncontact method for measuring macroscopic and microscopic surface characteristics [5]. In general, optical interferometry method determines how much interference fringes depart from going straight and equally spaced. A computer analysis is used to characterize a surface.

5.2.4 Laser Surface Profilometry

In this noncontact type method, a laser light of a suitable wavelength is focused onto the surface, and the reflected light is directed to a detector. The moving position of the light is converted to an electrical signal with the help of a transducer. The electrical signal is further processed to obtain a two-dimensional or three-dimensional profile of the surface.

5.2.5 Scanning Electron Microscopy

Electron microscopy can reveal microscopic and macroscopic surface features [6]. The surface profiles are generated by processing scattered electron signals (SES) in a scanning electron microscope (SEM). Three-dimensional measurements of roughness are possible by making several scans. In a similar manner, stereomicroscopy can also be used where the stereo effects can be obtained by preparing two images of the same surface with slightly different angular views (typically less than 10°). The difference in the elevation between the feature and the reference point provides a parallax shift, which can be extracted to obtain height information. However, these methods require vacuum and conducting coating on ceramic surfaces. Also, these methods take long time and represent a limited field of view of few asperities.

Among the several methods of roughness measurements, the selection of the particular one is largely dependent on lateral and vertical resolution of tomography, the need for quantitative information, availability, etc.

5.3 Bearing Area Curve and Cumulative Distribution Function

Traditionally, bearing area is defined as a measure of the relative cross-sectional area of a plane passing through the measured surface, from the highest peak to the lowest valley, would encounter. The bearing area curve (BAC) is also known as the Abbott–Firestone curve.

The BAC represents deviation of any point on the surface from the mean line. For higher bearing area, the surface deviation is higher than that is covered by a very small fraction of the area. A thorough understanding of the BAC concept is necessary to understand the difference between different types of commercial lubricants. The change in BAC of Ti_3SiC_2 ceramics' surface as a result of fretting against steel ball is shown in Figure 5.5 [7].

The bearing area is related to the cumulative distribution function (CDF). It can be analyzed as a fraction of heights above a given height z (i.e. the bearing ratio at height h) and the CDF is represented as

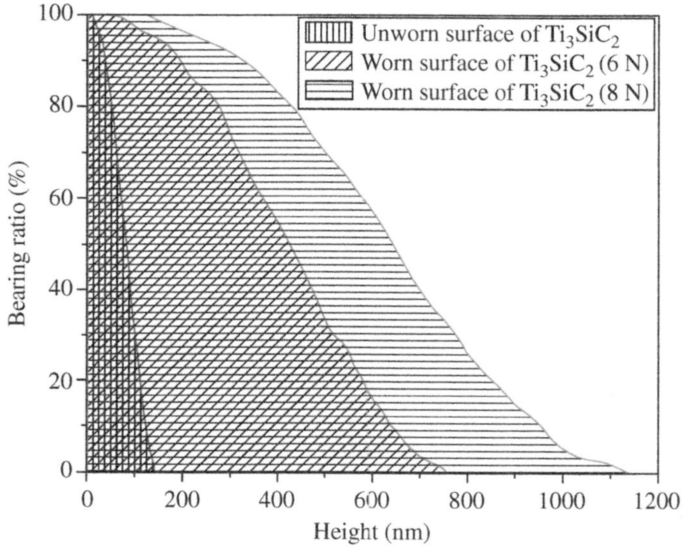

Figure 5.5 Representative bearing area of Ti_3SiC_2 surface before and after fretting against steel ball at 6 and 8 N load [7]. *Source:* Reproduced with permission of Elsevier.

$$p(z \geq h) = \int_h^{\infty} p(z)\,\mathrm{d}z \tag{5.4}$$

where $p(z)$ is the CDF at $z \leq h$. The bearing ratio histograph at height h is the progressive addition of all the values of $p(z)$ starting at the highest point and working down to the height $z = h$. The cumulative sum multiplied by the class interval z can be expressed as

$$p(z \geq h) = \Delta z \sum_{z=h}^{\infty} p(z) \tag{5.5}$$

The relationship between bearing ratio and the fractional real area of contact is highly approximate, as the material is sliced off in the construction of BAC and the material deformation is not taken into account. The BAC is complementary to the CDF and physical meaning of the BAC is that it represents a material ratio, when sliding at the surface at a certain height (Figure 5.6).

5.4 Nominal Versus Real Contact Area

However, it is considerately essential to outline from the previous two sections that the surfaces are not flat! There always exist some heights above and valleys below the surface mean line, respectively. Thus, when the two surfaces draw

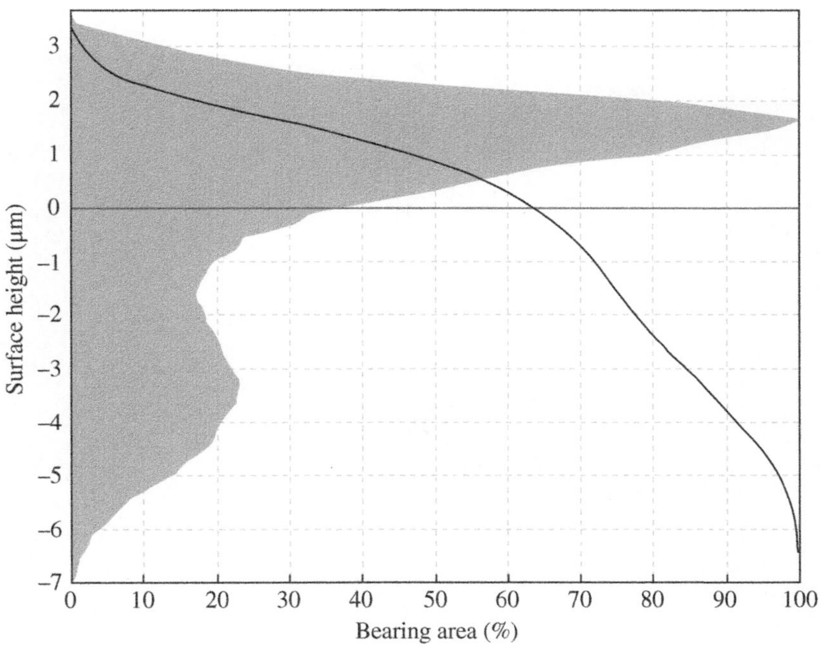

Figure 5.6 Bearing area curve and cumulative distribution function.

toward each other and ultimately come in contact, this contact exists only on the asperities that touch first and not on the entire surface of the samples.

The area of asperity junctions in contact is called real contact area.

In reality, it depends on the roughness, material mechanical properties (elastic modulus, hardness, etc.), type of material deformation (brittle, plastic), load, and so on. Figure 5.7a illustrates a schematic representation of contacts of a single-asperity against a precise smooth flat surface. As such "n" number of asperities are contained by the real engineering surfaces, the real contact area (A_r) can be determined as the total sum of the individual contact area (A_i) of these asperities:

$$A_r = \sum_{i=1}^{n} A_i \tag{5.6}$$

However, there is an abiding quest in determining of the real contact area. Though many attempts in its modeling and measuring have been made in the past, accordingly there would be many responses to this quest. Nevertheless, it is

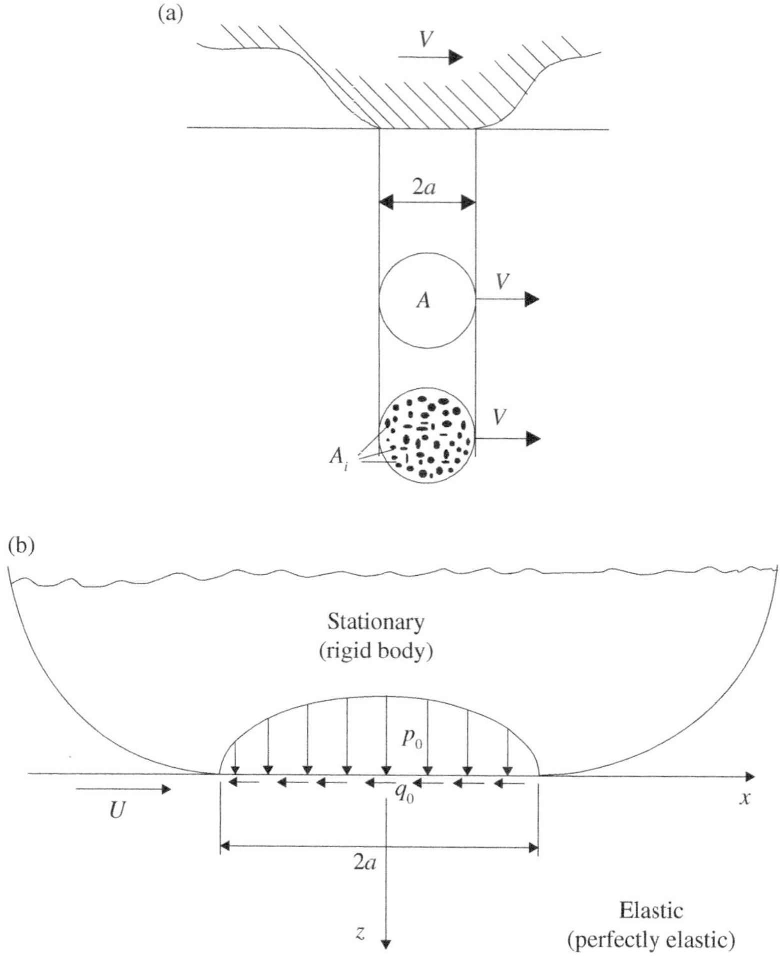

Figure 5.7 (a) Schematic illustration of the contact of a single asperity with a nominally flat surface. (b) Hertzian contact geometry and contact pressure distribution (p_0) over contact diameter ($2a$).

generally understood that the values of real contact area are significantly much less than the nominal contact area and this leads to larger than nominal contact pressure at asperity/asperity contacts on an engineering surface.

Furthermore, due to this, the ductile/brittle materials undergo deformation and/or fracture, respectively. Also, the partitioning of the heat generated due to frictional losses in the material is predominantly via conduction, rather than convection or radiation. Therefore, the surface temperatures reach a higher value, than generally expected.

5.5 Hertzian Contact Stress

Once a tribological surface is sufficiently characterized using different quantitative parameters, the next question arises as how those surfaces experience contact stresses at a loaded tribocontact. The researchers employ the well-accepted Hertzian theory, which can be implemented for various geometries in contact, while assuming the elastic contact conditions [8]. The use of test samples of predefined geometries, materials, and loads is more suitable in tribology of materials, and a detailed explanation is presented in the book by Johnson [4]. A sphere-on-flat contact geometry is often used to understand the Hertzian pressure distribution. The numerous adaptations of Hertzian theory for specific parameters and conditions are reported in the literature, thus its in-depth discussion is not be provided here, apart from the simple ball-on-flat case. The distribution of the contact pressure $p(r)$ consisting of a parabolic shape is shown in Figure 5.7b, and its distribution can be calculated as

$$p(r) = p_0\sqrt{1-\frac{r^2}{a^2}} \tag{5.7}$$

where $r^2 = x^2 + y^2$.

The contact pressure at the contact edges increases from zero to a maximum p_0 at the center of the contact. According to this distribution, a generally valid solution for the contact area and maximum contact pressure can be derived as

$$p_0 = \frac{3W}{2\pi a^2} \tag{5.8}$$

In the above expression, the radius of the contact area is

$$a = \sqrt[3]{\frac{3WR^*}{4E^*}}$$

Also, W represents the normal contact load, and E^* and R^* are the equivalent elastic modulus and radius of contact bodies, respectively. These parameters can be further calculated using the following expressions:

$$\frac{1}{E^*} = \frac{\left(1-\upsilon_1^2\right)}{E_1} + \frac{\left(1-\upsilon_2^2\right)}{E_2} \tag{5.9}$$

$$\frac{1}{R^*} = \frac{1}{R_1} + \frac{1}{R_2} \tag{5.10}$$

In Eqs. (5.9) and (5.10), suffixes 1 and 2 refer to two mating solids and υ is the Poisson's ratio.

In order to determine the resistance against motion at a loaded tribological contact, it is necessary to evaluate the stresses that are developed at the surface and subsurface. Two most crucial conditions for static Hertzian contacts are to be considered. As far as the surface radial stresses are concerned, it undergoes a transition from compressive to tensile at the edge of a contact (Figure 5.8a). This often explains the experimental observation of cracking at the edges of a Hertzian contact.

Another crucial outcome, pertinent particularly for the long-term loading, is elaborated in the delamination theory of wear, referring to subsurface shear stresses (Figure 5.8b).

> The shear stress has a maximum value somewhere below the surface – in this particular case of point static contact for steel (i.e. = $\sqrt{3}$), at the depth of *0.48a*.

This indicates that the cracks arising in ductile materials due to relatively low loads may extend beneath the surface.

Further, while considering friction force with the addition of a tangential component to the load, the loading conditions and stress distribution change. Moreover, the stresses and deformations under static loading conditions tend to be symmetric. Also, the stresses depend on the direction of sliding (tangential friction force) and become asymmetric, during sliding conditions [7].

> The normal stress in front of the contact will only be compressive, whereas tensile stresses will appear at the tail of the contact.

The occurrence of the crack opening and/or surface spallation will form inevitably with the movement of the slider after a certain point. Also, the shear stresses are moved further toward the surface by the tangential load. Accordingly, it is desired to evaluate the stresses for any application or experimental design in order to prevent maximum values at critical locations and also to restrict the loads for a long-term suitable contact performance.

5.6 Closure

Considering the importance of surface characteristics in tribological studies, this chapter discusses the qualitative and quantitative nature of surface topography. To quantify the surface roughness, various roughness parameters (R_a, R_q, etc.) are

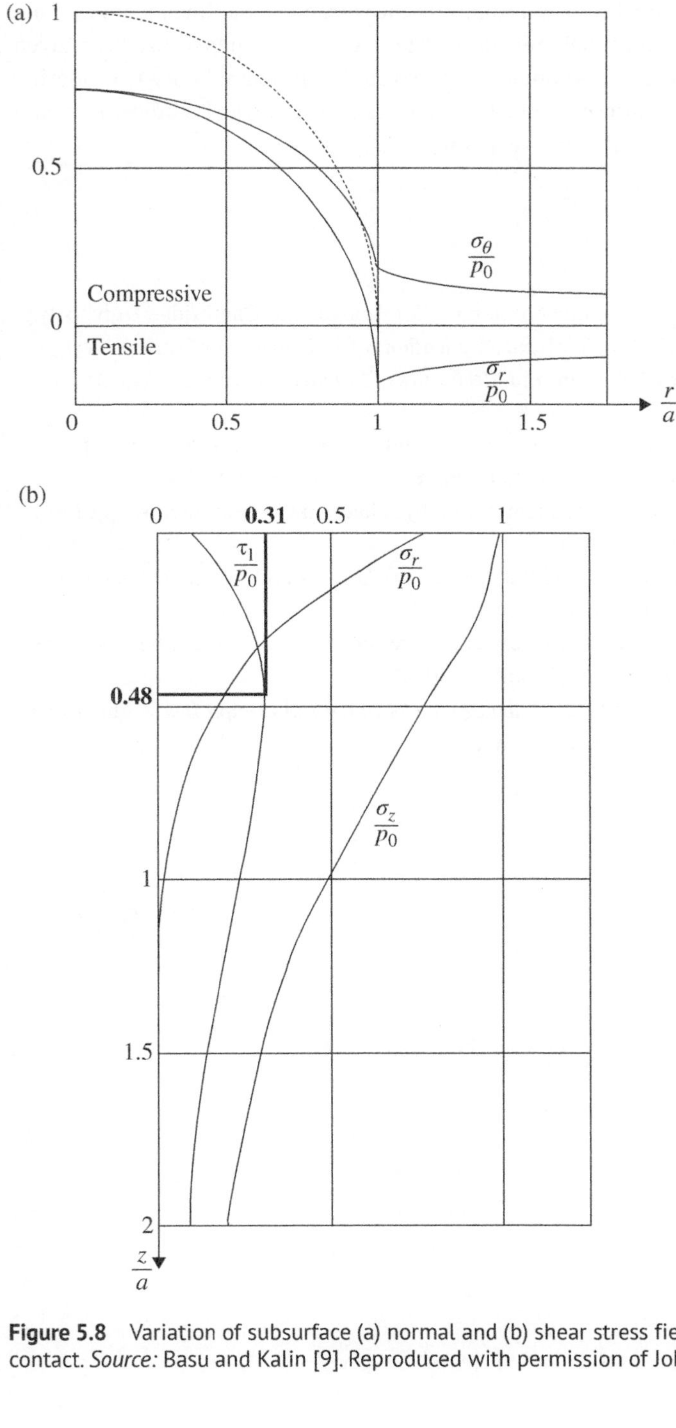

Figure 5.8 Variation of subsurface (a) normal and (b) shear stress field of a Hertzian contact. *Source:* Basu and Kalin [9]. Reproduced with permission of John Wiley & Sons.

described and also some of the roughness measurement methods were emphasized. Also, the BAC and CDF are emphasized. A major emphasis has been given on the nominal and real area contact, and also on the Hertzian theory. Considering the importance of common contact-mechanics in tribological studies, Hertzian contact stress topic is elaborately explained.

References

1 Johnson, K.L. (1985). *Contact Mechanics*. Cambridge, UK: Cambridge University.
2 Bhushan, B. (1999). *Principles and Applications of Tribology*. New York: Wiley.
3 Bhushan, B. (1996). *Tribology and Mechanics of Magnetic Storage Devices*, 2e. New York: Springer.
4 Thomas, T.R. (1999). *Rough Surfaces*, 2e. London, UK: Imperial College Press.
5 Tolansky, S. (1973). *Introduction to Interferometers*. New York: Wiley.
6 Halliday, J.S. (1955). Surface examination by reflection electron microscopy. *Proc. Inst. Mech. Eng.* 109: 777–781.
7 Sarkar, D., Kumar, B.V.M., and Basu, B. (2006). Understanding the fretting wear of Ti_3SiC_2. *J. Eur. Ceram. Soc.* 26: 2441–2452.
8 Johnson, K.L. (1955). Surface interaction between elastically loaded bodies under tangential forces. *Proc. R. Soc. Lond. A* 230: 531.
9 Basu, B. and Kalin, M. (2012). *Tribology of Ceramics and Composites: A Materials Science Perspective*. Hoboken, NJ: Wiley.

6

Friction and Interface Temperature

Friction mechanisms, friction laws, and key factors affecting friction in tribological contacts are discussed in this chapter. Moreover, the contact temperature of the engineering surfaces at the multiple-asperity level is discussed. Key aspects of the heat generation and the consequent contact temperature rise are presented and some ready-to-use models to calculate the contact temperature are provided.

6.1 Theory of Friction

The theory of friction is taught at the school and a few laws, discussed below, are also known to the young students. The friction force acting on sliding and rolling bodies is shown in Figure 6.1. Work is required to overcome the friction force, which results in a loss of energy. This loss of energy is manifested as noise, vibration, frictional heat generation, and the energy needed to deform materials. Moreover, tribochemical reactions can be altered due to friction.

> Friction is the resistance to sliding or rolling in the tangential direction when two bodies are in contact and in relative motion with respect to each other. The tangential force acting opposite to the direction of sliding is called the *friction force*.

In the vast majority of applications, friction is considered to be an undesirable phenomenon, although there are cases where friction is required and is thus necessary. As an example, in daily life, friction is needed for walking, for the grip of automobile tires on the road, achieving enough purchase to pick up objects, etc. Moreover, there are engineering applications where maximizing friction is

Friction and Wear of Ceramics: Principles and Case Studies, First Edition.
Bikramjit Basu, Mitjan Kalin, and B. Venkata Manoj Kumar.

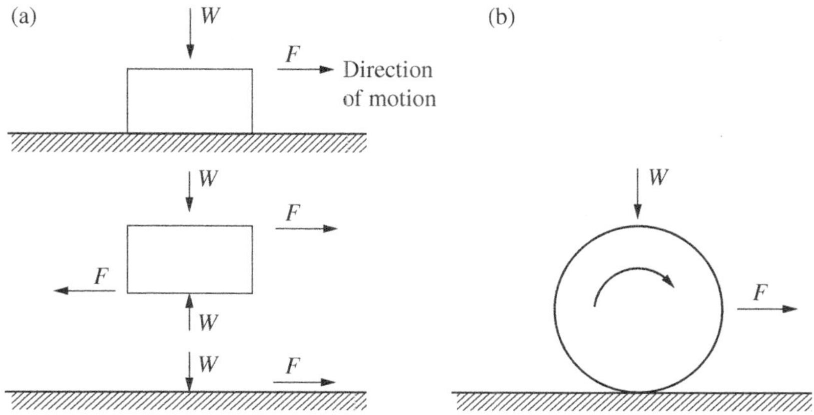

Figure 6.1 Schematic of (a) a body sliding on a surface (free-body diagram) and (b) a body rolling on a horizontal surface. *W* is the normal load and *F* is the friction.

required, such as vehicle brakes, the frictional transmission of power (such as belt drives), and clutches. Nevertheless, most applications indeed tend to reduce friction, such as seals, bearings, gears, and piston rings. In these cases, friction increases the loss of energy and is thus undesirable, and moreover, often alters the wear of contacting surfaces.

6.1.1 Friction Laws and Mechanisms

As a first approximation, friction is described by the three "laws of friction." Two of these are also known as Amonton's law [1, 2], after the first person to give them a clear and straightforward definition, although it is now known in the community that a similar phenomenon was realized and conceived at an earlier date by Da Vinci; however, his notes were unfortunately found much later.

> Amonton's first law defines the frictional force in proportion to the normal load.

Amonton's first law takes the form (6.1):

$$F = \mu W \tag{6.1}$$

where F, W, and μ are the friction force, the normal force, and the coefficient of friction (COF), respectively. The COF is a constant in this linear relationship. This equation is also the simplest and most commonly used definition for friction.

Friction can also be defined or divided by the type of motion, such as common sliding friction and rolling friction. The rolling friction, which is characteristic for ball bearings, is much lower than the sliding friction. In this section, it is primarily sliding friction that is discussed.

Amonton's second law suggests that the friction force is independent of the apparent area of the contact between the contacting bodies.

The second law of friction states that a certain body, irrespective of the size and shape of the contacting surface, results in the same frictional force when moving over a certain counter-body. This can be easily understood, if we think of moving a simple rectangular box over a flat on sides with different sizes: the force that is required to move the box will always be the same.

Third law of friction is known as Coulomb's law.

Coulomb's law defines the friction force (or COF) as a parameter that does not depend on the sliding velocity once the motion has started.

With this [3], a definition of static friction and the kinetic friction is anticipated and a distinction between the two is described.

It has been well established that the force required to start the motion of a body is larger than the force required to continue same motion. This is due to the differences at the interfaces when they are standing still, or moving, all being related to several physical and chemical phenomena, such as the adhesion, roughness, asperity interlocking, and many others, which are described in the following sections. However, it is important to note that the coefficient of static friction is much higher than the kinetic value, which can reach differences up to 20–30% – and even more in some special cases. An illustration of the frictional response describing the differences between the static and the kinetic friction is presented in Figure 6.2.

However, even with very limited experience when it comes to measuring friction, it becomes clear that the COF will not remain constant with the velocity change. There are many interface parameters influencing this change, which include oxidation, the formation of tribochemical layers, the heat, and the increase in contact temperature, which can affect material softening, various asperity transitions, etc.

As described earlier, these three laws are completely empirical; if these laws are not followed in some situations, they do not imply any violation of the more fundamental laws of nature. Actually, friction seldom follows any of these three laws exactly, since it also typically depends on many other parameters: surface morphology, material properties, and environmental conditions being some of the most important. Accordingly, it is difficult to make a prediction for the friction

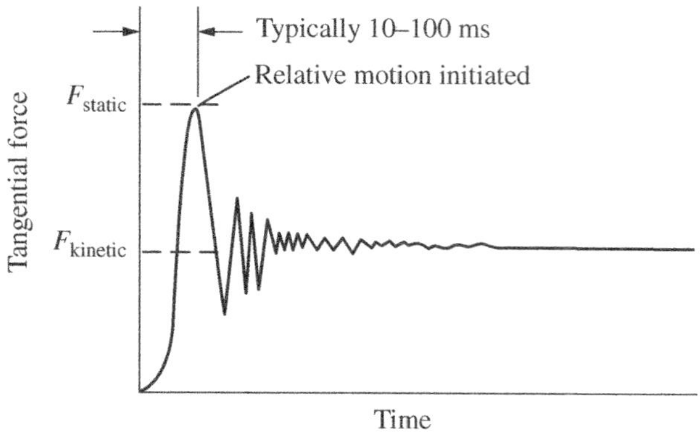

Figure 6.2 Tangential force as a function of time or displacement; F_{static} is the static friction force necessary to initiate motion and $F_{kinetic}$ is the kinetic friction force necessary to sustain motion [1]. *Source:* Reproduced with permission of John Wiley & Sons.

behavior when knowing just the materials and some of the operating conditions [4, 5]. While a rough estimation is of course feasible, the details and accurate values are almost impossible to anticipate. However, it should be noted that friction is not a material property, it is a strong system-response property. This depends on the materials in contact, their properties, the operating conditions, the environment, the tribochemistry and the in-situ-developed interface films, and the structural behavior of the whole system, such as stiffness, for example. If we consider all these, it is obvious that these parameters and all their inter-relations change over the time, and as a result so does the friction.

When two surfaces are in relative motion in the contact, the surface asperities are first touching and so the load is carried at these discrete contact spots. Accordingly, in conditions without lubricants, so-called "dry" contacts, the complete friction force is the sum of all the tangential frictional forces that are opposing the motion on these asperity spots. The conditions at these asperity spots vary a lot due to the wide variation in the topographic features of realistic surfaces, which furthermore influences the friction at these junctions. In general, more than one mechanism acts simultaneously, but most probably some of those will dominate the behavior and define the overall friction behavior, and possibly affect most of the total amount of friction. This suggests that it is very important to recognize the key mechanisms of friction that need to be controlled under specific contact conditions.

Under nonlubricated conditions, friction arises as a result of three mechanisms: adhesion, deformation, and plowing/cutting.

The adhesion term results from the force, which is needed to shear the adhesive bonds that form at the real contact area's asperity interactions. In hard materials, like most metals, in particular steel, ceramics, and others, the deformation term represents the force needed to deform the surface asperities. Plowing and/or cutting is, however, the contribution term when the asperities of a harder material in the contact then groove or cut the surface of the mating material. Plowing can also occur as a result of wear particles trapped in the contact interface. In cases when the contacting material is a viscoelastic (rubbery) material, the friction is largely due to the deformation term caused by elastic deformation hysteresis.

In terms of friction behavior, the above-described mechanisms can be discussed as follows. When two surfaces slide over each other, changes in the interface properties occur, which affect the friction and the wear. Initially, the friction force tends to be very high, due to the nature of the static friction before the motion starts, associated with the asperities' interlocking. The duration of this period, known as "running-in period" can vary and sometimes it is very short, only a few passes, and thus negligible. However, this phase can span over much longer timescale for an inappropriate mating couple.

During the running-in period, some of the highest asperities at the surface will be deformed or fractured, and so wear occurs. The two surfaces adapt to a more compliant form and the real contact area will increase. However, also the tribochemical changes occur at the interface, where the initial surface films from oxides and other contaminants might be worn, and re-formed, and new boundary films can be generated, depending on the surrounding atmosphere and its species, and some material structural changes that might occur.

This run-in period causes the friction to increase or decrease, and is very critical for the whole component's lifetime. Namely, an incorrect run-in can result in severe damage to the surfaces and so early failures. In contrast, a proper run-in can lead to long-lasting, beneficial surface conditions. In practice, surface engineering techniques, such as phosphating or functionally graded multilayers are adopted to facilitate the beneficial running-in period. Nevertheless, the friction force is reported to stabilize to what is known as *steady-state* sliding, after the run-in period.

> Depending on various parameters, including the run-in, there are different scenarios for the behavior and the resulting changes of the COF. These can lead to low or high friction.

It is common that after steady-state sliding and stable friction, the friction increases again and reaches another plateau, as shown by the S-shaped curve in Figure 6.3a. During sliding, even in the steady state, changes to the interface might occur gradually, for example, surface fatigue, roughening, trapping of wear

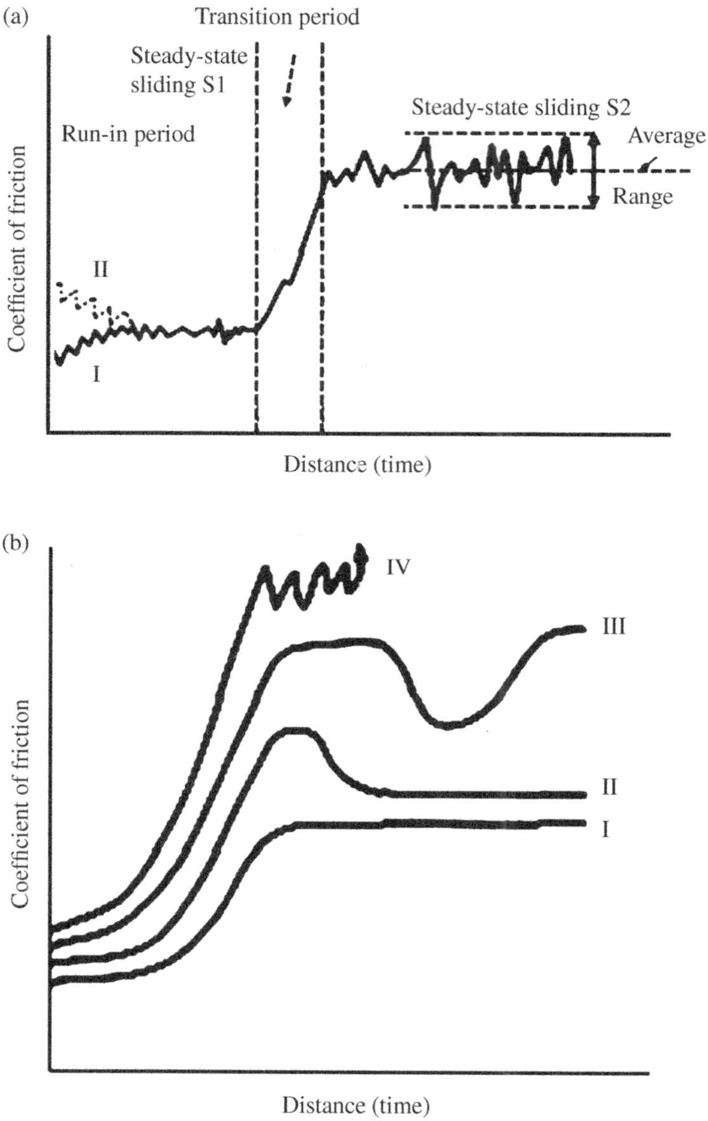

Figure 6.3 COF as a function of sliding distance with a (a) typical S-shaped curve showing running-in period and (b) four hypothetical cases [1]. *Source:* Reproduced with permission of John Wiley & Sons.

particles, and many tribochemical and structural changes, which might lead to an increase in the friction. This process can repeat and the friction can reach several plateaus. If there is a significant amount of wear present, the surfaces will eventually become rough and the friction could become very high. It is worthwhile to mention that the evolution or shape of the friction curve largely depends on interface materials and operating conditions.

Friction can increase in different ways, for example: (i) the friction might remain at its initial value for some time and slowly increase to another steady-state value; (ii) the friction might remain at its initial value for some time and then increase to a high value, followed by a leveling off at a lower value, which is higher than the initial friction; (iii) the friction might increase immediately to a high value, level off at this value, later drop to a lower value, and increase again to a high value; (iv) friction might also change in a completely random, non-repeatable way, as presented in Figure 6.3b. In a number of tribological conditions, the value of COF reaches a high value after a certain period of sliding. Identical metals in contacts exhibit the behavior shown in case (I); the increase is associated with the plowing because of the roughening and the trapped wear particles. For smooth surfaces involving elastic deformation, the occurrence of adhesive friction increases the COF. This is reported to be associated with surface smoothing, due to dominance of adhesive friction [6].

In Figure 6.3b, the decrease of the COF in case (II) is related to smoothening of the two hard surfaces, which undergo notable plastic deformation. This leads to a reduced plowing component of the friction. In cases where there is an adhesive component of the friction, dominant, roughening and/or trapped wear particles, or transfer layers reduce the contact area, and this in fact reduces the adhesive friction term. The friction decrease in the case of (III) in plastic contacts relates to the ejection of wear particles, and a subsequent increase is related to the generation and entrapment of wear particles. A large, unacceptable increase to very high levels of friction in a short period of sliding in case (IV) relates to the wrong selection of contacting materials, where the total friction comes from several sources.

6.2 Types of Friction

6.2.1 Static and Kinetic Friction

It was mentioned earlier that static friction is distinct from kinetic or dynamic friction. While the static friction is characteristic for the contact surfaces just prior to the motion, the kinetic friction occurs when surfaces are already in relative motion. The difference in friction exists because of several reasons. First, the surfaces in contact without any motion interlock at the asperity interactions at the interface. When motion is to begin, these asperities need to be overcome, and this requires some level of their deformation or brakeage, which oppose the motion and thus results in high friction, at least initially at higher value than when in motion. Another major reason of static friction may be the adhesive term. When surfaces rest in contact for longer time, the adhesive interactions may become stronger due to various surface interactions that could occur during this time, and their kinetics, which often takes some time and may not be present in relative motion. In this view, it is reasonable that static friction is higher than when the motion already started and surface therefore slide over each other.

6.2.2 Slip-Stick Friction

Macroscopic stick-slip is a quite complex phenomenon not well described at the macroscale level. In general, in this case we can refer to static and kinetic friction situations that spontaneously occur repeatedly. This causes vibrations and noise from the contact surfaces. In a classical mechanics, it can be described as a mass sliding over the surface, which has some elastic and stiffness properties. If the static friction is large enough, the interface may standing still – "stick" – in spite the part of the body will already move – this is due to its elastic deformation. However, at certain value of tangential load (spring force), this will overcome the static friction and the contact body (mass) will "slip" with high velocity in a sense of damping–stopping motion until the static friction component prevails over the tangential force and again stops the motion, causing stick again. This cycle then repeats.

Stick-slip can be prevented by ensuring situations, where static and kinetic friction coefficients are very similar, designing systems that are more rigid, increasing the inertia-damping properties of the systems, and producing more smooth and more massive contacts.

6.2.3 Rolling Friction

Rolling friction is friction associated with the components in rolling motion, such as rolling bearings, for example. However, there is some controversy in this name. Namely, rolling is by definition situation of contacting bodies, when their relative sliding equals to zero. Friction, on the other hand, is the force that is due to relative motion and opposing it. So, where the rolling friction arises? In fact, in contacting rolling components there is always some elastic deformation, which causes Reynolds slip, namely the relative slip between the surface that is increasing, and one that is decreasing its circumference due to this deformation. This causes micro-sliding in the contacts and so the friction. Another source is Heathcote slip, which, however, is a consequence of kinematic conditions and differences in translatory velocity at the circumference with different radius. Again, rolling is not possible at most of the contacting spots, where the slip occurs and causes friction. Finally, elastic hysteresis of the contacting bodies due to high stresses in rolling is another possible source of rolling friction.

6.3 Friction of Engineering Material Classes

> Friction is not directly related to a particular material, rather it is a property of the tribological system. As a result, the amount of friction cannot be known only from the materials in contact, it is determined by the mating materials, the operating conditions, the stiffness associated with the system and the various contacts, and the conditions of the surface, e.g. the roughness and the topography, and the environmental conditions and any boundary films that are present.

Combinations of materials produce different types of films at the boundaries and show certain types of behavior when in contact. This is a consequence of their various properties that have an effect on the friction. For instance, many materials are characteristically hard; others are brittle and maintain their roughness because they fail in a brittle manner; some materials tend to smear, deform, and become smoother, dramatically increasing the real contact area; some lack chemical stability, some oxidize easily, while others do not. As a consequence, if the conditions are not extreme, we usually have a relatively accurate approximation in terms of the value of the COF. In the literature, these values are used to describe the friction between materials. While this is not strictly the case, they can be useful for understanding the principles of how materials mate and can be used to compare various pairings. They also provide information about the friction losses.

Typical values for a wide range of different materials are listed in Tables 6.1–6.3. The steady-state COF of metallic materials (self-mated and with mild steel, under ambient, non-lubricated conditions) are reported in Table 6.1. How the COF varies depends on the normal load and the sliding velocity. Added to this, a COF of less than 0.3 is impossible to achieve with normal engineering materials. Comparing Table 6.1 with Table 6.2 shows us that many self-mated ceramics have a COF of 0.3 or less. The COFs of ceramics range from 0.3 to 0.7, although COF values as low as 0.1 can result from the use of diamond-based materials.

Clean metallic surfaces are often strongly adhesive and so exhibit high friction and wear when self-mated, which means that if there are no oxides, lubricants, etc., or the materials are in a high vacuum, the steady-state COF can be much

Table 6.1 Typical values for the COF of non-lubricated metals and alloys either in a self-mating contact or with mild steel at room temperature in air [1].

	Coefficient of friction	
Material	**Self-mated**	**On mild steel**
Pure metals		
Precious metals		
Au, Pt, Ag	1–1.5	0.3–0.5
Soft metals		
In, Pb, Sn	0.8–2	0.5–0.8
Metals		
Al	0.8–1.2	0.5–0.6
Co	0.5–0.6	0.4–0.5
Cu	0.8–1.2	0.6–0.7
Fe	0.8–1.5	0.8–1.5
Mo	0.5–0.6	0.4–0.6
Ni	0.7–0.9	0.6–0.9
Ti	0.5–0.6	0.4–0.6
W	0.7–0.9	—
Alloys		
Leaded brass (Cu, Zn, Pb)	—	0.2–0.4
Gray cast iron	0.8–1	0.3–0.5
Mild steel	0.7–0.9	—
Intermetallic alloys		
Co-based alloy	0.3–0.5	—
Ni-based alloys	0.6–0.9	—

Source: Reproduced with permission of John Wiley & Sons.

higher than found in Table 6.1. This is because strong bonds form at the interface, and so the exchange of metal from one contact to the other occurs when there is relative movement.

Plastically deforming materials in a frictional contact can cause the adhesion to increase rapidly; this is the result of junction growth and a larger contact area that has to shear.

Table 6.2 Typical values for the COF of non-lubricated ceramics in a self-mating contact at room temperature in air [1].

Material	Coefficient of friction
Al_2O_3	0.3–0.6
BN	0.25
Cr_2O_3	0.25
SiC	0.3
Si_3N_4	0.25
TiC	0.3
WC	0.3
TiN	0.25
Diamond	0.1

Source: Reproduced with permission of John Wiley & Sons.

Table 6.3 Typical values of the COF for polymers and solid lubricants on a hard surface at room temperature in air [1].

Material	Coefficient of friction
Polymer-plastics	
Acetal	0.2–0.3
Polyamide (nylon)	0.15–0.3
High-density polyethylene (HDPE)	0.15–0.3
PTFE (Teflon)	0.05–0.10
Polymers-elastomers	
Natural and synthetic rubber	0.3–0.6
Silicone rubber	0.2–0.6
Solid lubricants	
Layer-lattice solids	
MoS_2	0.05–0.10
Graphite	0.05–0.15
Non-layer-lattice solids	
Fullerenes (C_{60})	0.05–0.10

Source: Reproduced with permission of John Wiley & Sons.

However, any form of contamination reduces the metal-to-metal contact and the level of adhesion, so lowers the friction [7]. Ultra-fine-grain and nano-grain materials have helped in the understanding of both friction and wear on various length scales. A good example is the paper by Chen et al., who reported low friction in a gradient, nanograined material that was produced by a grinding process [8]. In dry sliding conditions, applying a ball-on-plate tribometer with WC-Co balls, a Cu-5%Ag alloy having a gradient, nanograined surface had a COF equal to 0.29, much lower than a Cu alloy with coarse grains (COF of approximately 0.64). It was suggested that such a large reduction in the COF for a 90 N load could be due to the lowering of the sliding-induced roughening of the surface and some delamination. This supports the idea that nanograined materials can accommodate large plastic strains in their elastically homogeneous but plastically gradient subsurface structures.

In contrast to metals and alloys, the structural ceramics exhibit different mechanical behavior as a result of the different types of interatomic forces associated with the covalent/ionic bonds. The presence of covalent/ionic bonds exhibit restricted plastic flow and less ductility at room temperature. Although adhesive forces – covalent, ionic, or van der Waals – are present at the contact between ceramic materials, the small contact area means relatively small values for the adhesive component of friction compared with the same situation with metallic materials. Furthermore, in a contamination-free situation, the COFs for pairs of ceramic materials do not compare with the much higher values seen for clean metals in an ultrahigh vacuum or in the absence of oxygen [9]. Having said this, ceramic pairs could produce a lot of friction because of the mechanical components of friction, e.g. asperity interlocking and/or asperity fracture. The way in which a surface is prepared and finished is very important if we want to have the right surface conditions for low friction. In addition, humidity is critical for the friction associated with ceramics and the wear. Sometimes it can mean extremely smooth, low-friction tribochemical layers, while in others, it could lead to a hydrothermal change and the danger of a fractured surface as a result of the high friction.

In comparison to metals (and certain ceramics), the majority of polymers have a much better frictional response, which means a smaller steady-state COF. Table 6.3 reveals that a lot of polymers have a COF of 0.2 or less in the unlubricated state. For an explanation, we need to look more closely at the structure of the material. Polymers have a macromolecular, chain-like structure. The chains are made of C, H, O, and N, with covalent bonds being found along the chain, and van der Waals bonds going across the chain. When subjected to stresses, the neighboring chains slide over each other, contributing to the time-dependent viscoelastic deformation observed with polymers. Polymers are of three main types: (i) thermoplastics, like high-density polyethylene (HDPE) and polytetrafluoroethylene (PTFE), that easily deform when heated; (ii) thermosets, like polyoxymethylene and polyamide, that do not deform after they are shaped at elevated temperature; and (iii) elastomers, like rubber, that can deform up to 1000% without fracturing.

Many of these polymers have low friction and in many applications are used as a type of solid lubricant, where they are sometimes referred to as self-lubricating materials when conventional lubricants are added to the polymer matrix. PTFE, polyamide, and HDPE are among the world's most widely used materials, because of their low friction and high capacity for carrying loads. Of the various polymers, PTFE has the smallest friction, as low as 0.05. PTFE is also able to operate well in a wide range of different environments.

The friction and wear associated with biomaterials used for replacement joints often determine the implant's lifespan. Any loss of bone in the region of a prosthesis can result in periprosthetic osteolysis, which reduces the useful lifetime of joints. This medical situation can be caused by bone-resorptive responses to wear debris on the contacting surfaces. In an attempt to more wear-resistant implants, Moro et al. looked at the surface functionalization of polyethylene with respect to the friction and wear observed in contacts with CoCrMo alloys. The experiment used a 12-station hip-joint simulator with repetitive motion and loading (a maximum force of 280 kg and a fixed frequency of 1 Hz) in water-based lubricant with bovine calf serum (25%) [10]. The surfaces of the polyethylene had grafts of the biocompatible phospholipid polymer 2-methacryloyloxyethyl phosphorylcholine (MPC). The friction torque measured for the MPC polyethylene that was grafted and cross-linked was more than six times less than for polyethylene implants that did not have a modified surface. In the same way, a reduction of two orders of magnitude in terms of wear loss was reported for MPC-grafted polyethylene, in comparison with an as-received polyethylene after 3×10^6 cycles (Figure 6.4). Significantly, the wear particles from the MPC-grafted polyethylene were inert in terms of the resorptive responses of the bone when measured using the cytokine expression, RANKL induction, and osteoclastogenesis in a mouse model.

Plastic flow at low loads and temperatures observed with polymers greatly affects their behavior in terms of friction. This is in addition to their well-known time-dependency, which means that their mechanical properties are not constant for a particular time period, especially with respect to high temperatures and large deformations. As a result, the laws associated with friction (Amonton, Coulomb) cannot be easily applied to polymers, because they are very dependent on factors such as load, speed, temperature, and the modification of certain mechanical properties over time. This means that the data in Table 6.3 should be treated with a degree of skepticism. Even though low COFs are possible with polymers, their high wear rate, low melting point, or glass-transition temperature greatly restricts their use in many areas. However, composites formed from polymers and ceramics can have better wear resistances.

Layered-structure materials like MoS_2 and graphite can have a COF as small as 0.05. Such materials are often employed as solid-form lubricants. The much lower friction observed in so-called layered structures results from their basal planes being covalently bonded, which implies very strong bonds, while the bonds between these layers are relatively weak, being the result of van der Waals forces. The result is a material that shears relatively easily.

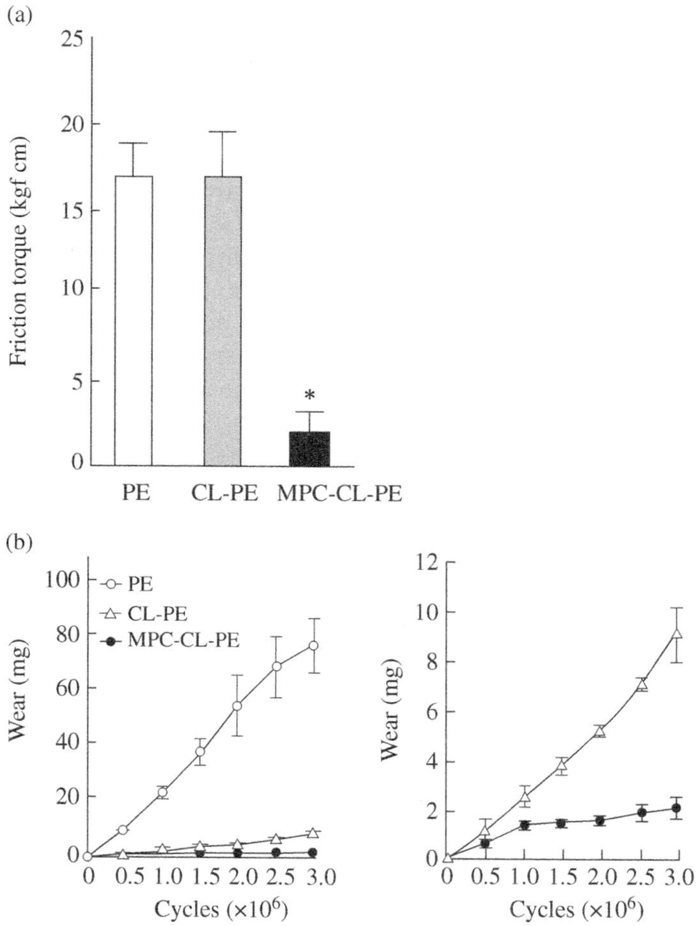

Figure 6.4 Friction torque and the amount of wear in a hip-joint simulator with PE, Cl-PE, and MPC-CL-PE liners. (a) Friction torque against the femoral heads measured before the loading test. (b) Wear quantity as a function of loading cycles. (c) Wear quantity for every 5×10^5 cycles. (d) Typical SEM images of wear particles isolated from lubricants of the simulators with CL-PE and MPC-CL-PE liners. The size distribution of particles is also shown [10]. *Source:* Reproduced with permission of Springer Nature.

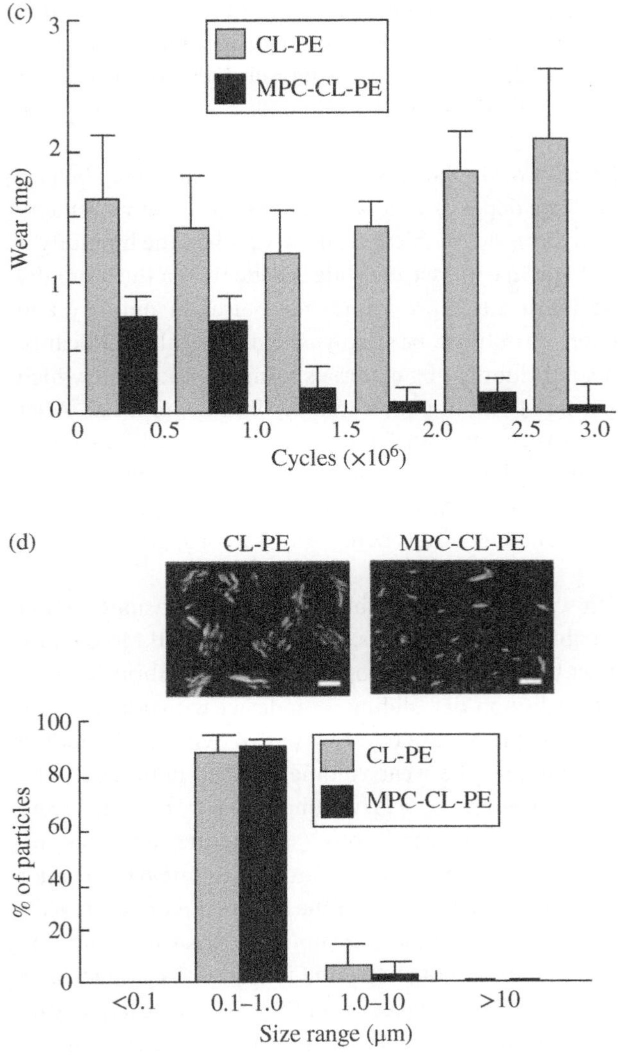

Figure 6.4 (Continued)

While explaining the origin of friction in layered materials, Koren et al. reported on measuring the line tension and friction in graphite with its layered form [11, 12]. Their results suggested friction's stochastic basis, originating in the interactions between the lattices of incommensurate interfaces. The adhesion energy (227 mJ/m^2) was attributed to the layers in materials like graphite, which reduce stiction forces and the amount of friction due to sliding. The interesting phenomenon of super lubricity derives from the mismatch between surfaces that are flat at the atomic level. These have the ability to revert spontaneously to the original equilibrium state, immediately once any externally force is taken away. All these small-scale tests and theories were based on scanning probe microscopy.

Although graphite and MoS_2 are similar in structural terms, how they behave can very often be different. This depends to a large extent on the environment. MoS_2 is the better material in terms of low friction and wear when the humidity is low and the vacuum is ultrahigh. In contrast, graphite is better when the humidity is high. As a result, these lubricants, have found many uses in industry and beyond. They can be powders, thin films, or simply added to regular lubricants. Most of the so-called "soft-solid" lubricating materials form a surface film, which means that after a brief period of running-in, a transfer film is produced. After this, a real contact forms between the transfer film and the solid lubricant. A material such as MoS_2 can be produced as nanoparticles, in the shape of fullerenes, nanowires, or nanotubes. Currently, a number of practical applications are being investigated, with these nanomaterials being added to oils, greases, and a variety of coatings.

Problems associated with water-lubricated rubber stern tube bearings restrict the effectiveness of submersibles. To address this, Dong et al. looked MoS_2 nanoparticles and how they affect the tribological properties of water-lubricated rubbers [13]. The tests employed a disc-on-disc sliding-wear device with a speed of up to 3.3 m/s under a normal contact pressure of 0.3 MPa. With or without the adding of MoS_2 to the rubber, the COF and the wear volume reduced nonlinearly the speed dropped until a COF of 0.06 and a wear volume of 0.07×10^{-3} cm^3 were achieved in hydrodynamic lubrication conditions (see Figures 6.5a and b). Lubrication models using an SEM were also put forward (see Figure 6.6) to better comprehend how the MoS_2 affected the tribology of the system. Figure 6.7 reveals MoS_2 nanoflakes bulging on the worn surface, forming a contact with the ring disc. Furthermore, the nanospheres do not enable the forming of a contact with the ring disc. It is clear that the MoS_2 nanoflakes lower the COF; however, they do not affect the damping or the mechanical properties. In contrast, the MoS_2 nanospheres increase the tribological and mechanical properties.

WS_2 also has potential as an additive for lowering friction. Despite a lack of clarity regarding the mechanisms involved, the formation of sheet-like, low-shear layers and the ball-bearing-type properties of the nanoparticles are commonly proposed [15].

Friction occurs as a result of several non-equilibrium processes that involve the detaching and re-attaching of many macro-contacts at the surfaces that mate in the contact. To better comprehend friction on the nanoscale, atomic force microscopy (AFM) can be used to reveal the many single-asperity contact points

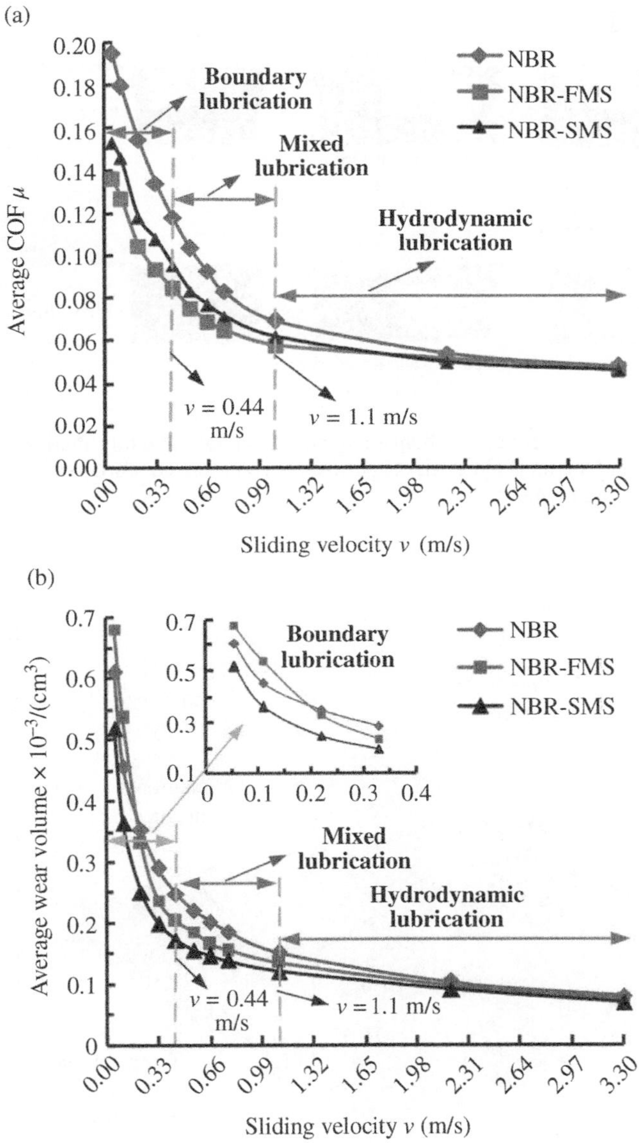

Figure 6.5 (a) Average COF as a function of sliding velocity for a rubber and $ZCuSn_{10}Zn_2$ ring-disc system with different MoS_2 nanoparticles. (b) Average wear volume of rubber ring disc as a function of sliding velocity with different MoS_2 nanoparticles. *Source:* Conglin Dong et al. [13]. https://www.nature.com/articles/srep35023. Licensed under CC BY 4.0.

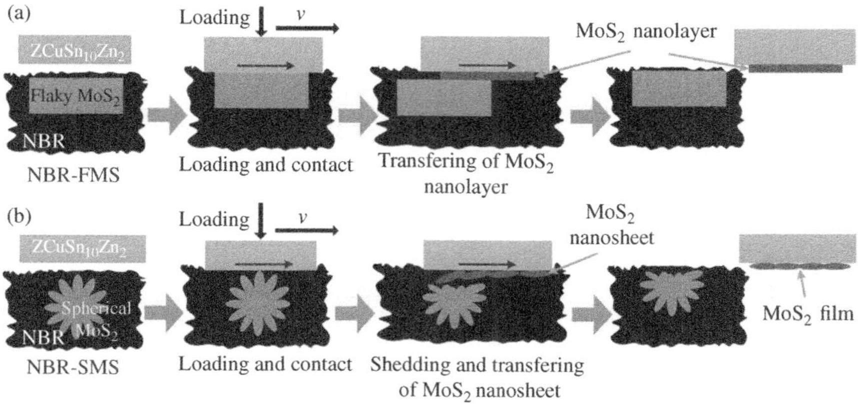

Figure 6.6 Schematic of the lubrication mechanism model for (a) flaky and (b) spherical MoS_2 nanoparticles. *Source:* Conglin Dong et al. [13]. https://www.nature.com/articles/srep35023. Licensed under CC BY 4.0.

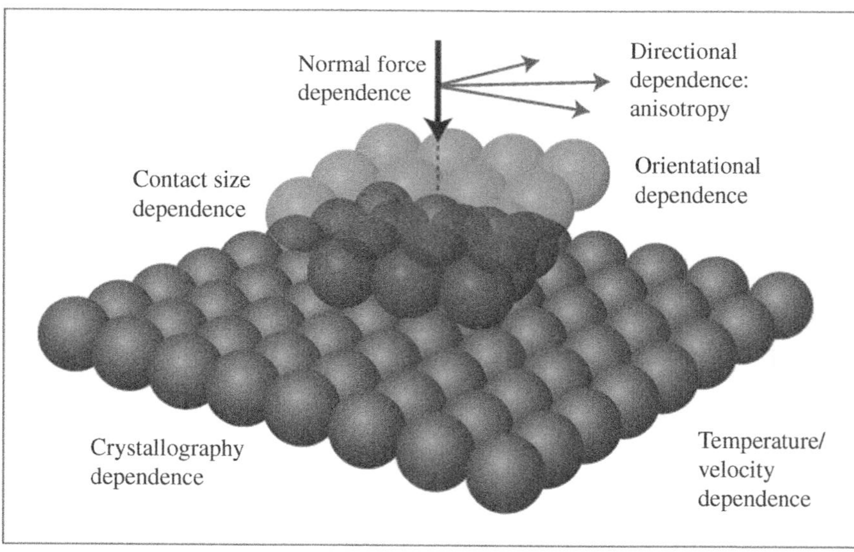

Figure 6.7 Scheme of a typical AFM experiment to study friction on the nanoscale [14]. *Source:* Reproduced with permission of Springer Nature.

(see Figure 6.7) that are subjected to GPa-level stresses [14], in comparison to macroscopic contacts for ceramics or metals, which are in the MPa range, and polymers, which are in the KPa range for self-mated contacts. In addition, the amplitudes observed during sliding in AFM testing for single asperities are in the range of just a few micrometers. Contrast this with the Hertzian contact diameter in a macroscopic contact, which ranges from hundreds of micrometers to millimeters. These differences are also seen very clearly in the values recorded for the friction. An additional key feature is that the friction is not just the sum of single-asperity effects, rather it is the summation of the temporal and spatial dynamics of the various processes that take place in the contacts.

Molecular dynamics-type simulations were widely used to comprehend both wear and friction in the area of tribology. As an example, Pastewka et al. looked at micron-scale effects occurring when polishing with diamond paste aided by classic molecular dynamics simulations [16]. Sliding at 30 m/s and a 10 GPa contact stress were applied, and showed how an amorphous layer developed at the interface as the result of a sp^3-to-sp^2 transition of the order–disorder type. The amorphous layer between the diamond surfaces becomes thicker, depending on the direction of the sliding and the orientation surface, which points toward the importance of anisotropic amorphization in the case of such diamond materials. Basically, this kind of anisotropy is connected to the dissociation of the covalently formed bonds as a result of the occurring stresses. The same authors also put forward a pilot-atom-based idea in order to provide an explanation for the microscale origin of the anisotropy. Following this idea, a single atom is displaced from the crystal lattice to create a bond with any of the available amorphous carbon (a-C) atoms in the sliding process. The atom has two possibilities that depend on the slope maxima of the interatomic potential wells; it can connect with the a-C or it can remain at the original place in the lattice.

Only a few tribological studies have been carried out close to the melting point of metals. In Zykova-Timan et al., the authors presented molecular dynamics results that enabled them to investigate the high-temperature nanoscale friction of a NaCl (100) surface [17]. A rigid apex diamond tip was employed for the modeling, a sharp conical tip was attached to see how the plowing contributed to the friction. Finally, a blunt, flat 200-atom tip was employed to determine the grazing contribution to the overall friction. Figure 6.8 shows how the friction sharply drops rapidly at $T/T_m \sim 0.7$ and so-called skating is the dominant feature until the friction practically disappears at $T = T_m$. The results of the simulations (Figure 6.9) showed that this loss of friction is related to the forming of a local thin cloud of Na^+ and Cl^- ions that have liquid-like mobilities. This also makes it possible to have a skating motion with tip. In addition, any scratches that form at a lower temperature ($T/T_m < 0.5$) can then heal rapidly at the elevated temperature ($T/T_m > 0.8$). This involves a timescale of picoseconds at $T \sim T_m$. Contrastingly,

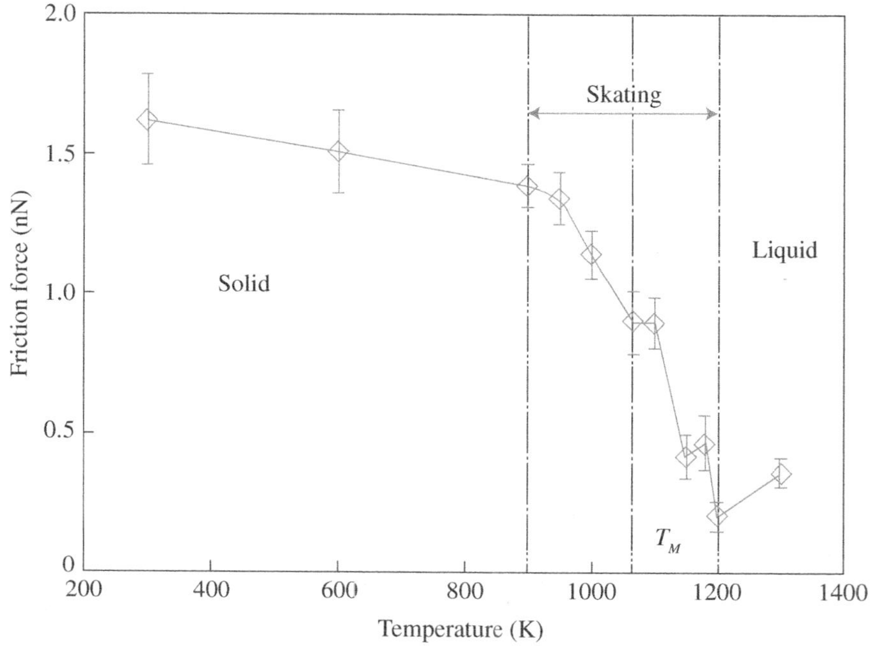

Figure 6.8 Temperature dependence of the plowing frictional force [17]. *Source:* Reproduced with permission of Springer Nature.

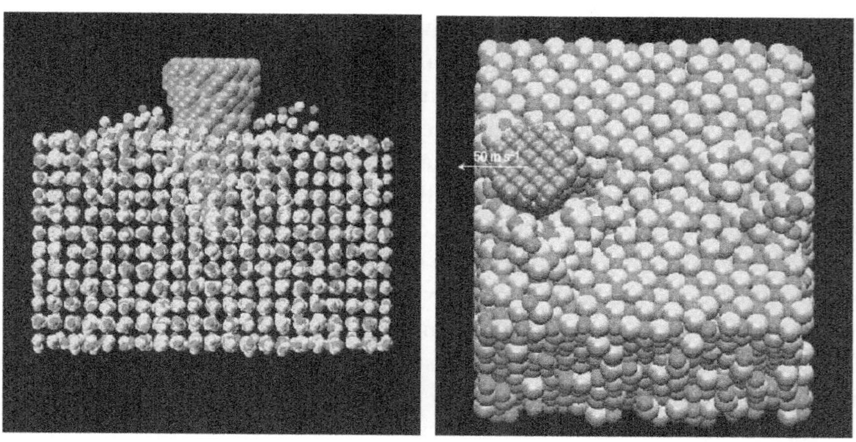

Figure 6.9 MD simulation of a hard tip plowing through solid NaCl (100) surrounded by a local liquid cloud at T_m [17]. *Source:* Reproduced with permission of Springer Nature.

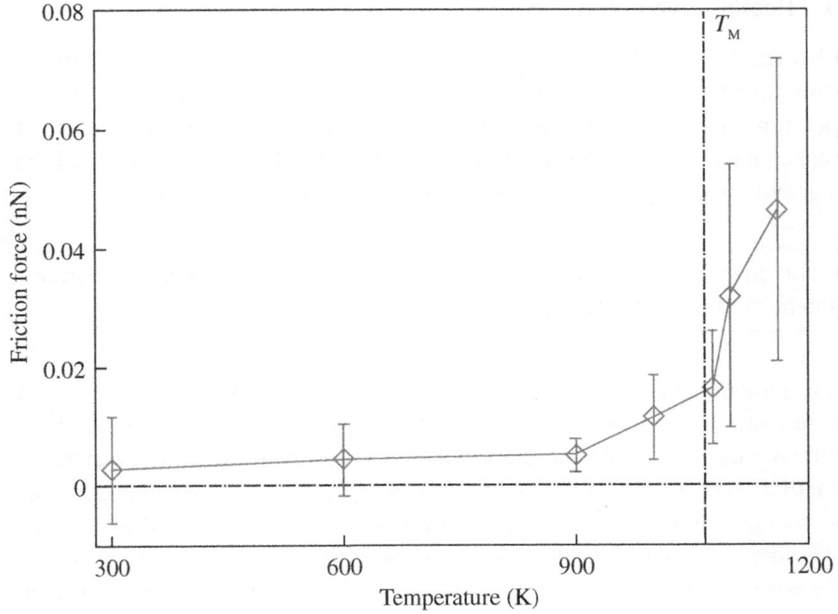

Figure 6.10 Temperature dependence of the adhesion friction of a grazing tip on NaCl (100) [17]. *Source:* Reproduced with permission of Springer Nature.

the "wear-less" form of friction observed with a large flat tip that grazes the surface of the NaCl (100) in a zero-load makes contact with the substrate only in terms of adhesion, which means a high-temperature frictional peal close to T_m (Figure 6.10). Despite the fact that Zykova-Timan suggested that such simulations could be carried out for other low-melting-point systems, there have been no more reports on this subject.

6.4 Frictional Heating and Temperature at the Interface

Many materials have been looked at in terms of their tribological potential and how the material is actually removed from the contact interfaces. However, a better sense of what is actually happening needs an investigation of the physicochemical and thermomechanical interactions that take place at the interface. Here, there have been many such studies [18–32]. This section discusses heating due to friction and the temperature at the interfaces of contacting surfaces during sliding.

6.4.1 Heating Due to Friction

Two bodies that are in contact and moving relative to each other do work by overcoming friction. Such work is dissipated primarily in the form of heat. The temperature of the contacting surfaces very much depends on the dimensions of the real contact area, together with the COF, the applied load, the velocity of the sliding, and the thermal characteristics of the bodies in the contact.

> Heating due to friction, which only occurs in real contacts, produces an abrupt gradient in the subsurface layer's temperature.

Such gradients can reach hundreds of degrees over a micrometer, reaching down just a few micrometers, see Figure 6.11. The distribution of the subsurface temperature is related to the thermal characteristics, for example, the heat capacity, the conductivity, and the diffusivity, and to the sliding velocity and the applied loads.

The friction and wear when contacts are sliding very much depend on the friction-based energy that is able to dissipate at the contact. The friction from sliding bodies raises the temperature in the contact and creates energy, which leads to deformation, cracking, or tribochemical reactions.

The energy that is dissipated is as follows:

$$E_{\mathrm{d}} = \mu W v t \tag{6.2}$$

where μ is the COF, W is the load applied normally, v is the fretting velocity, and t is the time of the testing. The energy dissipated is calculated from Eq. (6.2). The energy can trigger a variety of mechanisms associated with wear. This energy plays a role in the mechanical transformations, the tribochemical effects, the third-body changes, and the thermal effects [17].

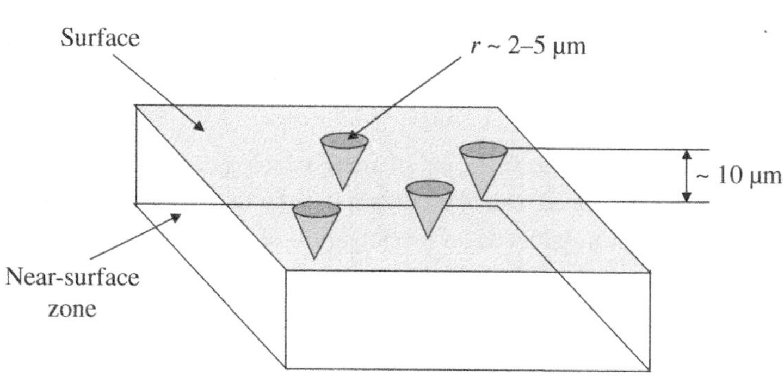

Figure 6.11 Schematic of phase-transformed material formation at the asperity contacts due to the extreme surface temperatures and the extreme thermal gradients, as per the evidence presented in Refs. [33, 34]. *Source:* Adapted from Vodopivec et al. [33] and Kalin [34].

6.4.2 Understanding the Temperature in the Contact: The Bulk and Flash Temperatures

The forces in a contact are normally shift from a body in motion to a body at rest, or vice versa [33, 34] (Figure 6.11). For sliding contacts, a transversal force only acts across the asperities, while the surfaces in the contact are subject to elastic or plastic deformations when subjected to the load, of course depending on the characteristics of the material (E, H) or the roughness of the surface (asperity size and shape). The contact stresses and how they are distributed are also related to the friction and the configuration of the contact (this means circular in the case of a sphere versus spherical or rectangular in the case of crossed cylinders). The temperature in the contact affects the oxidation-related wear for metallic materials [35, 36], and the tribo-wear for conventional materials, including ceramics [37–42] and composites [39, 40, 43]. The reason for this is that at the atomic level, chemical and mechanical properties depend on the energy accessible to initiate the processes [44]. The friction and wear are very much affected by how the microstructure, mechanical, and physical characteristics of the materials in the contact are affected by the temperature. A lot of materials degrade in terms of the mechanical and lubricating properties as the temperature at the interface increases.

> The interface temperature is the result of the flash temperature and the bulk temperature.
>
> The contact-area temperature, which is not related to the instantaneous generation of heat from friction, is referred to as the bulk temperature. In contrast, the temperature increase actually in the contact as a result of friction is referred to as the flash temperature.

Of course, the challenge here is to define the area of the contact, because the majority of the heat dissipates from the contact to the bulk via the contact area. However, the area of the contact is not as big as the apparent contact; it is smaller, and is equal to real contact area. To provide an illustration of this effect, the surface interactions on the microscale are highlighted in Figure 6.12. Such an interaction must be defined by the multi-asperity interactions between two rough surfaces. So, the primary question is: how big is the real contact area in the sliding contact? This is not generally appreciated, but it does greatly affect any calculations of the contact temperature [34].

> The models used to measure the real contact area are different and there is no one generally accepted and properly validated model.

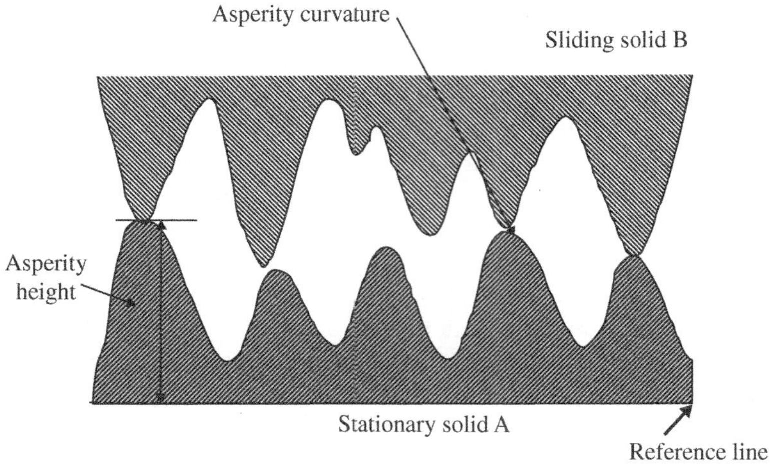

Figure 6.12 Schematic of multi-asperity contacts at the tribological interface.

Furthermore, some models include a contact area, so it is hard to make a comparison between different models with the same inputs. In addition, there are many questions surrounding how the "contact" temperature is different for the asperities in the contact. As a result, in many publications the calculations of the temperature can vary a great deal. Finally, the problem is made worse by the fact some people refer to apparent contacts, while other people use a real contact area.

> For a real contact point, the true contact pressure come from dividing the nominal load by the ratio of the real to the apparent contact area.

This is illustrated in Figure 6.13. All of the two-dimensional contact points are asperities that could be loaded (or not). This relates to how the asperities are distributed, which is hard to calculate. All of these problems have been comprehensively addressed by Kalin and coworkers [30, 34, 45].

> For a body at rest in contact with a body in motion, the peak flash temperature is related to the normal load, the sliding speed, the COF, and the thermal characteristics.

The temperatures in the contact area, as a result, dependent on the supply of heat, the size and velocity of the source of the heat, and the thermal characteristics of the materials being investigated. The friction at an interface generates heat

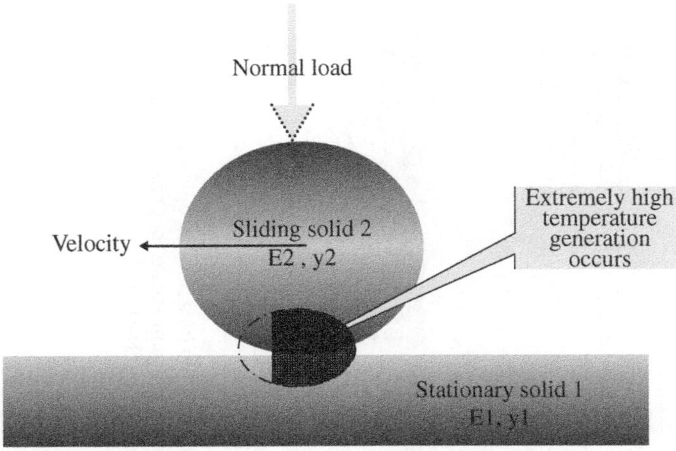

Figure 6.13 Schematic of a sphere making relative movement on a nominally flat surface. *E* and *v* are the elastic modulus and Poisson's ratio, respectively, of the solids in contact.

and transports the heat between the contacts, and depends on the transmission probability of the heat-carrying phonons that are across the contact. Gotsmann and Lantz studied these effects experimentally [46]. They put forward a model that considers the pressure dependence for the conductance of the thermal interface (see Figure 6.14) and suggested that the conductance at the thermal interface is reduced at a pressure of 100 MPa. The experiments were additionally related to the concept of quantized thermal conductance across contacts that are atomically smooth.

Bhushan [1] put forward a new model that considered both the transient and steady-state flow of heat, and the contact's thermal resistance with respect to the slider and the clamp. They initiated temperature maps so as to more easily envisage the model. These temperature maps are diagrams having the normal load and the velocity on each axis, with plots of the temperature contours. The temperature is then determined for each load–velocity combination using the map. They calculated the temperatures for a number of the contacts (so-called material pairs), by taking into consideration the pressure and the normalized velocity. This can be convenient for estimating the temperatures in the contact; however, it is important to remember that the model has assumptions that need to be considered. If not, large differences and misinterpretations of the results can occur [30, 34, 45]. A characteristic temperature map of a ceramic–ceramic contact can be seen in Figure 6.14. It is clear that increases in temperature occur when the velocity or the load are increased.

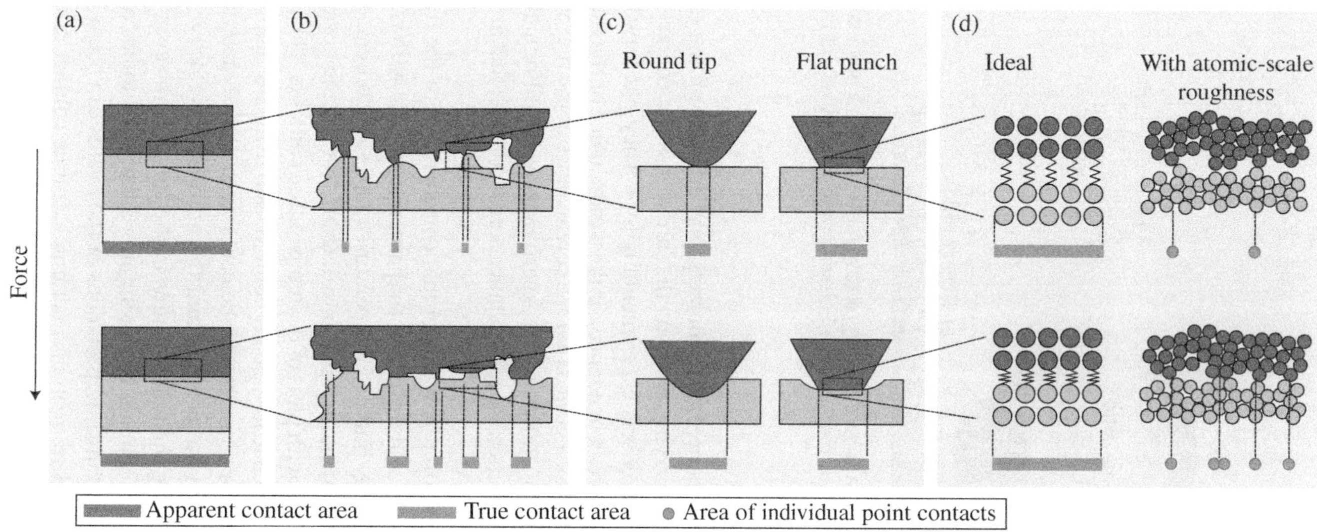

Figure 6.14 Schematic illustration of the pressure dependence of the contact between two bodies in (a) a macroscopic contact, (b) a multi-asperity contact, (c) single asperity round tip and flat tip contacts, and (d) atomistic ideal and atomic-scale roughness contacts. Also shown are the corresponding contact areas [46]. *Source:* Reproduced with permission of Springer Nature.

6.5 Analytical Models Used to Predict the Temperatures in the Contact

High temperatures occur in multi-asperity contacts (see Figure 6.12) for brief periods of time (~10^{-3} seconds or less, depending on sliding speed) and over short distances (~10^{-4} m, depending on the distribution of the load and the size of the asperities).

It is hard to determine the temperature in a contact with any degree of accuracy. As a consequence, it is usual to estimate the temperature with Blok's theory. When these models were first developed, the outcomes were very inconsistent. Some took into account mainly the physics of the contacts and the transfer of heat, while others were more focused on the material factors and experimental input. There has been many experiment- and theory-based investigations, resulting in a large number of papers and reports. However, the difficulty is in accurately knowing how the contact temperature relates to the real contact area (which is unknown), the real contact "geometry" (also unknown), the temperature-dependent material properties, and the distribution of the friction at the various asperities [30, 34, 45].

All the developed models assume ideal behavior on the contact. In a tribocontact, an asperity of the solid B slides harmonically on the at-rest surface of solid A (see Figures 6.12 and 6.13). At such a contact, the normal force is transmitted from the oscillating surface to the at-rest surface across a single asperity. When deriving the model, it makes no difference whether it is solid A or solid B that is moving. On this basis, theoretical and maximum flash temperatures were calculated by a number of researchers. Tian–Kennedy's average and maximum flash-temperature model [47], Greenwood–Greiner's average flash-temperature model for fretting [48], and Ashby–Abulawi–Kong's bulk and flash temperature model [49]. These models are different in terms of their assumptions regarding the physical, dynamic, and geometrical characteristics [30, 36]. For example, the geometry used by Ashby et al. can be seen in Figure 6.15. By convention, one body is considered to be at rest, while the other is moving.

In the different analytical models for computing the temperatures of the tribological interfaces, the following are assumed:

1) There is an elastic Hertzian contact between the solids.
2) The heat generated in the contact is only the result of dissipated frictional energy.
3) The siding of two rough surfaces is considered to be the same as sliding of one rough surface with the equivalent or composite roughness over a smooth surface.
4) There is a finite temperature gradient perpendicular to the contacting surface.

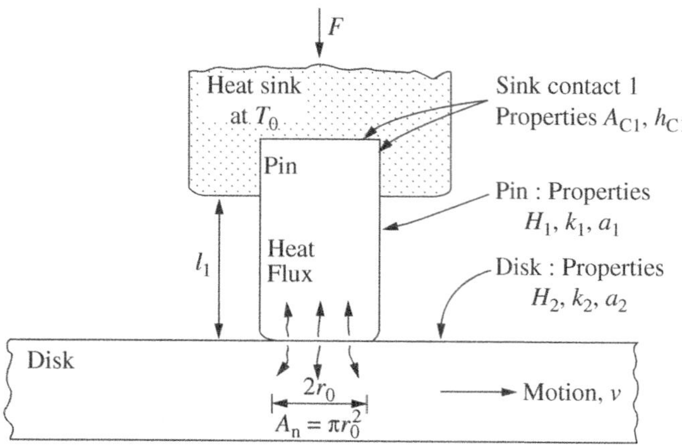

Figure 6.15 Schematic of frictional heat generation in a pin-on-disk configuration [49]. *Source:* Reproduced with permission of Taylor & Francis.

5) The maximum temperature in the contact point is the same for the two solids.
6) In a fretting contact the heat dissipation to the surroundings can be neglected.
7) The flash temperature is the average of the maximum temperature in an asperity–asperity contact.
8) There is a steady-state condition for heat generation.

6.6 Implications of the Important Contact Temperature Models

From the preceding discussion in this chapter, it is perceivable that the sliding between two solid surfaces generates heat at the interface and this can be extracted from the contact zone by one or combination of the mechanisms, conduction, convection and radiation. Which one of these dominates depends on the tribological environment. In case of dry sliding, the heat would dissipate by conduction to both the mating solids. In contrast, the heat dissipates by convection or evaporation in a cryogenic environment (for example, liquid nitrogen) [50, 51]. For the simplistic ball-on-flat configuration, the energy dissipation can be determined as [1],

$$q = \mu P_a V \left(\frac{A_a}{A_r} \right) \tag{6.3}$$

where q is the heat generated per unit area (J/m^2), μ is the COF, P_a is the load per unit apparent area (P/A_a) (N/m^2), P is the load (N), A_a is the apparent area of contact (m^2), A_r is the real area of contact (m^2), and V is the sliding velocity (m/s).

From the above equation, it should be evident that more heat will be generated at the interface experiencing higher COF. Such situation will lead to a higher flash temperature.

6.6.1 Archard Model

Based on the assumption that heat is generated at true contact area and thereafter conducted away into the bulk of the mating solids, Archard [21] proposed a model to quantify the contact temperature (Figure 6.16). In his model, it was further assumed that the peak temperature occurs when the heaviest load is concentrated on the smallest area, i.e. when the applied load causes plastic deformation on a single area.

The flash temperatures are calculated assuming that the heat comes from the area of true contact and that this heat is conducted away to the bulk of the materials in the sliding contact.

The sliding system approximates to a finite but long band heat source with a uniform distribution of the heat (Figure 6.16).

Using the Peclet number ($L = Va/2\chi$), the criterion for heat sources moving in different velocity ranges can be defined, and the following expressions for the flash temperature are the result:

$$\theta_{\mathrm{m}} = 0.25NL \qquad \text{at low speeds; } L < 0.1 \tag{6.4}$$

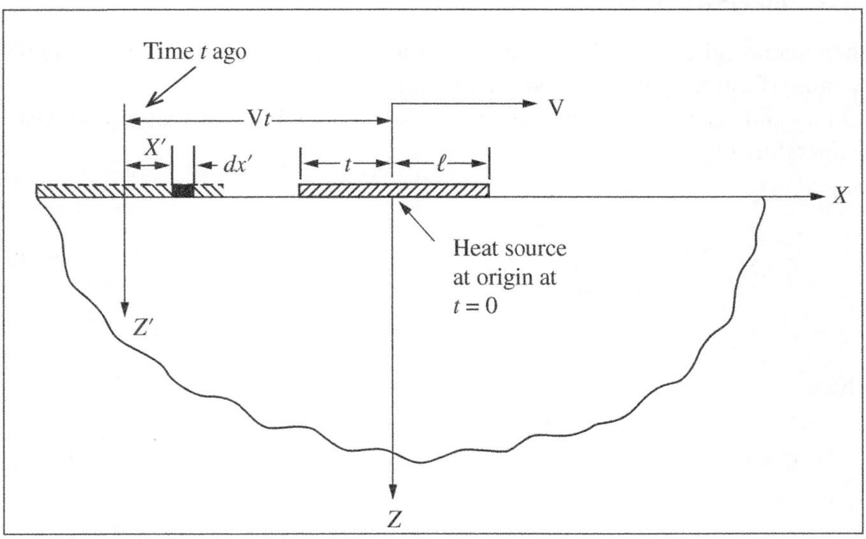

Figure 6.16 Schematic of a moving band heat source (length "2l") on a semi-infinite solid body (stationary), illustrating the concept used in determining the contact temperature [1]. *Source:* Reproduced with permission of John Wiley & Sons.

$$\theta_m = 0.25\beta NL \quad \text{at moderately low speeds; } 0.1 < L < 5; \tag{6.5}$$

$$\theta_m = 0.345NL^{1/2} \quad \text{at high speeds; } L > 100; \tag{6.6}$$

$$\theta_m = 0.435\gamma NL^{1/2} \quad \text{at moderately high speeds; } 5 < L < 100; \tag{6.7}$$

where

$$N = \frac{\mu g}{J}\frac{\pi p_m}{\rho s} \quad \text{and} \quad L = \frac{P^{1/2}v}{2\chi(\pi p_m)^{1/2}} \tag{6.8}$$

and where μ is the COF, g is the acceleration due to gravity, J is the mechanical equivalent of heat, p_m is the flow or yield pressure, s is the specific heat, ρ is the density, P is the normal load, v is the sliding speed, χ is the thermal diffusivity, β is a constant that ranges from approximately 0.95 at $L = 0.1$ to approximately 0.5 at $L = 5$, and γ is a constant that varies from approximately 0.72 at $L = 5$ to approximately 0.92 at $L = 100$.

The flash temperature increases in a linear fashion with the sliding velocity and is proportional to (load) 0.5 at low velocities, while it is proportional to (velocity) 0.5 and (load) 0.25 at higher velocities.

6.6.2 Kong–Ashby Model

Phenomenologically speaking, a high sliding velocity increases the contact temperature (flash temperature) and causes oxidation.

Kong and Ashby [52] proposed an alternative model for calculating the flash temperature (T_f):

$$T_f - T_b' = \frac{\mu P v}{A_r}\left[\frac{1}{\frac{k_1}{l_{1f}} + \frac{k_2}{l_{2f}}}\right] \tag{6.9}$$

where

$$T_b' = T_b - \frac{A_r}{A_n}(T_b - T_o) \tag{6.10}$$

$$T_b = T_o + \frac{\mu F v}{A_n}\left[\frac{1}{\frac{k_1}{l_{1b}} + \frac{k_2}{l_{2b}}}\right] \tag{6.11}$$

$$\frac{A_r}{A_n} = \frac{P}{P_s} \tag{6.12}$$

$$l_{1f} = \frac{r_j}{\pi^{1/2}} \tan^{-1}\left[\frac{n2\pi\chi_1}{r_j v}\right]^{1/2} \tag{6.13}$$

$$l_{2f} = \frac{r_j}{\pi^{1/2}} \tan^{-1}\left[\frac{n2\pi\chi_2}{r_j v}\right]^{1/2} \tag{6.14}$$

$$r_j = r_0 \left\{\left(1 - \frac{P}{P_s}\right)\left(\frac{r_0}{r_a}\right)^2 + 1\right\}^{-1/2} \tag{6.15}$$

$$P_s = \frac{A_n H_0}{\left(1 + 12\mu^2\right)^{1/2}} \tag{6.16}$$

$$H_0 = \text{Least of}\left(H_1, H_2\right) \tag{6.17}$$

In the above, T_f is the flash temperature (K); A_n is the nominal contact area; A_r is the real contact area; χ_1 and χ_2 are the thermal diffusivities associated with 1 and 2, P is the normal load; P_s is the seizure load; H is the hardness, k_1 and k_2 are thermal conductivities of 1 and 2; l_{1f} and l_{2f} are the equivalent linear heat-diffusion distances for surfaces 1 and 2 for flash heating; l_{1b}, l_{2b} are the equivalent linear heat-diffusion distances for two surfaces in contact, bulk heating (m); n is the lifetime of a contacting asperity; r_a is the radius of a junction; r_j is the radius of a contact junction (that can be made up of many unit asperities); r_0 is the radius of the nominal contact area; T_b' is the effective bulk temperature for bulk heating; v is the sliding velocity; and μ is the COF. In the Kong–Ashby model, the flash temperature is in proportion to the (load) 0.5; however, its relationship with the sliding velocity is complex.

> Archard's model and the Kong–Ashby model predict a similar dependence on load at slow velocities, but they deviate at higher sliding velocities and in terms of the sliding dependence of the flash temperature.

6.7 Closure

It is clear that friction is a property that depends on the complete system, i.e. friction does not only depend on the material or materials but also depends on the environment and the operating characteristics. Similarly, the friction for similar operating and environmental conditions could be very different for a different pair of materials.

In other words, to say the "friction of iron is 0.5" has no meaning, unless the operating conditions and the other material in the contact are specified.

Since friction is only influenced by the surface, applying a solid lubricant to an otherwise high-friction surface can give us a system with very low friction. Typically, the friction of a material combination has two stages: the running-in time and the steady state. Since the transition from one to the other happens quickly after the start of the asperity–asperity interactions, it is the steady-state response that is more important for tribology-associated applications. It is not correct to group the friction and wear properties for a given system, because it is possible to have low friction and a high wear rate, and high friction and a low wear rate. Atomic-force-microscopy- and molecular-dynamics-based simulations can help to understand the complexity in friction and wear properties that involve single-asperity contacts.

Theoretical models are available to estimate the contact temperature at an interface. But the model has to be chosen in accordance with the test parameters (load, sliding velocity, etc.). Also, the material data for the temperature window of the contact temperatures must be employed. This is vital because thermal properties and mechanical properties depend on the temperature, and so room-temperature data are not necessarily going to be very useful. It is also important that the flash temperature is well estimated in the context of any transition in the applicable wear mechanisms.

References

1 Bhushan, B. (1999). *Principles and Applications of Tribology*. New York: Wiley.

2 Amontons, G. (1699). De la resitance cauisee dans len Machines. *Memoires de l'Academie Royale A*: 257–282.

3 Coulomb, C.A. (1785). Théorie des Machines Simples, en ayant regard au Frottement de leurs parties, et á la Roideur des Cordages. In: *Mémoire de Mathématique et de Physique de l'Académie Royale*, 161–342.

4 Guerra, R., Tartaglino, U., Vanossi, A., and Tosatti, E. (2010). Ballistic nanofabrication. *Nature Materials* 9: 634–637.

5 Kisiel, M., Gnecco, E., Gysin, U. et al. Suppression of electronic friction on Nb films in the superconducting state. *Nature Materials* 10: 199–122.

6 Bhushan, B. and Kulkarni, A.V. (1996). Effect of normal load on microscale friction measurements. *Thin Solid Films* 278: 49–56. 293, 333.

7 Buckley, D.H. (1981). *Surface Effects in Adhesion, Friction, Wear and Lubrication.* Amsterdam: Elsevier.

8 Chen, X., Han, Z., Li, X., and Lu, K. (2016). Lowering coefficient of friction in Cu alloys with stable gradient nanostructures. *Science Advances* 2 (12): e1601942. https://doi.org/10.1126/sciadv.1601942.

9 Hutchings, I.M. (1992). *Tribology: Friction and Wear of Engineering Materials.* Boca Raton, FL: CRC Press.

10 Moro, T., Takator, Y., Ishihara, K. et al. Surface grafting of artificial joints with a biocompatabile polymer for preventing periprosthetic osteolysis. *Nature Materials* 3: 829–836.

11 Liechti, K.M. Understanding friction in layered materials. *Science* 348: 632.

12 Koren, E., Lorstscher, E., Rawlings, C. et al. Adhesion and friction in mesoscopic graphite contacts. *Science* 348: 679–683.

13 Dong, C., Yuan, C., Wamg, L. et al. Tribological properties of water lubricated rubber materials after modification by MoS_2 nanoparticles. *Scientific Reports* 6: 35023. https://doi.org/10.1038/srep35023, 1–12.

14 Urbakh, M. and Meyer, E. Nanotribology: the renaissance of friction. *Nature Materials* 9 (1): 8–10.

15 Wahl, K.J. (2012). Positively "negative" friction. *Nature Materials* 11: 1004–1005.

16 Pastewka, L., Moser, S., Gumbsch, P., and Moseler, M. Anisotropic mechanical amorphization drives wear in diamond. *Nature* 10: 34–38.

17 Zykova-Timan, T., Ceresoli, D., and Tosatti, E. Peak effect versus skating in high-temperature nanofabrication. *Nature Materials* 6: 230–234.

18 Blok, H. The flash temperature concept. *Wear* 6 (963): 483–494.

19 Archard, J.F. (1958). The temperature of rubbing surfaces. *Wear* 2: 438–455.

20 Griffiths, B.J. (1985). White layer formations at machined surfaces and their relationship to white layer formations at worn surfaces. *ASME Journal of Tribology* 107: 165–171.

21 Archard, J.F. (1988). The temperature of rubbing surfaces: part 2, the distribution of temperatures. *Wear* 128: 1–17.

22 Pezdirnik, J., Vizintin, J., and Podgornik, B. (1999). Temperature at the surface and inside an oscillatory sliding microcontact – theoretical part. *Tribology International* 32: 481–489.

23 Vick, B., Furey, M.J., and Iskandar, K. (2000). Theoretical surface temperatures generated from sliding contact of pure metallic elements. *Tribology International* 33: 265–271.

24 Cameron, A., Gordon, A.N., and Symm, G.T. (1964). Contact temperatures in rolling/sliding surfaces. *Proceedings of the Royal Society London A* 286: 45–61.

25 Ling, F.F. (1973). *Surface Mechanics.* New York: Wiley Interscience.

26 Ling, F.F. and Pu, S. (1964). Probable interface temperature of solids in sliding contact. *Wear* 7 (9): 23–34.

27 Francis, H.A. (1970). Interfacial temperature distribution within a sliding Hertzian contact. *ASLE Transactions* 14: 41–54.

28 Greenwood, J.A. (1991). An interpolation formula for flash temperatures. *Wear* 150: 153–158.

29 Liu, G., Wang, Q., and Ao, Y. (2002). Convenient formulas for modeling three-dimensional thermo-mechanical asperity contacts. *Tribology International* 35: 411–423.

30 Kalin, M. and Vizintin, J. (2001). Comparison of different theoretical models for flash temperature calculation under fretting conditions. *Tribology International* 34: 831–839.

31 Vick, B. and Furey, M.J. (2001). A basic theoretical study of the temperature rise in sliding contact with multiple contacts. *Tribology International* 34: 823–829.

32 McCauley, R.A. (2004). *Corrosion of Ceramic and Composite Materials*, 2e. New York: Marcel Dekker.

33 Vodopivec, F., Vižintin, J., and Šuštaršič, B. (1996). Effect of fretting amplitude on microstructure of IC-1.5Cr steel. *Materials Science and Technology* 12: 355–360.

34 Kalin, M. (2004). Influence of flash temperatures on the tribological behaviour in low-speed sliding: a review. *Materials Science and Engineering A* 374: 390–397.

35 Surender, M., Basu, B., and Balasubramaniam, R. (2004). Wear characterization of electrodeposited Ni-WC composite coatings. *Tribology International* 37: 743–749.

36 Choubey, A., Basu, B., and Balasubramaniam, R. (2004). Tribological behaviour of Ti-based alloys in simulated body fluid solution at fretting contacts. *Materials Science and Engineering A* 379: 234–239.

37 Basu, B., Vitchev, R.G., Vleugels, J. et al. (2000). Influence of humidity on the fretting wear of self-mated tetragonal zirconia ceramics. *Acta Materialia* 48: 2461–2471.

38 Basu, B., Vleugels, J., and Van Der Biest, O. (2001). Fretting wear behaviour of TiB_2-based materials against bearing steel under water and oil lubrication. *Wear* 250: 631–641.

39 Vleugels, J., Basu, B., Kumar, K.C.H. et al. (2002). Unlubricated fretting wear of TiB_2 containing composites against bearing steel. *Metallurgical and Materials Transactions A* 33 (12): 3847–3859.

40 Basu, B., Vleugels, J., and Van Der Biest, O. (2003). Fretting wear behaviour of advanced ceramics and cermet against alumina. *Journal of Materials Research* 18 (6): 1314–1324.

41 Basu, B., Vleugels, J., Kalin, M., and Van Der Biest, O. (2003). Friction and wear mechanism of SiAlON ceramics under fretting contacts. *Materials Science and Engineering: A* 359: 228–236.

42 Sarkar, D., Ahn, S., Kang, S., and Basu, B. (2003). Fretting wear of TiCN-Ni cermet: influence of secondary carbide content. *P/M Science and Technology Briefs* 5 (2): 5–11.

43 Kumar, B.V.M., Basu, B., Murthy, V.S.R., and Gupta, M. (2005). The role of tribochemistry on fretting wear of Mg-SiC particulate composites. *Composites: Part A* 36: 13–23.

44 Bhushan, B., Israelachvill, J.N., and Landman, U. (1995). Nanotribology: friction, wear and lubrication at the atomic scale. *Nature Materials* 374: 607–616.

45 Kalin, M. (2001). High temperature phase transformations under fretting conditions. *Wear* 249: 172–181.

46 Gotsmann, B. and Lantz, M.A. (2013). Quantized thermal transport across contact of rough surfaces. *Nature Materials* 12: 59–65.

47 Tian, X. and Kennedy, F.E. Jr. (1994). Maximum and average flash temperatures in sliding contacts. *ASME Journal of Tribology* 116: 167–174.

48 Greenwood, J.A. and Alliston-Greener, A.F. (1992). Surface temperatures in a fretting contact. *Wear* 155: 269–275.

49 Ashby, M.F., Abulawi, J., and Kong, H.S. (1991). Temperature maps for frictional heating in dry sliding. *Tribology Transactions* 34: 577–587.

50 Khanna, R. and Basu, B. (2007). Sliding wear properties of self-mated yttria-stabilised tetragonal zirconia ceramics in cryogenic environment. *Journal of American Ceramic Society* 90 (8): 2525–2534.

51 Khanna, R. and Basu, B. (2006). Low friction and severe wear of alumina in cryogenic environment: a first report. *Journal of Materials Research* 21 (4): 832–843.

52 Kong, H.S. and Ashby, M.F. (1991). Friction-heating maps and their applications. *Materials Research Society Bulletin* 16 (10): 41–48.

7

Wear of Ceramics and Lubrication

The previous chapters emphasized basic concepts of surface characteristics, friction, and frictional heat generation. Against this backdrop, the present chapter is dedicated to discuss primary wear mechanisms of ceramic materials, which are explained in terms of physico-chemical changes at tribocontacts, leading to (i) tribomechanical wear and (ii) tribochemical wear. The quantitative understanding of the material damage or removal as a consequence of variation in material parameters and test conditions is also explained. The last part of the chapter consists of basic concepts of lubrication with respect to different regimes and the Stribeck curve.

7.1 Introduction

When two solid surfaces come in physical contact, one cannot avoid material damage, whether measurable or not. A number of parameters influence the extent of such damage, which often leads to material removal from contacting interfaces.

> The progressive material loss due to relative movement with contacting substance is defined as wear.

Though the wear is generally referred to as material loss due to relative movement, the damage due to displacement of material without net weight loss or volume loss shall also constitute wear.

At the microscopic length scale, the material removal occurs through interactions of surface asperities. It is often observed that a large amount of material is

Friction and Wear of Ceramics: Principles and Case Studies, First Edition.
Bikramjit Basu, Mitjan Kalin, and B. Venkata Manoj Kumar.

removed in contacts having high friction. However, this is not always true and neither physically justified. It is preferred to deal with friction and wear as separate phenomena in understanding the overall tribological performance of an engineering component. While manufacturing processes like polishing, grinding, cutting, milling, etc., necessarily need controlled wear of the material, wear is strictly undesirable for the long-lasting performance of the majority of engineering components. The immediate consequence of wear is to replace the engineering component, and the involved cost, whereas the long-term and essential consequence is the increased cost associated with the maintenance of the system and manufacturing process. Besides, the worn components cause vibrations and frictional heating at the contact, which result in poor product quality and process efficiency.

It is a general consensus that suitable lubrication at the contact results in a reduction in friction and wear. Lubricant separates the contacting surfaces by forming low-shear films. While detailed overlook on the basics of lubrication can be found in a later part of the present chapter, it is worthwhile to note here that the lubricated surfaces significantly affect the wear behavior. In material aspects, adsorbed layers of oxides, hydroxides, or other species from the surrounding environment act as lubricants and affect the properties of the contact surface.

Sliding and rolling are two major wear types and these are distinguished on the basis of the difference in relative motion of the contacting surfaces.

It should be noted that varying contact kinematic conditions in the majority of the contact surfaces in practical applications like gears, bearings, etc., can be better represented by the simultaneous actions of sliding and rolling. It is also worthwhile to note that pure rolling is restricted to some points or a line during initial contacts, where micro-sliding dominates during the majority of the contact period.

In general, wear is influenced by operating conditions such as load, temperature, counterbody, environment, etc., and material parameters such as surface roughness, toughness, hardness, microstructure, interfacial elements, geometry, etc. Wear represents a complex physico-mechanical phenomenon and referred to as a "system property."

Wear conditions and material parameters influence material removal mechanisms such as adhesion, abrasion, erosion, tribochemistry, etc. On the other hand, an improved understanding of underlying wear mechanisms and material properties, together with technological parameters, can lead to substantial improvement against and tailoring of material damage processes. In other words, the diagnosis of wear mechanisms helps to a large extent in selecting proper action for suppressing material removal [1].

7.2 Testing Methods and Quantification of Wear of Materials

The quantification of wear of materials is strongly influenced by the test variables like load, velocity, etc., in sliding wear conditions and impingement angle, particles velocity, etc., in erosive wear conditions, and wear test configuration and specimen geometry. Therefore, a controlled wear testing is necessary for estimating the material loss. In this regard, important geometries in sliding contacts are pin-on-disc, pin-on-plate, block-on-ring, twin disc, and ring-on-ring tribotesters (Figure 7.1) [2]. Test methods generally used for studying erosive wear behavior are the jet impingement method, recirculating loop, centrifugal accelerator, and whirling arm rig method (Figure 7.2).

After performing a particular wear test on a given material, it is necessary to quantify the wear loss of material for performance evaluation. Depending on the amount of wear loss, one can comment on the severity of wear in terms of mild or severe wear. Generally, the weight loss method is used for the wear volume quantification for most of the metals and polymers. In the case of ceramics (low density and superior mechanical properties), wear loss is less as compared with metals. In this regard, the weight loss method is not suitable to quantify the wear loss for ceramics. Therefore, it is a common practice to assess the wear of ceramic materials by profilometry. For this, the depth and width of the wear scars are measured using non-contact (laser) profilometry on multiple tracks. Results of average depth and width of wear scars are used for determining the volumetric wear as per the following equation:

$$V = 2\pi rwd \tag{7.1}$$

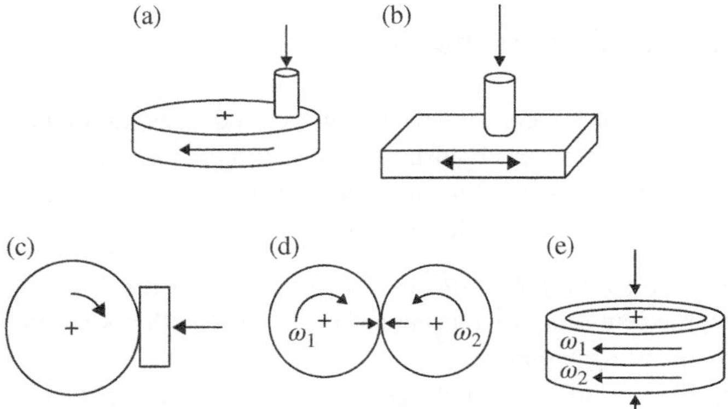

Figure 7.1 Schematic illustration of sliding wear test geometries: (a) pin-on-disc, (b) pin-on-plate, (c) block-on-ring, (d) twin disc, and (e) ring-on-ring [2].
Source: Reproduced with permission of Elsevier.

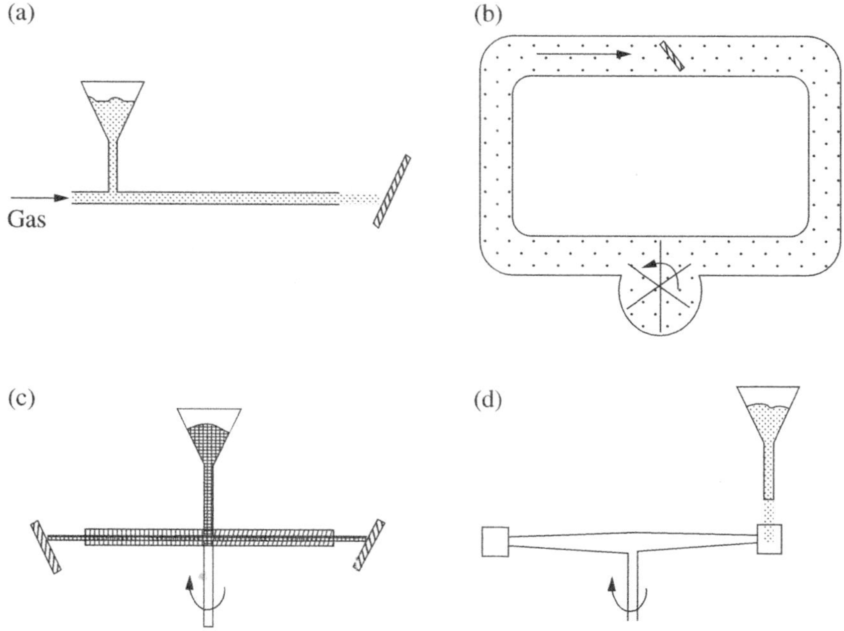

Figure 7.2 Schematic illustration of testing methods for erosive wear: (a) jet impingement method, (b) recirculating loop, (c) centrifugal accelerator, and (d) whirling arm rig [2]. *Source:* Reproduced with permission of Elsevier.

where *r* is the radius of the circular wear scar, and *d* and *w* are the depth and width of the wear scar, respectively.

7.3 Classification of Wear Mechanisms

In general, the mechanisms of material removal in wear conditions are highly complex. Often, more than a single mechanism operates at the mating surfaces. Therefore, it is imperative to review the fundamental wear mechanisms. Such a review would enable the reader to realize the vital issues related to

1) Significance of the microstructural parameters on wear mechanisms.
2) Correlation of wear mechanisms with mechanical properties (hardness, elastic modulus, etc.) and operating parameters.
3) Guiding parameters for developing materials systems with better wear resistance.

Majority of available literature on tribology explains the wear mechanisms in terms of conventional manner such as adhesive wear, abrasive wear, fretting wear, erosive wear, etc. In order to understand the multifactorial nature of wear, wear

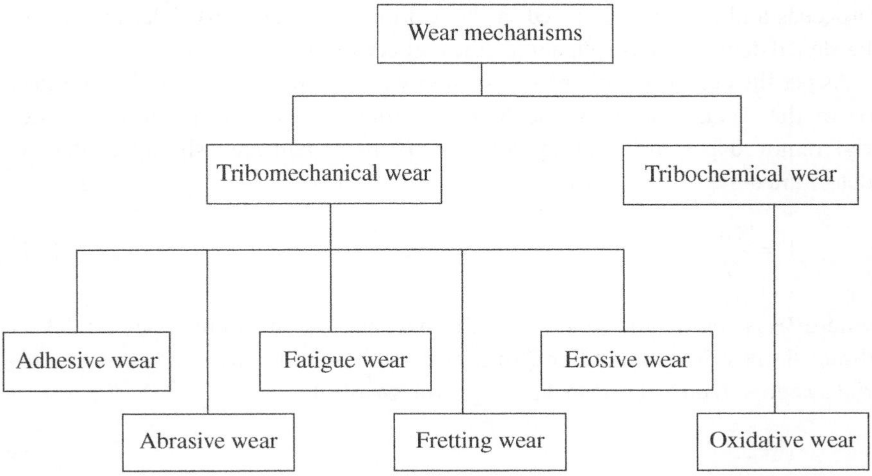

Figure 7.3 Some major types of wear mechanisms of ceramics.

mechanisms of ceramic materials in this book are explicated in a rationalized manner in terms of physicochemical changes at tribocontacts leading to (i) tribomechanical wear and (ii) tribochemical wear. Tribomechanical wear is generally observed in wear conditions of adhesion, abrasion, fatigue, fretting, and erosion, whereas oxidative wear with complex reaction products at the tribological surface is a representation of tribochemical wear in case of ceramics (Figure 7.3).

7.3.1 Tribomechanical Wear

Tribomechanical wear is the removal of material due to dominant mechanical forces/stresses acting at the tribological interface. This can take place by one or a combination of the following wear mechanisms.

7.3.1.1 Adhesive Wear

This type of wear is largely reported for self-mated metals. However, for the sake of providing a knowledge base of key wear mechanisms in general, adhesive wear is discussed below.

> The wear caused by the asperity junction's welding and its removal from the sliding contact is referred to as adhesive wear.

When the asperity junction is stronger than the interatomic bonds in bulk, the material from the weaker surface is detached and remains attached to the countersurface in the form of a layer. The transfer layer may grow as the sliding proceeds and is finally removed in the form of wear debris particle. This work-hardened debris may as well abrade the contact surface.

As per the classical Archard's wear theory [3], shearing occurs at the interface or in the weakest part in one of the sliding bodies. The amount of material removed per unit sliding distance (V) in a deformed sliding contact is determined by:

$$V = \frac{KW}{H} \tag{7.2}$$

where W is the normal load, H is the hardness of the softer body, and K is dimensionless wear coefficient. To compare the wear behavior of varying material systems, Lancaster proposed a specific wear rate k as

$$k = \frac{V}{W} \tag{7.3}$$

When the breaking strength surrounding the local contacting asperity region is more than the interfacial adhesion strength, shearing at the interface results in rupture (path 1 in Figure 7.4) without wear. In small fractions of contacts, rupture occurs at one of the surfaces resulting in the removal of a fragment (path 2 in Figure 7.4) [4].

In an important work, Kayaba and Kato [5] accounted for plastic shearing of successive layers to adhesive wear, in conjunction with the propagation of a shear crack on a slip line field. Sasada [6] further proposed that the adhesive wear leads to the debris formation [7]. According to Sasada, surface asperities experience plastic deformation and strain hardening.

The adhesion primarily occurs in self-mated material contacts or contacts between materials with large mutual solubility. The adhesion is also pronounced in contacts of metals possessing high toughness and a limited tendency for oxide formation. Further, the adhesion can be prevented by the use of dissimilar materials, which provide a small real area of contact or by the use of surface contaminants like low surface energy coatings, lubricants, etc.

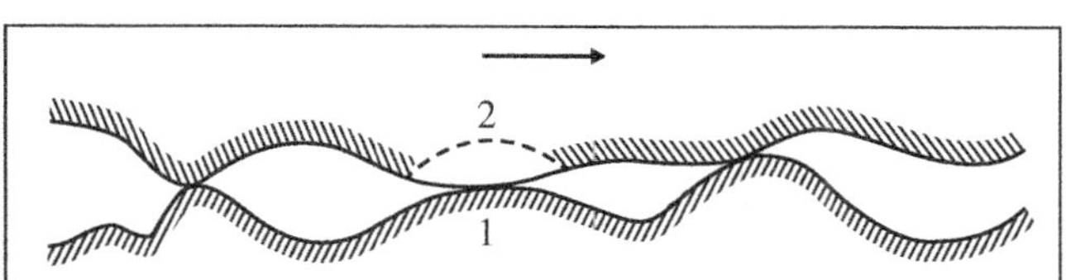

Figure 7.4 Schematic diagram showing the rupture of adhesive bonds via paths 1 and 2 during shearing of an interface [3]. *Source:* Reproduced with permission of AIP Publishing.

7.3.1.2 Abrasive Wear

Abrasion is associated with the multiple indenting of sharp and hard asperities or particles into the softer surface. The hard particles may result from inclusions during the process or as a result of interactions of two surfaces with a large difference in their hardness or reaction of the debris with atmosphere during sliding.

> Abrasion is generally defined as a mode of material removal that is caused by hard asperities or hard particles sliding over a relatively softer surface.

While abrasion by hard asperities closely resembles cutting of a series of teeth of a tool surface at the macroscopic scale, the mechanisms of material removal in abrasion can involve both plastic flow and brittle fracture at the microscopic scale (Figure 7.5). Micro-cutting occurs when sharp grits or asperities abrade over the softer surface, releasing debris of chip or ribbon-type morphology (Figure 7.5a). Moreover, plowing is also a similar mechanism, but differs only in material removal efficiency. Plowing is less efficient than the cutting and often results in the formation of grooves (Figure 7.5b), instead of ejecting material debris. The repeated sliding action results in plastic deformation and strain hardening of the surface, where plowing is dominant. It is noted that both cutting and plowing are typically observed in abrasion of deformable materials. On the other hand, irregular-shaped sharp wear particles are generated by lateral crack propagation from ideally brittle materials (Figure 7.5c). At high contact stress condition, tensile stresses across the vertical midplane initiate a median crack. During unloading, the median crack closes, and the residual stresses, due to relaxation of the deformed material around the contact region, resulting in lateral cracks. Both plastic deformation and brittle fracture often occur together, even in brittle materials.

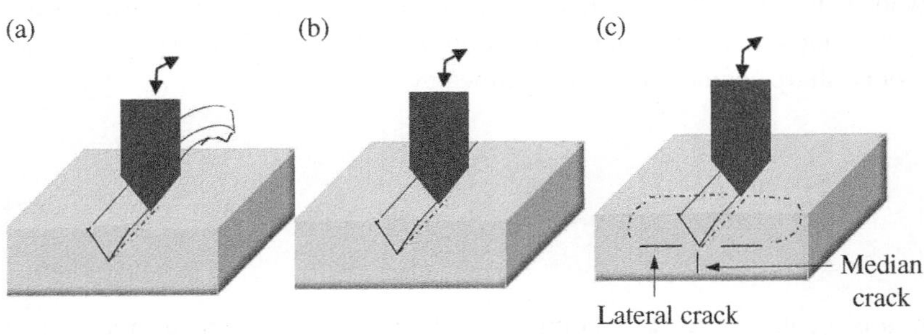

Figure 7.5 Schematic representation of microscopic mechanisms of abrasive wear: (a) cutting, (b) plowing, and (c) cracking.

Abrasive damage at the interface of two sliding materials with a large difference in hardness is referred to as two-body abrasion. Hard particles act as third-body elements at the contacting interface. This accounts for wear on one or both the surfaces and the mechanism of material removal is termed as three-body abrasion.

> The material removal in three-body abrasion is lower than that in two-body abrasion due to the smaller angle of attack in the former.

The total amount of material loss (V) of the displaced material by associated conical asperities in sliding a distance x can be determined as [8]:

$$V = \frac{2Px\overline{\tan \Phi}}{\pi H} \tag{7.4}$$

In Eq. (7.4) (equivalent to Archard's wear equation in case of plowing), Φ is the attack angle of conical asperity, P is the applied load, and H is the hardness of the displaced material.

Under repeated stress cycles at the contact, fatigue facilitates easy propagation of deeper cracks into the subsurface and lateral cracks to cover small local regions. When multiple asperity interlocks hard or large grains or phases, crack propagation is favorable through weaker grain boundaries or interfaces of different phases of ceramics and ceramic composites, resulting in grain pull-outs or surface pits. As mentioned above, the mechanism of abrasive wear involves both plastic flow and brittle fracture. In high-stress conditions of abrasion, sharp asperities of brittle materials follow deformation at low load. However, abrasion at high load leads to dominant cracking. In other words, plastic deformation-induced abrasive wear at low load changes to brittle fracture-induced abrasive wear at high load. As per the indentation fracture mechanics, the minimum load required to produce fracture from a sharp contact can be determined by

$$P^* = \frac{54.47\beta}{\pi\eta^2 g^4}\left(\frac{K_{IC}}{H}\right)^3 K_{IC} \tag{7.5}$$

where β is a constant (2.16) relating to the hardness and diagonal in a Vickers indentation. H is hardness and K_{IC} is fracture toughness of the material indented. g and η are constants. The above discussion provides us a basis as why material parameters (H, K_{IC}) are to be tailored to shift the critical fracture load to higher load wear regime in tribomechanical wear regime.

Taking the above discussion further, the amount of tribomechanical wear in abrasion conditions can be estimated from the occurrence of the cone and radial-lateral cracks.

While a more detailed discussion on these two major crack types is provided in a preceding chapter, it can be reiterated that in case of a static ceramic surface, the indentation by a sharp indenter causes cone cracks. In contrast, radial-lateral cracking develops as a consequence of loading–unloading response against a blunt indenter. As shown in Figure 7.6, when a blunt indenter is pressed with a sufficiently large load against a brittle surface, surface ring cracks are generated that propagate downward upon repeated sliding and develop Hertzian cone crack to cause material pull-out [9].

Based on the above mechanistic description of tribomechanical wear, Tewari et al. formulated a simple analytical model to determine wear volume for brittle materials in ball-on-disk fretting wear contact of blunt indenter (V_B) as per the following equation: [10]

$$V_B = \frac{2\pi\beta^2 s}{3}\left(\frac{3PR}{4E}\right)^{2/3}\sqrt{\left[\left(\frac{4\chi^2 EP}{3\beta^3 K_{IC}^2 R}\right)-1\right]} \qquad (7.6)$$

where χ and β are constants, R is the radius of the ball, and s is the stroke length in fretting contact.

On the other hand, when a sharp indenter is pressed against a brittle material, the material removal occurs by the formation and propagation of lateral cracks. Marshall et al. determined the wear volume of the brittle solid in ball-on-disk

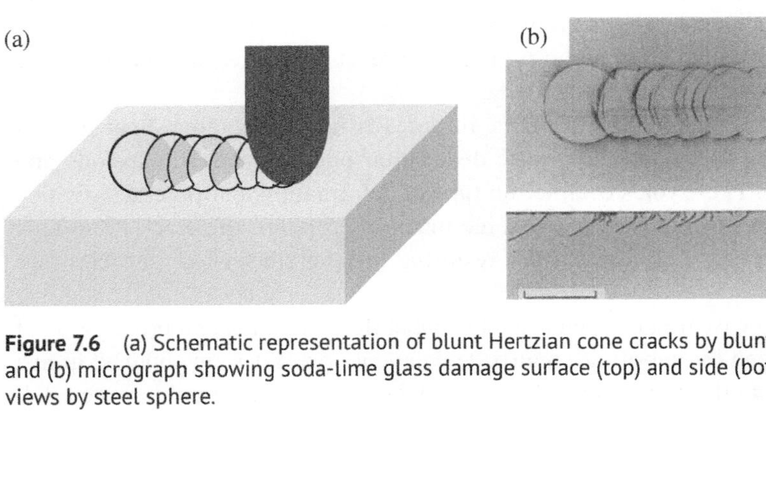

Figure 7.6 (a) Schematic representation of blunt Hertzian cone cracks by blunt indenter and (b) micrograph showing soda-lime glass damage surface (top) and side (bottom) views by steel sphere.

wear contact of sharp indenter (V_S) at a given applied sliding load (P) and total sliding distance (S), using the following equation [11]:

$$V_S = \alpha \frac{P^{9/8}}{K_{IC}^{1/2} H^{5/8}} \left(\frac{E}{H} \right)^{4/5} S \tag{7.7}$$

where α is a material-independent constant; K_{IC}, H, and E are, respectively, fracture toughness, hardness, and elastic modulus of abraded brittle material. In general, the ratio E/H for a brittle material does not vary to a significant extent. In this regard, the amount of material loss is strongly dependent on the sliding load, fracture toughness, and hardness. For brittle materials like magnesium oxide, zirconium oxide, silicon nitride, and boron carbide, the wear resistance is high with a large value of parameter $[W^{9/8}/K_{IC}{}^{1/2}H^{5/8}]$. Further, for a given load, abrasive wear is influenced neither by hardness nor fracture toughness but influenced by a combination of hardness and fracture toughness. The significance of this parameter and the microstructural characteristics will be discussed in a number of case studies in this book.

7.3.1.3 Fatigue Wear

The removal of the material due to repeated loading–unloading cycles is called fatigue wear.

The material loss in fatigue conditions is contributed by the initiation and propagation of surface cracks and subsurface cracks. A primary crack initiates at some weak point at the contacting surface and propagates downward along weak planes (slip planes or dislocation cell boundaries), and result in secondary cracks. Alternatively, the primary crack connects with an existing subsurface crack, which upon propagation reaches the free surface to release wear particles.

As per Shu's theory of subsurface-initiated fatigue wear, large shear stresses beneath the sliding surface cause dislocation pileup at hard inclusions and form voids. These voids coalesce on further deformation to form cracks, which grow parallel to the surface. At some unspecified point, the crack eventually turns upward to the free surface resulting in long ribbon-like particles (see Figure 7.7).

The fatigue wear rate depends on the rate of the generation and propagation of crack through the subsurface matrix. Low fatigue wear rates are obtained from a material with the minimum inclusions and flaws. The effect of unlubricated slid-

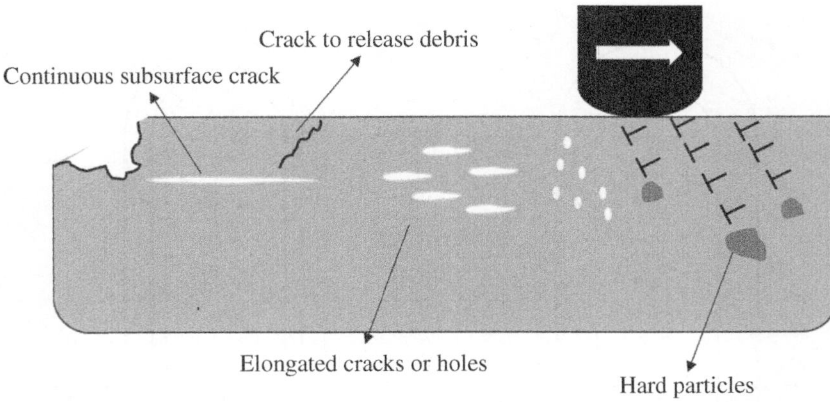

Figure 7.7 Schematic representation of the release of debris particles during fatigue wear.

ing wear on fatigue failure of bulk material is often shrouded by other wear processes, such as adhesion, abrasion, tribochemical wear, etc., which produce wear debris faster than the cracks owing to the delamination process.

The total amount of material removal in fatigue conditions is contributed by plastic strain accumulation both at the leading and trailing edges of asperities, as per the following relation [12]:

$$\frac{V}{x} = c\left(\frac{\overline{\varphi}}{\overline{\varphi}_{\mathrm{f}}}\right)^{m}\frac{F_{\mathrm{N}}}{H} \tag{7.8}$$

where x is the sliding distance, H is the hardness, F_{N} is the normal load, and c is a tribosystem-dependent constant. $\overline{\varphi}_{\mathrm{f}}$ is the average plastic strain to failure by a fracture in one loading cycle and $\overline{\varphi}$ is the average strain in the static condition. The value of coefficient m is about 2–3.

7.3.1.4 Fretting Wear

> Fretting is the surface degradation caused by small-amplitude relative motion that may occur between contacting surfaces.

Fretting is a variant of reciprocated sliding in the small amplitude of displacement. The critical difference between fretting wear and sliding wear is that the debris remains in the contact area (due to the small displacement amplitude) in

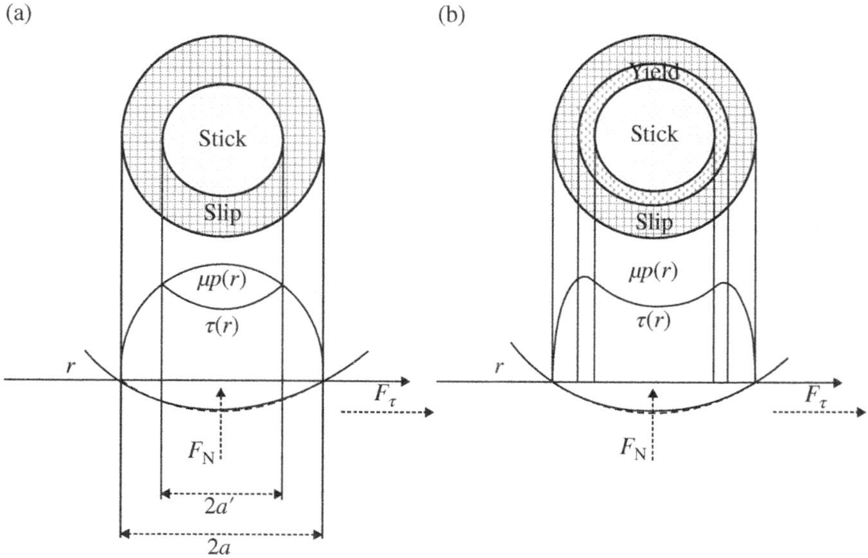

Figure 7.8 (a) Elastic contact and (b) elastic–plastic contact conditions in a fretting contact.

the former, and this influences the progression of wear. A comparison of the wear behavior of ceramics in reciprocated sliding and unidirectional sliding conditions will be discussed in a later chapter of the book.

A stress–strain field can be defined at the contact of an ideally smooth ball and a flat surface. In case of elastic contact, it is often assumed that relative displacement contains micro-slip and elastic deformation at the contacting solid surfaces. In elastic conditions, the contact stress reaches the maximum at the center of the contact circle and falls to zero at the edges (Figure 7.8a) [13]. The contact radius is given by

$$a = \sqrt[3]{\frac{3\left(1 - \nu^2\right) F_{\mathrm{NR}}}{2E}} \tag{7.9}$$

In elastic–plastic contacts, asperities are elastically deformed in a central stick zone and plastically yielded (not fractured) in a surrounded yield annulus (Figure 7.8b). The asperities are subjected to shear fracture in the slip annulus, which surrounds the yield annulus. When fretting is combined with corrosion, the wear mode is known as fretting corrosion. Shrink fits, riveted joints, bolted flanges, and keys are some of the examples, where fretting wear is commonly observed.

7.3.1.5 Erosive Wear

The material damage in erosion is possible by liquid impingement, slurry erosion, cavitation erosion, and solid particle erosion [14–16]. The material damage by erosive wear is predominant for engineering components like grit blasting nozzles, coal turbines, pipelines used for coal transportation, cutting tools, soil erosion, etc. Erosive wear is helpful in sandblasting, abrasive deburring, and the erosive drilling of hard materials.

As solid particle erosion is largely relevant for ceramics and composites [17, 18], the emphasis is given on solid particle erosion in this chapter. A schematic of solid particle erosion is shown in Figure 7.9.

It is to note that the erosion differs from three-body abrasion, mainly with respect to the origin of the forces between the particles and the surface. In particular, contact stresses in erosion result from the kinetic energy of erodent particles, while impinging on the material surface.

In erosion, eroding particle is subjected to a variety of forces, including gravitational force due to weight of the particle, contact forces exerted by the surface, interparticle contact forces in a stream of particles, and drag force due to flowing fluid (Figure 7.10). As abrasion also involves hard particles movement along the surface under contact stress, it should be noted that the material loss depends on the normal load and the distance traveled, and size and shape of the particles. On the other hand, the transfer of kinetic energy of the eroding particles on to the target surface significantly contributes to the material removal in erosion.

According to Finnie [19], three critical factors that influence the erosive wear are: (i) conditions of fluid flow like nature and temperature of fluid, and concentration, velocity, rotation, and impingement angle of particles; (ii) particle characteristics like shape, size, and properties like strength and hardness; and (iii) target surface properties like temperature, stress level, residual stresses, strain rate and fracture toughness, hardness, elastic modulus, and microstructure. For better

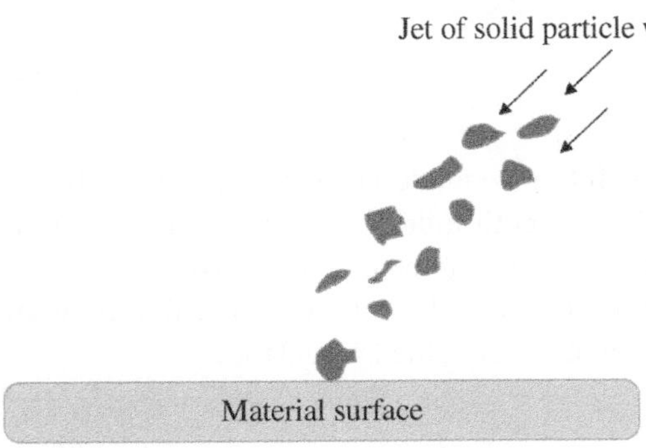

Figure 7.9 Schematic of solid particle erosion.

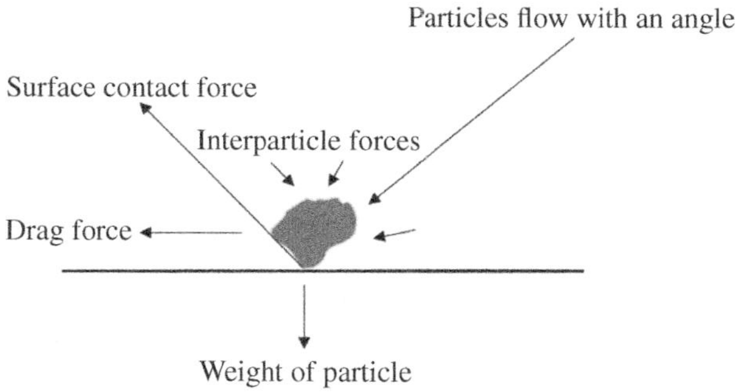

Figure 7.10 Free-body diagram representing dominant forces on erosion particle while contacting the surface.

understanding the significance and response of various materials, the erosion phenomenon is explained below for both ductile and brittle materials.

Several investigations reported the erosive wear behavior of ductile metals [20–22]. A large number of abrasive particles strike the surface during erosion. As it is not possible that every particle cuts the striking surface, the motion of a single rigid abrasive particle is believed to result in displacement or cutting away of the material from the surface.

For a given total mass of particles (M) impinged on the surface of hardness H at a velocity v, the volume (V) of the eroded material can be determined by

$$V = \frac{k_e M v^2}{2H} \tag{7.10}$$

where k_e is the factor (10^{-5} to 10^{-1}) related to the amount of displaced material involving the effect of impingement angle, shape, and size of the particles. In general, the ratio of the mass of eroded material (of density ρ) to the mass of erodent particles is expressed as erosive wear rate (E_r),

$$E_r = \frac{\rho V}{M} \tag{7.11}$$

It is to note that the erosive wear volume equation (7.10) is similar to that of abrasion wear volume equation (7.4), as both indicate inverse relation of wear volume with hardness, but different as the mv^2 (m being the mass of erodent particle) in erosion wear volume equation is replaced by normal load in abrasive wear volume equation. Hutchings [23] proposed a modified model for E_r as

$$E_r = \frac{g_1 \rho v^n}{H f(\alpha)} \tag{7.12}$$

where g_1 is a constant related to particle geometry and fraction of the particles actually cutting the surface in an idealized manner. $f(\alpha)$ is a function of the angle of impingement, α. The value of exponent n lies between 2 and 2.5. The steeper impingement angles result in higher values of n. However, the experimental studies on pure metals indicate that the erosive wear rate is related to the velocity of eroded particles as

$$E_r \propto v^n \tag{7.13}$$

The exponent n in Eq. (7.13) varies between 2.3 and 3. It is understood that the eroded particles dig deeper at a higher velocity, and the resultant force on the particles change their position and leads to excess loss of material.

In ductile materials, the maximum erosion occurs at shallow angles (generally 20°–30°), which decreases to the minimum at the normal impingement (Figure 7.11) [24]. With respect to material removal mechanisms, cutting and plowing are two underlying mechanisms in the erosion of ductile materials.

While impingement of angular particles results in crater formation by cutting, the impingement of rounded particles results in plowing by deformation. Single particle impingement displaces the material in the form of crater lips, whereas

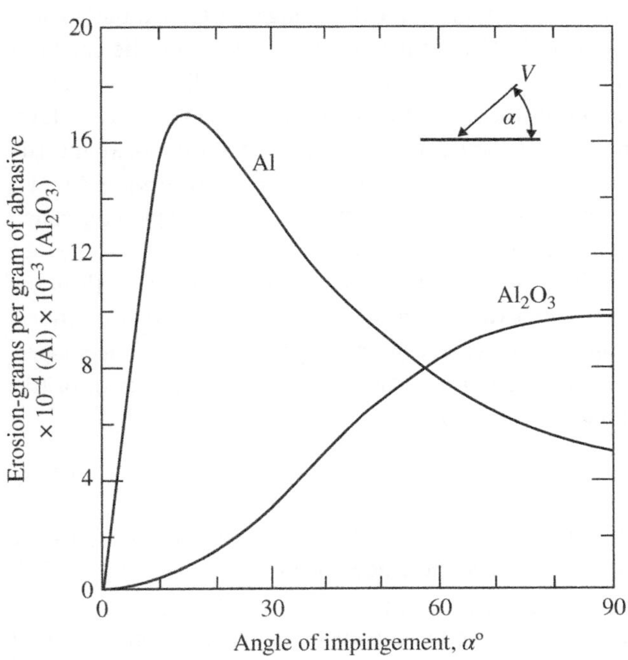

Figure 7.11 Erosion loss of aluminum and aluminum oxide when impinged by silicon carbide particles [2]. *Source:* Reproduced with permission of Elsevier.

multiple particle impingement results in the removal of highly stressed material in the form of thin slices. Similar to abrasion mechanisms, cutting dominates at shallow angles and deformation dominates at normal impingement.

It is to note that hardness of the eroded surface is essential in determining the erosive wear as the surface becomes work-hardened by impingement of multiple particles. Also, the degree to which plastic flow is localized around the impact zone of each particle site is essential, as this represents susceptibility for removal of the displaced material [25].

> The impingement-induced surface and subsurface cracks and their propagation account for the erosive wear of brittle materials.

In relation to the effect of impingement angle, severe cracking occurs when a particle indents the brittle solid at normal impingement (Figure 7.11). This can be understood that the kinetic energy associated with eroding particle at normal impingement is completely transferred to the material surface to result in extensive cracking, whereas only a fraction of the kinetic energy is transferred to the material at shallow angles.

Since the formation and propagation of cracks strongly influence the erosive wear behavior of ceramic materials, it is imperative to understand the cracking characteristics. As explained in the case of abrasive wear conditions, two crack systems are primarily found in erosion with the change in particle velocity. In low-velocity erosion conditions, (i) conical cracks by blunt erodent [26–28] and (ii) lateral and median cracks by sharp erodent dominate the wear process [26, 29–31]. In case of high-velocity erosion conditions, chipping, cracking, and plowing cause material loss dominantly [16, 26, 27].

At high load, static (such as in sliding) or dynamic indentations with blunt indenter generate surface ring cracking and result in material pull-out. Continued particle impingement causes ring cracks to propagate to form cone-shaped fracture surface and result in material removal. The erosive wear rate of brittle materials can be determined by the relation:

$$E_{\mathrm{r}} \propto v^{m} r^{n} \tag{7.14}$$

where the values of exponents m (0.1) and n (3.0) are related to the statistical distribution of fracture stress raisers on the eroded surface [4].

> As per elastic–plastic indentation fracture theory, sharp particle impingement results in intersection of lateral cracks among themselves and those with the free surface of brittle solids.

The lateral-median crack formation and the subsequent material removal in erosion are similar to that occurring in abrasion by sharp asperities in sliding. In erosion, the residual stresses, generated under the plastic zone, originate small lateral cracks that propagate beneath the impact sites and further interact with the free surface, resulting in material removal in the form of chips or lips. In practical erosion conditions, particles impinged at lower angles slide on the surface for a finite distance and cause longer and broader grooves.

For given erosion parameters and erodent, the erosive wear of ceramic material can be expressed as [32]

$$E = \alpha H^m K_{IC}^n \tag{7.15}$$

where α is the proportionality constant. The m and n exponents are 0.11 and −1.33 as per the dynamic model, and −0.25 and −1.33 as per the quasi-static impact model. However, Wiederhorn and Hockey [26] reported the values of m as 0.48 and n as −1.9 in dimensional analysis, while fitting the erosive wear data of a wide range of ceramic materials. However, none of these models considered the vital contribution from a local microstructure in assessing the erosion loss.

7.3.2 Tribochemical Wear

The dominant effect of tribochemical wear is generally observed in race and shaft contacts of a roller bearing, bevel wheels, riveted joints, clutches, links of chains, plate springs, etc. The products are formed from the chemical reactions of contacting surfaces with the environment. These reactions are found to be not feasible under the same static load and temperature conditions. In fact, such reactions at the contact are triggered even at a lower temperature due to the stimulated high-stress conditions. Thus, the kinetics of these reactions in the contacting zone vary from the static thermochemical reactions.

> The material removal as a result of *in situ* reactions from the contacting zone is referred to as tribochemical wear.

Further, it is evident that stimulated tribological conditions weaken the intermolecular bonds at the contacting surface due to the effect of chemical reactions and cause material removal. Thus, tribochemical wear is a synergistic effect of mechanical action and surface chemistry. For example, the molecular bonds at

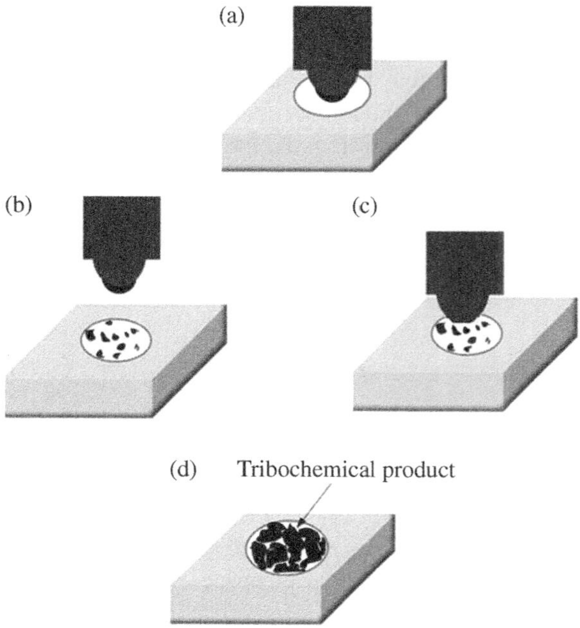

Figure 7.12 Schematic representation for the mechanism of the tribolayer formation: (a, b, and d) in case tribochemical reaction initiates on the ball and (a, c, and d) in case tribochemical reaction initiates on a flat surface in a ball-on-flat configuration.

the surface of dental enamel are weakened by the dietary acids or gastric reflux and then rubbed away by the mechanical rubbing actions [33].

The material loss associated with the tribochemical wear in a ball-on-flat configuration can be understood as per the following (see Figure 7.12). During initial running-in-period contact, the mechanically interlocked asperity junctions are deformed and fractured, accounting to the dominant tribomechanical type of wear. Subsequently, the wear debris particles are compacted to form a layer under the effect of suitable tribochemistry conditions. In this regard, the following three situations may arise with respect to the initiation of the tribochemical reactions and the subsequent formation of tribochemical layer.

1) When tribochemical reaction initiates at the ball surface, the high speed and loading conditions lead to the thick layer that transfers and spreads on to the flat surface, protecting it from further wear. The process sequence for the tribochemical layer formation, initiated at the ball surface, is shown through Figure 7.12a, b, and d.
2) In case of initiation of tribochemical reaction at the flat surface, the tribochemical layer on the flat surface avoids direct contact with the atmosphere.

Thus, the progress of tribochemical reaction is possible only through a diffusion mechanism. Figure 7.12a–c represents schematic progress of the event for tribochemical layer formation, initiated at the flat surface.

3) In the case of tribochemical reaction initiating at the interface between the ball and flat surfaces, the material loss is accounted by either reaction-controlled tribochemical reaction or diffusion-controlled tribochemical reaction or a combination of both.

It is a general observation that debris particles resulting from the wear process act as abrasives, which roughen the contacting surfaces and react with the environment to form a fresh layer. Therefore, the tribochemistry-induced wear is associated with the continuous formation and removal of the layer. In this regard, two essential conditions exist with respect to the protectiveness or stability of the layer.

1) If the rate of formation is more than the rate of removal of tribochemical layer, then the tribochemical layer is considered to be stable, and material removal is primarily reduced by the presence of tribochemical layer.
2) Conversely, the tribochemical layer is unstable when the rate of removal is more than the rate of formation of tribochemical layer due to which the material removal is accelerated.

With regard to protectiveness and stability, tribochemical layers with low shear strength and high load-bearing ability protect the surface from further friction and wear, whereas less viscous and unstable layers are vulnerable for spalling, exposing the surface underneath and cause increased loss of material.

> The overall effect of tribochemical wear depends on both formation kinetics and nature and properties of tribolayers.

A qualitative analysis of tribochemical wear is provided by Kalin et al. [34] However, it is difficult to determine the effect of rates of tribochemical reaction and the formation of an associated layer on the mating materials in the given wear condition. Kitaoka et al. [35] proposed a wear volume calculation model for the non-oxide ceramics with the unprotective tribochemical layer in high temperature and high-pressure water conditions.

However, it is important to emphasize that the tribochemical layer formation is affected by many factors, including physics and chemistry of the counterbody materials, the formation of clean reactive surfaces, shear, environment, and friction-induced contact temperature. In several studies, the role of carbonaceous reinforcements like graphene or carbon nanotubes (CNTs) on the wear resistance or structural materials is reported. For example, Yazdani et al. reported the ball on disc wear of alumina reinforced with graphene nanoplates and CNTs against silicon nitride ball at a sliding speed of 10 mm/s at different loads of up to 35 N for the sliding duration of two hours [36]. When compared with unreinforced alumina, the composites with 0.3% graphene and 1% CNT exhibited 86% reduction in wear rate. The authors attributed such impressive wear performance toward the formation of tribofilm, mediated by exfoliation of graphene nanoplatelets. The addition of CNT is reported to improve fracture toughness and reduce grain pull-out during the sliding test.

7.3.2.1 Oxidative Wear

The most general type of tribochemical wear in many non-oxide ceramic components is the oxidative wear, which represents the material loss in dry unlubricated materials in the presence of air or oxygen. The atmospheric oxygen radically changes the friction and wear rates of the dry sliding materials. Furthermore, the oxidative wear shows a moderate and stable coefficient of friction (COF) in comparison with much larger fluctuating values in severe wear. The oxidative wear is prominently observed in situations, where a high-process temperature results in accelerated oxidation at the contact.

The suitable combination of wear conditions like humidity, temperature, speed, load, partial pressures, etc., results in the formation of tribochemical oxide layers at the sliding contact. On the other hand, the structure, nature, and properties of oxides affect the wear behavior of the mating materials. For example, among oxides layers of iron, the layers rich in magnetite (Fe_3O_4) provide less wear than the layers rich in hematite (Fe_2O_3). In a classical work, Hong and coworkers [37] proposed a quantifying model for oxidative wear volume (V), as per reaction (7.16):

$$V = \frac{A\left[A_1 \exp^{-\left(\frac{Q}{R_g T_f}\right)}\right]}{\rho_o f_o} t_s \tag{7.16}$$

In the above reaction, t_s is the total traversed time, T_f is the flash temperature, R_g is the molar gas constant, Q is the activation energy for oxidation, f_o is the mass fraction of the oxide that is oxygen, ρ_o is the average density of the oxide in contact, and A_l is Arrhenius constant.

Moreover, oxidative wear is also observed in erosion conditions. In a study, the effect of high temperature on the solid particle erosive wear of SiC-Si_3N_4 composites exhibiting a transition from brittle fracture with easy stripping of aggregates of composites at a lower temperature (up to 800 °C) to surface oxidation along

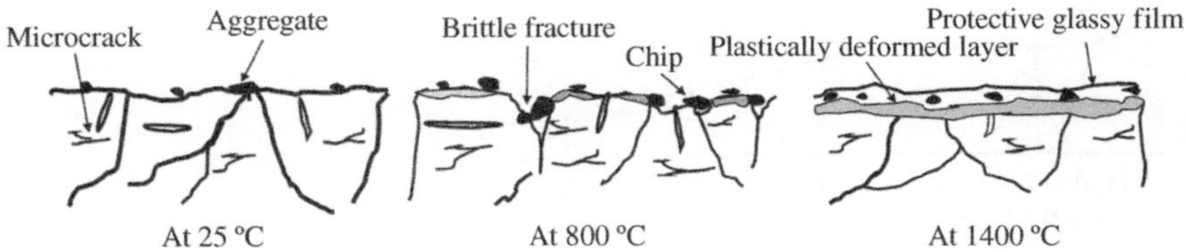

Figure 7.13 Schematic illustration of material removal mechanisms in the erosion of SiC-Si_3N_4 composites at different temperatures [38]. *Source:* Reproduced with permission of Elsevier.

with plastic deformation of the surface at higher temperature (beyond 800 °C) is notably observed [38] (Figure 7.13). Also, a thin dense layer of SiO_2 glassy film is generated on the material surface as a result of oxidation of Si_3N_4 and SiC at 1400 °C, which further prevented direct contact of the material surface with erodent and resulted in less material loss.

7.4 Lubrication

Lubrication is a general practice to reduce friction and wear of mating materials. Lubrication can be achieved through the addition of lubricant oil film at the contact or by generating molecular-level boundary films at the contacting surfaces that act as low-shear and/or wear-resistant protective layers. Lubricants prevent direct contact between the mating surfaces. Conventionally used lubricants are compounds consisting of base lubricant oil and additives. The use of inorganic nanoparticles and ionic liquids in lubricants not only reduces friction and wear but also makes the lubricant environmentally friendly [39, 40].

7.4.1 Regimes of Lubrication and the Stribeck Curve

A schematic representation of contacts with different regimes of lubrication in fluid lubrication is displayed in Figure 7.14. The Stribeck curve represents the COF as a function of (Stribeck) parameter, which is proportional to viscosity (η) and velocity (N) and inversely proportional to load (P). The lubrication parameter (ratio of lubrication film of a thickness, h, and surface roughness, σ) and the COF are plotted as a function of Stribeck parameter ($\eta N/P$) in Figure 7.15.

Hydrostatic lubrication (HL) refers to a thick layer of lubricant maintained between the mating surfaces with little or no relative motion. For example, a thick film of fluid supports hydrostatic bearings by creating an adequate hydrodynamic pressure to support the load.

Both load-bearing capacity (h/σ) and COF increase with an increase in the bearing capacity in the HL region (Figure 7.15). As there is no direct contact of the

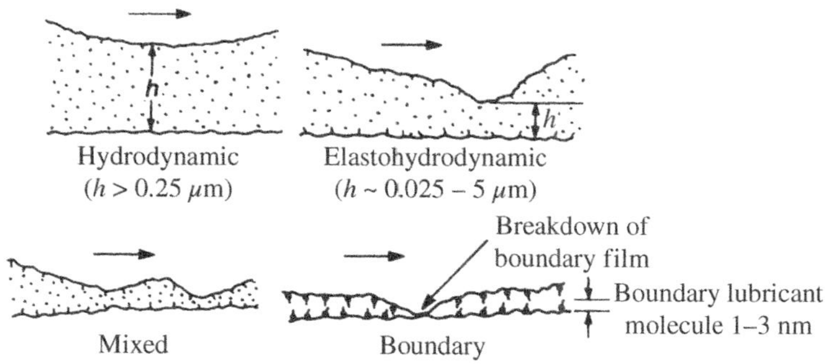

Figure 7.14 A schematic representation of contacts with different lubrication regimes [4]. *Source:* Reproduced with permission of John Wiley & Sons.

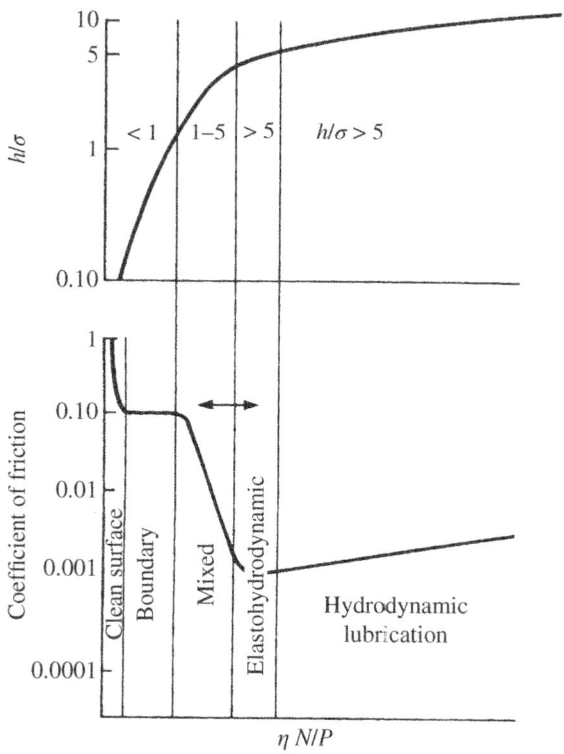

Figure 7.15 The lubrication film parameter (h/σ) and the coefficient of friction as a function of Stribeck parameter ($\eta N/P$) representing lubrication regimes in fluid lubrication [4]. *Source:* Reproduced with permission of John Wiley & Sons.

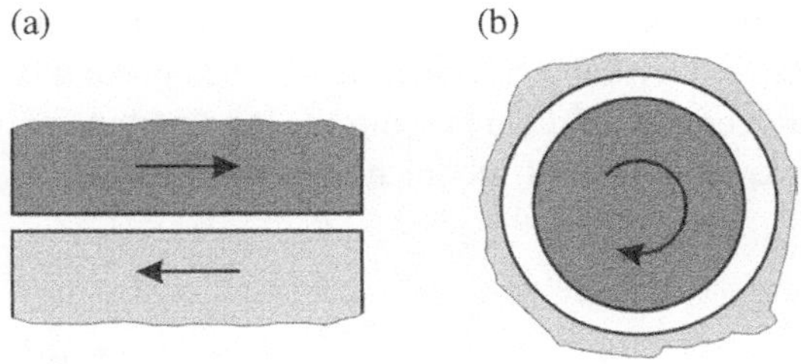

Figure 7.16 Typical conformal contacts where the hydrodynamic regime is possible. (a) Sliding of two contacts and (b) rotational contact.

mating surfaces in the presence of thick lubricant, the contact behavior in the HL regime is governed by lubricant's physical properties, like viscosity, as frictional characteristics arise purely from the shearing of viscous lubricant. The HL regime is possible in conformal contacts, like flat-on-flat (Figure 7.16a) or shaft-in-a-bore contacts (Figure 7.16b). Relatively high loads can safely be applied to the contacts as the high area of contact keeps the contact pressure low.

Elastohydrodynamic (EHD) lubrication is considered as a subset of HL. This lubrication is most readily induced in contacts where high loads act over small contact areas, such as the point contacts in the gear teeth and ball bearings [41]. In EHD, the elastic deformation of the contacting surface and lubricant viscosity play significant roles. As shown in the Stribeck curve (Figure 7.15), the lubricant film in EHD location is thinner than that in HL.

In the boundary lubrication (BL) condition, solid surface asperities and monomolecular or multimolecular films of lubricant dominate the contact. This situation arises due to the increase in speed, load, and decrease in fluid velocity, and results in an increase in COF (see Figure 7.15). The BL regime fails due to adhesive wear or chemical wear. Moreover, the transition between the HL/EHD and BL is referred to as mixed lubrication (ML) regime. In the ML regime, two lubrication mechanisms may occur simultaneously. One possibility is that the solid–solid contacts lead to adhesion, metal transfer, debris formation, and finally, closure. On the other hand, adhesion is prevented in areas of contact where liquid films bear the load. The COF sharply increases, and the h/σ ratio sharply decreases in the ML regime (see Figure 7.15).

Hamrock–Dowson equation is mostly used to calculate the minimum oil film thickness in EHD conditions [42]

$$h_{\min} = 3.63 R_x U^{0.68} G^{0.49} W^{-0.073} \left(1 - e^{-0.68k}\right) \tag{7.17}$$

where U, W, G, and k are non-dimensional parameters for speed, load, materials, and contact geometry, respectively. R_x is the effective radius in the x-direction.

However, the critical issues, like non-ideal smooth surfaces, vibration-induced temperature and viscosity changes, surface damages, wear debris, etc., make the calculation of the oil film thickness complex. Tallian [43] introduced a parameter λ, which is a ratio between the thickness of oil film and the roughness of two moving surfaces,

$$\lambda = \frac{h_{\min}}{\sqrt{R_{q1}^2 + R_{q2}^2}} \tag{7.18}$$

where R_{q1} and R_{q2} are the root mean square (RMS) roughness values of two mating surfaces, respectively. The parameter λ indicates the ability of the oil film to avoid direct contact of surfaces. Based on the parameter λ, the lubrication regimes are largely accepted as follows: (i) if $\lambda \geq 3$, the contact is in HL/EHD lubrication and life of the component is not reduced because of no wear; (ii) if $3 \geq \lambda \geq 1$, the contact is in ML with mild wear; (iii) if $\lambda \leq 1$, the contact is in the BL regime with severe wear and substantial reduction in component life.

7.4.2 Influence of Lubricant Composition, Contact Pressure, and Temperature on Lubrication

In an effort to understand the tribological properties of nanostructured coatings, Erdemir et al. conducted a series of experiments on MoN_x copper-coated steel and uncoated steel in PAO 10 oil and fully formulated 5W30 oil [44]. The instantaneous formation of carbon-based tribofilms on catalytically active coated steel surfaces were reported to cause negligible wear and reduced friction (COF ~ 0.02; Figure 7.17a) in self-mated sliding contact at 1.3 GPa pressure. The reduction in wear of the ball (see Figure 7.17b) was extremely significant. Importantly, Erdemir et al., based on *ab initio* and reactive molecular dynamics simulation results, proposed that the catalytic property of the nanocomposite coatings facilitated dehydrogenization of linear olefins in PAO 10 oil and a random session of C–C backbone chain. This tribochemical phenomenon leads to the growth of a compact, amorphous lubricating tribofilm on nanocomposite coating, which was deposited on martensitic steel substrates. The authors commented that the above mechanism could be operative in case of metal-on-metal articulating joints, as biological proteins can be the primary carbon source.

It is to note that compounds in the lubricant are supposed to have different roles [45]. For example, the aim of using base oil is also to carry the additives, to remove debris particles from the contact, to remove heat from the contact, to cool the

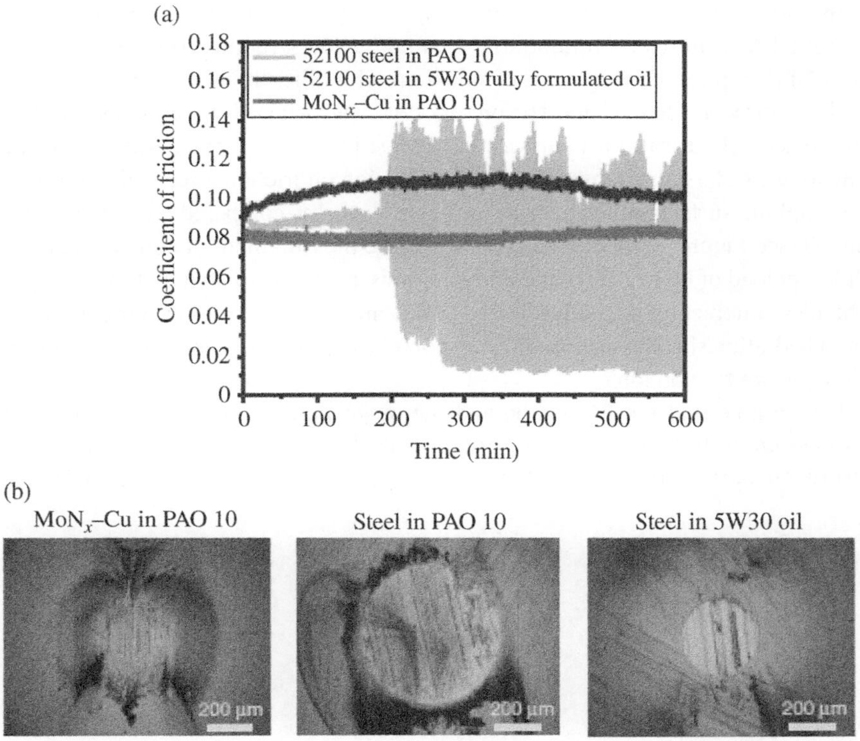

Figure 7.17 Comparative analysis of (a) the coefficient of friction and (b) ball wear scars of MoN_x–Cu-coated steel ball in PAO 10 oil, uncoated steel ball in PAO 10, and uncoated steel ball in fully formulated 5W30 oils [44]. *Source:* Reproduced with permission of Springer Nature.

system, and to reduce noise or vibrations. However, base oils also have the ability to reduce wear and friction and/or to separate the contacting surfaces via hydrodynamic (HD) oil film.

In the field of lubrication, extensive efforts have been made to use different surfactants or additives to achieve ultralow friction properties. It is known for several years that in the BL regime, the mating surfaces are coated with molecular monolayers having good lubricity. From the chemistry point of view, such monolayers are characterized by amphiphilic surfactants with polar head groups [46]. Further, the process of BL is not only used for engineering components to reduce friction and wear but also for biomedical applications at articulating joints, where lubrication can be mediated by phospholipid films with hydrated layers. The latter mechanism is significant for effective clinical treatment of osteoarthritis.

In one of the seminar works, Briscoe et al. reported the effectiveness of *N,N*-dimethyl-*N,N*-diundecylammonium bromide with structure $(CH_3(CH_2)_{10})_2N^+(CH_3)_2Br^-$, typical of a large class of similar amphiphilic surfactants, both in dry air and in (surfactant-free) water. The presence of DDunAB leads to reduction in COF, in crossed-cylindrical mica surfaces [47]. Most importantly, the authors reported one to two orders of magnitude or more reduction on friction force with the use of amphiphilic surfactant layers of DDunAB in water, as compared with that in dry air/oil (see Figure 7.18). Based on the extensive measurement of contact area at a different load of up to 20 mN and other analysis, it was postulated that the physico-chemical interaction at the lubricated surfactant–substrate interface mediated by fluid hydration sheaths surrounding the polar head group contributed to the significantly lower friction force.

It is to note that the BL regime and the major part of the ML regime depend strongly on the formation and nature of the tribochemical layers, i.e. boundary films. In order to increase the protectiveness of lubricant, specific additives are added to the

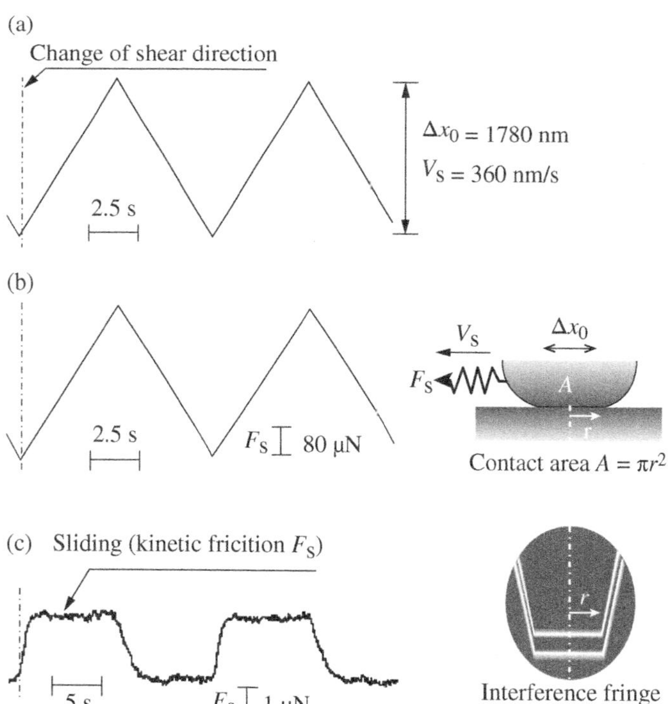

Figure 7.18 Friction characteristics between two DDunAB coated surfaces in adhesive contact [47]. (a) Asperity characteristics, (b) sliding conditions, and (c) sliding induced asperity damage. *Source:* Reproduced with permission of Springer Nature.

lubricants, coolants, or emulsions to control the chemistry in order to react with the surface and form tribochemical layers. The role of different additives in lubrication of surfaces in relative motion has been widely recognized. In one of the most interesting papers, Gosvami et al. reported the quantitative understanding of the contact pressure and temperature-dependent exponential growth rate of tribofilm at single asperity sliding nano contacts in lubricant medium containing zinc dialkyldithiophosphates (ZDDPs) [48]. Using extensive atomic force microscopy-based analysis, the authors reported the Arrhenius-type stress-activated growth rate of the tribofilm at various contact pressures up to 6.5 GPa. The similar exponential growth rate was also measured at temperatures up to 140 °C. Further, the activation energy of the tribofilm growth was found to be 0.8 eV, and this is similar to a single atomic bond breaking or formation process. Moreover, the authors also proposed the thermally activated and stress-assisted reaction rate model to explain the role of ZDDP on the tribological properties. Such a mechanism is consistent with covalent bond reaction pathway, and it was argued that repeated sliding at single asperity contact could facilitate abundant tribochemical reactions, which were triggered by the nucleation and growth of sacrificial tribofilm. Nevertheless, the in-situ measurements also suggested that ZTTPs' anti-wear property can be attributed to the mechanical protection by the tribofilm, which is weaker and has a complaint graded structure.

7.5 Closure

In view of the critical aspect of degradation and failure of ceramic components in engineering applications, the fundamentals of wear and lubrication are discussed in this chapter. In this aspect, the material removal mechanisms for given operating conditions (e.g. type of contact, load, temperature, humidity, etc.) and material properties (e.g. hardness, toughness, elastic modulus, etc.) are emphasized in the two categories of (i) tribomechanical and (ii) tribochemical wear. Under tribomechanical wear category, adhesion, abrasion, fatigue, fretting, and erosion are elaborated. Further, the reactive nature of the material in a given tribological condition leads to tribochemical or oxidative wear. The role of lubrication in reducing the wear is also emphasized. Summarizing the lubrication section, the Stribeck curve and lubrication regimes are affected by solid–liquid interface properties, such as wetting and/or surface energy. These effects shall be properly dealt in designing a lubricated contact system [49]. In a later chapter, the significance of wetting and surface energy parameters in understanding the relation between surface–oil interaction and friction in EHD contact are discussed. The detailed discussion on wear, wear mechanisms, and lubrication concepts is believed to provide a baseline knowledge needed for the development of the new generation of wear-resistant ceramic materials.

References

1 Zum Gahr, K.H. (1987). *Microstructure and Wear of Materials*. Oxford: Elsevier.
2 Hutchings, I. and Shipwa, P. (2017). *Tribology Friction and Wear of Engineering Materials*, 2e. Amsterdam: Butterworth-Heinemann.
3 Archard, J.F. (1953). Contact and rubbing of flat surfaces. *J. Appl. Phys.* 24: 981–988.
4 Bhushan, B. (1999). *Principles and Applications of Tribology*. New York: Wiley.
5 Kayaba, T. and Kato, K. Adhesive transfer of the slip-tongue and the wedge. *ASLE Trans.* 24: 164–174.
6 Sasada, T. (1984). Tibology. In: *Tibology in the 80s*, vol. I. NASA Conference Publication 2300, 197–218.
7 Rigney, D.A., Chen, H., Naylor, M.G.S., and Rosenfield, A.R. (1984). Wear processes in sliding systems. *Wear* 100: 195–219.
8 Rabinowicz, E. (1995). *Friction and Wear of Materials*. New York: Wiley.
9 Lawn, B. (1993). Indentation fracture. In: *Fracture of Brittle Solids*, 2e (eds. E.A. Davis and I.M. Ward), 249. Cambridge University Press.
10 Tewari, A., Basu, B., and Bordia, R.K. (2009). Model for fretting wear of brittle ceramics. *Acta Mater.* 57: 2080–2087.
11 Marshall, D.B., Lawn, B.R., and Evans, A.G. (1982). Elastic/plastic indentation damage in ceramics: the lateral crack system. *J. Am. Ceram. Soc.* 65 (11): 561–566.
12 Tavernelli, J.F. and Coffin, L.F. (1959). A compilation and interpretation of cyclic strain fatigue tests on metals. *Trans. ASM* 51: 438–450.
13 Kalin, M. (1999). Fretting wear mechanisms in contact of steel and silicon nitride ceramics. PhD thesis. University of Ljublijana.
14 Wang, D.F., She, J.H., and Ma, Z.Y. (1995). Effect of microstructure on erosive wear behavior of SiC ceramics. *Wear* 180: 35–41.
15 Medvedovski, E. (2010). Ballistic performance of armour ceramics: part 1. Influence of design and structure. *Ceram. Int.* 36: 2103–2115.
16 Evans, A.G. and Wilshaw, T.R. (1977). Dynamic solid particle damage in brittle materials: an appraisal. *J. Mater. Sci.* 12: 97–116.
17 Finnie, I. (1960). Erosion of surfaces by solid particles. *Wear* 3: 87–103.
18 Hutchings, I.M. (1987). Wear by particulates. *Chem. Eng. Sci.* 42 (4): 869–878.
19 Finnie, I. (1995). Some reflections on the past and future of erosion. *Wear* 186–187: 1–10.
20 Adler, W.F. (ed.) (1979). *Erosion: Prevention and Useful Applications*, 3e. ASTM STP 664. West Conshohocken, PA: ASTM International.
21 Shewman, P. and Sundararajan, G. (1983). The erosion of metals. *Ann. Rev. Mater. Sci.* 13: 301–318.

22 Finnie, I. (1958). The mechanism of erosion of ductile metals. In: *Proceedings of the 3rd US National Congress of Applied Mechanics*, 527–532. New York: American Society of Mechanical Engineers.
23 Hutchings, I.M. (1979). Mechanisms of the erosion of metals by solid particles. In: *Erosion: Prevention and Useful Applications*, 3e. ASTM STP 664 (ed. W.F. Adler), 59–76. West Conshohocken, PA: ASTM International.
24 Ritter, J.E. (ed.) (1992). *Erosion of Ceramic Materials*. Zurich, Switzerland: Trans Tech Publications.
25 Finnie, I., Wolak, J., and Kabil, Y. (1967). Erosion of metals by solid particles. *ASTM J. Mater.* 2: 682–700.
26 Wiederhorn, S.M. and Hockey, B.J. (1983). Effect of material parameters on the erosion resistance of brittle materials. *J. Mater. Sci.* 18: 766–780.
27 Knight, C.G., Swain, M.V., and Chaudhri, M.M. (1977). Impact of small steel spheres on glass surfaces. *J. Mater. Sci.* 12 (8): 1573–1586.
28 Lawn, B.R. and Marshall, D.B. (1978). Indentation fracture and strength degradation in ceramics. In: *Fracture Mechanics of Ceramics*, vol. 3 (eds. R.C. Bradt, D.P.H. Hasselman and F.F. Lange), 205–229. New York: Plenum Press.
29 Hockey, B.J. and Lawn, B.R. (1975). Electron microscopy of microcracking about indentations in aluminium oxide and silicon carbide. *J. Mater. Sci.* 10 (8): 1275–1284.
30 Chaudhri, M.M. and Brophy, P.A. (1980). Single particle impact damage of fused silica. *J. Mater. Sci.* 15 (2): 345–352.
31 Suh, M.S., Hinoki, T., and Kohyama, A. (2011). Erosive wear mechanism of new SiC/SiC composites by solid particles. *Tribol. Lett.* 41: 503–513.
32 Evans, A.G., Gulden, M.E., and Rosenblatt, M. (1978). Impact damage in brittle materials in the elastic-plastic response regime. *Proc. R. Soc. Lond. A* 361: 343–365.
33 Lambrechts, P., Debels, E., Landuyt, K.V. et al. (2006). How to simulate wear? Overview of existing methods. *Dent. Mater.* 22: 693–701.
34 Kalin, M. and Vizintin, J. (2001). A tentative explanation for the tribochemical effects in fretting wear. *Wear* 250: 681–689.
35 Kitaoka, S., Tsuji, T., Yamaguchi, Y., and Kashiwagi, K. (1997). Tribochemical wear theory of non-oxide ceramics in high temperature and high-pressure water. *Wear* 205: 40–46.
36 Yazdani, B., Xu, F., Ahmad, I. et al. (2015). Tribological performance of graphene/carbon nanotube hybrid reinforced Al_2O_3 composites. *Sci. Rep.* 5: 11579.
37 Hong, H., Huchman, R.F., and Quinn, T.F.J. (1988). A new approach to the oxidational theory of mild wear. *Tribol. Trans.* 31 (1): 71.

38 Li, X., Ding, H., Huang, Z. et al. (2014). Solid particle erosion-wear behavior of SiC–Si_3N_4 composite ceramic at elevated temperature. *Ceram. Int.* 40: 16201–16207.

39 Tenne, R. (2014). Recent advances in the research of inorganic nanotubes and fullerene-like nanoparticles. *Front. Physiol.* 9: 370–377.

40 Qu, J., Barnhill, W.C., Luo, H. et al. (2015). Synergistic effects between phosphonium-alkylphosphate ionic liquids and zinc dialkyldithiophosphate (ZDDP) as lubricant additives. *Adv. Mater.* 27: 4767–4774.

41 Hamrock, B.J. (1994). *Fundamentals of Fluid Film Lubrication*. New York: McGraw Hill.

42 Hamrock, B.J. and Dowson, D. (1981). *Ball Bearing Lubrication – The Elastohydrodynamics of Elliptical Contacts*. New York: Wiley-Interscience.

43 Tallian, T.E. (1967). On competing failure modes in rolling contact. *ASLE Trans.* 10: 418–439.

44 Erdemir, A., Ramirez, G., Eryilmaz, O.L. et al. (2015). Carbon-based tribofilms from lubricating oils. *Nature* 536: 67–71.

45 Hardy, W.B. (1920). Some problems of lubrication. *Nature* 106: 569–672.

46 Guo, F., Tian, Y., Liu, Y., and Wang, Y. (2017). Unexpected friction behaviours due to capillary and adhesion effects. *Sci. Rep.* 7 (1): 148.

47 Briscoe, W.H., Titmuss, S., Tiberg, F. et al. (2006). Boundary lubrication under water. *Nature* 444 (7116): 191–194.

48 Goswami, N.N., Bares, J.A., Mangolini, F. et al. (2015). Mechanisms of antiwear tribofilm growth revealed in situ by single-asperity sliding contacts. *Science* 348: 102–105.

49 Kalin, M., Velkavrh, I., and Vizintin, J. (2009). The Stribeck curve and lubrication design for non-fully wetted surfaces. *Wear* 267: 1232–1240.

Section III

Case Study: Sliding Wear of Ceramics

8

Sliding Wear of SiC Ceramics

Considering the importance of SiC-based composites in wear applications, this chapter summarizes author's published results on friction and wear behavior of SiC ceramics, sintered with different amount of sintering additive and sliding load. While discussing the dominant wear mechanisms, the lateral crack-induced wear model has been emphasized to explain the wear behavior of SiC ceramics.

8.1 Introduction

Owing to their superior properties like high strength, hardness, thermal conductivity, notable corrosion, and wear resistance, silicon carbide (SiC) ceramics are used for various structural and tribological applications including mechanical seals, bearings, nozzles, heat exchangers, and cutting tools [1–3]. Accordingly, research on friction and wear behavior of SiC ceramics is considerably reported [4–8]. However, only limited studies have been reported regarding sliding wear behavior of SiC ceramics sintered with small additive content. Ortiz et al. [9] prepared SiC ceramics with 3.6 vol% $Y_3Al_5O_{12}$ by colloidal processing and observed better sliding wear resistance in lubricated contacts. In other work [10], Kumar et al. prepared SiC ceramics with 3 wt% AlN-Sc_2O_3 via hot pressing and found improved wear resistance in dry sliding conditions. Nevertheless, wear resistance of SiC ceramics sintered with small additive content is not yet investigated probably because of difficulties involved in processing. Based on the above background, friction and wear properties of SiC ceramics sintered with 0.2 wt% yttria (Y_2O_3) and 3 wt% Y_2O_3 in dry sliding contacts [11] are discussed in the present chapter.

Friction and Wear of Ceramics: Principles and Case Studies, First Edition.
Bikramjit Basu, Mitjan Kalin, and B. Venkata Manoj Kumar.

8.2 Materials and Experiments

Powder batch mixtures of SiC and 3 wt% Y_2O_3 (designated as S3Y), SiC and 0.2 wt% Y_2O_3 (designated as S0.2Y) were hot-pressed in nitrogen atmosphere at 40 MPa, and 2000 °C for three hours and at 2200 °C for four hours, respectively. Sintered S3Y and S0.2Y ceramics exhibit relative densities of 97.4 and 99.9%, respectively. Typical SEM images of SiC ceramic surfaces are presented in Figure 8.1. The microstructure of sintered S0.2Y ceramics (Figure 8.1a) shows typical bimodal grain structure along with elongated grains (due to β–α transformation), whereas sintered S3Y ceramics (Figure 8.1b) reveals typical equiaxed grain structure with bright contrast secondary phase.

The S3Y ceramics showed lower hardness (26.1 GPa), lower strength (542 MPa), and higher fracture toughness (4.3 MPa·m$^{1/2}$) compared with that of S0.2Y (27.7 GPa, 561 MPa, and 4.0 MPa·m$^{1/2}$). Better hardness and strength for S0.2Y are possibly due to less amount of secondary phase. This, in fact, results in direct bonding between SiC grains and thereby increased transgranular fracture. Details of microstructural characteristics and mechanical behavior can be found elsewhere [11].

Systematic sliding wear tests were conducted in ambient and dry conditions to understand the tribological potential of SiC ceramics sintered with a small amount of yttria additive. The influence of applied normal load on tribological behavior of SiC ceramics was explored. The sliding wear tests were conducted against WC balls at 5, 10, and 20 N load. This resulted in the maximum (initial) Hertzian contact stresses varying from 1.3 to 2.07 GPa. The tangential frictional force was incessantly recorded to generate the coefficient of friction (COF). More details of sliding wear tests are reported elsewhere [11].

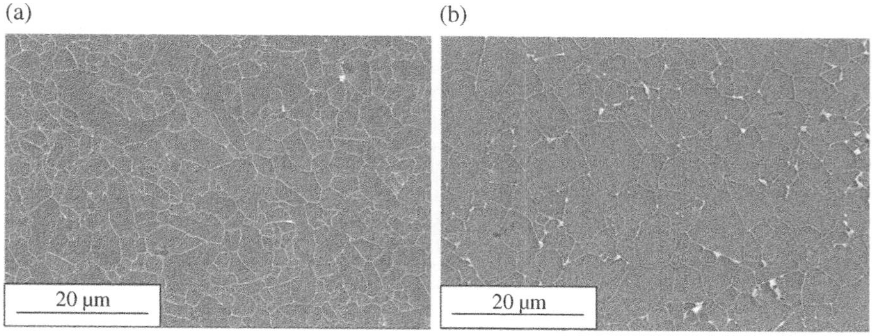

Figure 8.1 Representative SEM images of SiC ceramics sintered with (a) 0.2 wt% Y_2O_3 and (b) 3 wt% Y_2O_3 [11]. *Source:* Reproduced with permission of Elsevier.

8.3 Friction and Wear Behavior of SiC Ceramics Sintered with a Small Amount of Yttria Additive

Typical COF plots for the investigated S0.2Y and S3Y ceramics are shown in Figure 8.2. COF plots indicate observable fluctuations for both S0.2Y and S3Y ceramics, but considerable reduction in fluctuation is observed with an increase in load. The average COF in steady state ranged from 0.44 to 0.52 with the change in load and yttria content. Further, the insignificant variation in COF suggests negligible effect of yttria content on the frictional behavior, whereas COF is

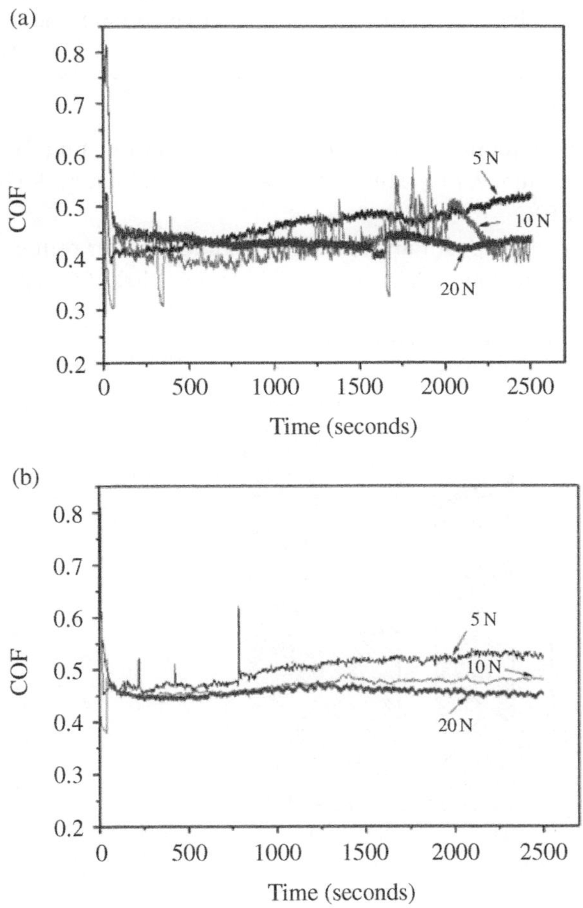

Figure 8.2 The coefficient of friction (COF) vs. time plots for (a) SiC ceramics sintered with 0.2 wt% Y_2O_3 and (b) SiC ceramics sintered with 3 wt% Y_2O_3 during sliding against WC balls at 5, 10, or 20 N load [11]. *Source:* Reproduced with permission of Elsevier.

observed to be less at high load. COF values for silicon carbide ceramics are reported to vary in a wide range from 0.2 to 0.8 with varying test or material parameters [4, 5, 7]. In most cases, high COF values are observed for abrasion- or fracture-induced rough surfaces, whereas low COF values are obtained for hydrated silica-rich smooth surfaces [4, 7, 12]. Accordingly, the higher side COF values observed in the current study suggests mechanical wear domination in sliding of the SiC ceramics.

The volumetric material loss in SiC ceramics is normalized by the product of applied load and total sliding distance and is shown as specific wear rate in Figure 8.3.

> The sliding wear considerably increased with sintering additive content and applied load.

With an increase in additive amount or applied load, specific wear rate ranged from 1.77×10^{-6} to 7.75×10^{-6} $mm^3/N{\cdot}m$. Minimum wear rate is obtained for S0.2Y ceramics at low load of 5 N.

It is known that fracture increases with an increase in sliding load for brittle solids [6]. At a given load, wear rate for S3Y is at least double than that of S0.2Y.

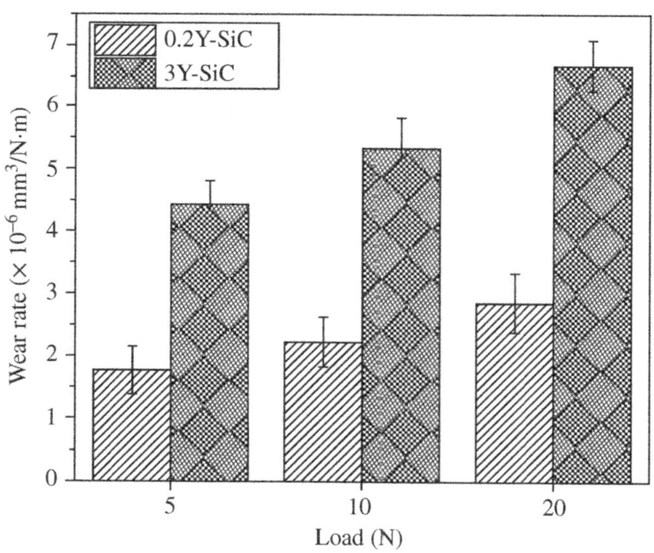

Figure 8.3 Wear rate of the investigated SiC ceramics as a function of additive content and sliding load. 0.2Y-SiC designates SiC ceramics sintered with 0.2 wt% Y_2O_3 and 3Y-SiC designates SiC ceramics sintered with 3 wt% Y_2O_3 [11]. *Source:* Reproduced with permission of Elsevier.

The inferior wear resistance in S3Y ceramics is attributed to the presence of a large fraction of weak amorphous yttria-rich phase.

In general, liquid-phase sintered SiC ceramics exhibit specific wear rate from 10^{-7} to 10^{-5} mm^3/N·m [7, 10, 12].

8.4 Influence of Mechanical Properties on Sliding Wear of SiC Ceramics

In this section, the correlation between material properties and wear property of SiC ceramics, prepared with a small amount of Y_2O_3, is analyzed. For both S0.2Y and S3Y ceramics, microcracking-induced fracture and pull-out were identified as major wear mechanisms when slid against WC ball. As per the lateral fracture model [13], the wear volume V_s of the brittle solids when slid against a sharp indenter can be determined by the relation:

$$V_S = \alpha \frac{P^{9/8}}{(K_{IC})^{1/2} H^{5/8}} \left(\frac{E}{H}\right)^{4/5} S \tag{8.1}$$

where P is sliding load, S is the total distance traveled, K_{IC} is the fracture toughness, H is the hardness, E is the elastic modulus, and α is a material constant. Wear volume for the investigated SiC ceramics is calculated and presented against experimentally determined wear volume in Figure 8.4.

The linear relation between the analytically calculated wear volume and experimentally determined wear volume suggests lateral crack-induced material removal in sliding of SiC ceramics sintered with yttria additive.

8.5 Wear Mechanisms

The worn surfaces of silicon carbide ceramics were subjected to SEM analysis and dominant mechanisms of material removal as a function of additive content and sliding load are studied. The worn surfaces of the investigated silicon carbide ceramics after sliding at different loads show grain fracture, grain pull-out, embedded debris, and deformation of yttria-rich phase. The worn surface of S0.2Y ceramics reveals only surface deformation at low load of 5 N (see inset (a) in Figure 8.4), while microcracking and embedded debris are observed at high load

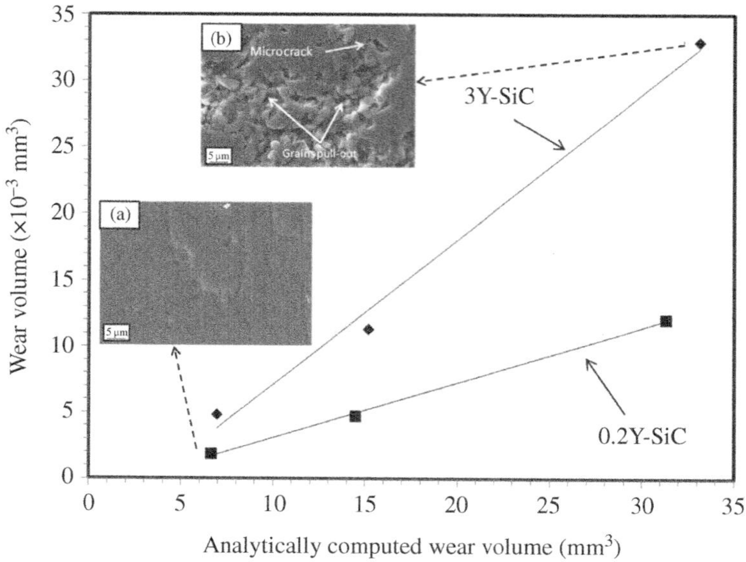

Figure 8.4 Linear relation between experimentally determined wear volume and analytically computed wear volume for the SiC ceramics. Insets (a) and (b), respectively, show worn surfaces of S0.2Y after sliding at 5 N load and S3Y after sliding at 20 N load. 0.2Y-SiC designates SiC ceramics sintered with 0.2 wt% Y_2O_3 and 3Y-SiC designates SiC ceramics sintered with 3 wt% Y_2O_3 [11]. *Source:* Reproduced with permission of Elsevier.

of 20 N (not shown in the figure). On the other hand, the severe microcracking, increased grain pull-out, and fractured SiC grains are noted in S3Y at a high load of 20 N (see inset (b) in Figure 8.4). Further, embedded debris particles are not present on worn S3Y surface as otherwise observed on worn S0.2Y surface. This is possibly due to the difficulty of retaining small oxide particles in sliding contact, where large grains can be easily fractured or pulled out in S3Y ceramics. The yttria additive settles in the SiC grain boundaries or triple junctions while sintering. The weaker yttria-rich phase in triple junctions or along the grain boundaries is susceptible to easy cracking and results in SiC grain pull-out.

8.6 Closure

The results discussed in this chapter clearly illustrate the critical role of sintering aids and load on the wear resistance of non-oxide ceramics. In particular, the wear of SiC ceramics increased with an increase in yttria content and load. The increased microfracture at high load caused a large amount of material removal, whereas the presence of increased amount of weaker Y_2O_3-rich second phase was responsible for high wear in silicon carbide ceramics sintered with 3 wt% Y_2O_3. Therefore, the discussion in the present chapter suggests that the sintering aid for non-oxide ceramics need to be carefully tailored for better wear resistance. More extensive analysis of friction and wear behavior of SiC ceramics, as reported in thic chapter, can be found in author's published paper [11].

References

1 Padture, N.P. (1994). In situ-toughened silicon carbide. *J. Am. Ceram. Soc.* 77: 519–523.

2 Kim, Y.W., Chun, Y.S., Nishimura, T. et al. (2007). High-temperature strength of silicon carbide ceramics sintered with rare-earth oxide and aluminium nitride. *Acta Mater.* 55: 727–736.

3 Lafon-Placette, S., Delbe, K., Denape, J., and Ferrato, M. (2015). Tribological characterization of silicon carbide and carbon materials. *J. Eur. Ceram. Soc.* 35: 1147–1159.

4 Cranmer, D.C. (1985). Friction and wear properties of monolithic silicon-based ceramics. *J. Mater. Sci.* 20: 2029–2037.

5 Dong, X., Jahanmir, S., and Ives, L.K. (1995). Wear transition diagram for silicon carbide. *Tribol. Int.* 28: 559–572.

6 Cho, S.J., Um, C.D., and Kim, S.S. (1996). Wear and wear transition in silicon carbide ceramics during sliding. *J. Am. Ceram. Soc.* 79: 1247–1251.

7 Gahr, K.H.Z., Blattner, R., Hwang, D.H., and Pohlmann, K. (2001). Micro and microtribologycal properties of SiC ceramics in sliding contact. *Wear* 250: 299–310.

8 Sharma, S.K., Kumar, B.V.M., and Kim, Y.W. (2015). Effect of WC addition on sliding wear behavior of SiC ceramics. *Ceram. Int.* 41: 3427–3437.

9 Ortiz, A.L., Borrero-Lopez, O., Quadir, M.Z., and Guiberteau, F. (2012). A route for the pressureless liquid-phase sintering of SiC with low additive content for improved sliding-wear resistance. *J. Eur. Ceram. Soc.* 32: 965–973.

10 Kumar, B.V.M., Kim, Y.W., Lim, D.S., and Seo, W.S. (2011). Influence of small amount of sintering additive on unlubricated sliding wear properties of SiC ceramics. *Ceram. Int.* 37: 3599–3608.

11 Gupta, S., Sharma, S.K., Kumar, B.V.M., and Kim, Y.W. (2015). Tribological characteristics of SiC ceramics sintered with a small amount of yttria. *Ceram. Int.* 41: 14780–14789.

12 Murthy, V.S.R., Kobayashi, H., Tsurekawa, S. et al. (2004). Influence of humidity and doping elements on the friction and wear of SiC in unlubricated sliding. *Tribol. Int.* 37: 353–364.

13 Evans, A.G. and Marshall, D.B. (1981). Wear mechanisms in ceramics. In: *Fundamentals of Frictional Wear of Materials* (ed. D.A. Rigney), 439–452. Materials Park, OH: ASM.

9

Sliding Wear of SiC-WC Composites

In this chapter, the tribological behavior of SiC-WC composites in unidirectional and reciprocated sliding wear conditions is discussed. In particular, an attempt has been made to answer the following important questions, using published results from author's research group: (i) What are the effects of WC content and mating material on friction and wear properties of SiC composites? (ii) Is there any effect of test temperature on wear properties of SiC-WC composites in reciprocated sliding conditions? (iii) What are the differences in mechanisms of material removal between unidirectional and reciprocated sliding wear conditions? It is perceived that the discussion in this chapter will provide guidelines to develop next-generation SiC-based composites for wear resistance applications.

9.1 Introduction

As mentioned earlier, silicon carbide (SiC) ceramics are recommended for tribological applications due to high hardness, strength, thermal conductivity, and superior resistance against corrosion or wear [1–7]. Literature indicates that the wear mechanisms of ceramics are influenced by the counterbody material. For example, a major transition in primary wear mechanism from tribomechanical wear to tribochemical wear was observed in SiC ceramics when worn against ZrO_2, Al_2O_3, WC-Co, and Si_3N_4 [4]. Some of the earlier studies focused on understanding the deformation and crack propagation underneath the worn surfaces, with an aim to understand the material removal mechanisms [8–14]. Many of those studies were conducted on the thermally or mechanically modified surfaces of the brittle ceramics. In the first part of this chapter, the tribological behavior of

Friction and Wear of Ceramics: Principles and Case Studies, First Edition.
Bikramjit Basu, Mitjan Kalin, and B. Venkata Manoj Kumar.

novel SiC-WC composites is investigated in dry sliding conditions against SiC ceramics SiC, WC-Co cermets, and steel balls [5, 6]. Especially, the influence of WC content (0–50 wt%) and counterbody on wear mechanisms of SiC ceramics is explained.

In applications such as turbine blades, rotary shaft bearings, hot abrasive materials' transportation medium, etc., understanding of reciprocated sliding wear resistance is needed. However, SiC-based composites are not considerably studied for their behavior; in this context, reciprocated sliding wear conditions. Therefore, tribological properties of SiC-WC composites in reciprocated sliding conditions against SiC ball were studied [7] and major results are discussed in the second part of the chapter. In particular, the effects of temperature: room temperature and 500 °C, and WC content: 0, 10, 30, and 50 wt% on the friction and wear properties of SiC ceramics are discussed. Tests were conducted with oscillations of 4.71 mm and a frequency of 8.3 Hz with a stroke of 4.71 mm for 1000 cycles. The influence of WC content on reciprocated sliding wear properties of SiC-WC composites at 19.3 N load at ambient temperature and elevated temperature (500 °C) conditions was studied.

9.2 Microstructure and Mechanical Properties of SiC-WC Composites

The materials' processing is similar to that of yttria-added SiC, as described in Chapter 8. In particular, powder mixtures containing 0, 10, 30, and 50 wt% WC were hot-pressed in two stages at 40 MPa in the argon atmosphere at 1800 °C for one hour. SiC-0 wt% WC, SiC-10 wt% WC, SiC-30 wt% WC, and SiC-50 wt% WC composites are designated, respectively, as SW0, SW10, SW30, and SW50. Typical fracture surfaces in Figure 9.1 show near-equiaxed SiC grains for SW0 and homogeneous distribution of WC particles inside and along the boundaries of SiC grains for SW30. Furthermore, the microstructural analysis of sintered composites indicates that the average grain size of SiC grains decreased from 835 nm for SW0 to 578 nm for SW50, and the average interparticle distance for WC decreased from 7 μm for SW10 to 2 μm for SW50 (Table 9.1). With respect to mechanical properties, hardness and fracture toughness ranged from 24 to 26 GPa, and 5.85 to 6.66 $MPa{\cdot}m^{1/2}$ with WC addition. SW30 exhibited maximum hardness, while SW50 exhibited maximum fracture toughness. More details of the microstructural and mechanical characteristics are explained elsewhere [15].

Figure 9.1 Representative fracture surfaces of SiC ceramics prepared with (a) 0% WC, (b) 10% WC, (c) 30% WC, and (d) 50% WC [15]. *Source:* Reproduced with permission of Elsevier.

9.3 Influence of Mating Material and WC Content on Tribological Properties

In this section, phenomenological understanding of the friction wear properties of SiC-WC composites in sliding wear contacts against SiC, WC-Co, and steel balls are discussed. The dominant material removal mechanisms are also elucidated.

9.3.1 Friction and Wear Behavior

The average steady-state coefficient of friction (COF) decreased from 0.49 to 0.39 against SiC ball and from 0.66 to 0.33 against WC-Co ball, while it increased from 0.34 to 0.43 against steel ball. The average steady-state COF values for SiC-based ceramics in dry sliding conditions are generally observed to vary in a wide range from 0.20 to 0.75 [16–19].

Table 9.1 SiC grain size, WC interparticle distance, hardness, and fracture toughness of sintered SiC-WC composites [15].

Sample composition (designation)	Average size of SiC grains (nm)	WC interparticle distance (μm)	Vickers hardness (GPa)	Indentation fracture toughness ($MPa\cdot m^{1/2}$)
SiC (SW0)	835 ± 73	—	23.95 ± 0.97	5.85 ± 0.30
SiC-10 wt% WC (SW10)	724 ± 65	6.805 ± 2.763	24.02 ± 1.12	6.36 ± 0.22
SiC-30 wt% WC (SW30)	637 ± 53	2.922 ± 0.667	26.33 ± 0.71	6.47 ± 0.13
SiC-50 wt% WC (SW50)	578 ± 62	2.303 ± 0.395	24.26 ± 1.12	6.66 ± 0.12

Source: Reproduced with permission of Elsevier.

For SiC-based ceramics, high friction is attributed to mechanical wear instances like abrasion, chipping, and fracture, whereas low friction results from the oxide layer formation at the contact.

The wear volume for the SiC-WC ceramic composites in dry sliding wear contacts as a function of counterbody and WC content at 20 N is shown in Figure 9.2. Wear volume is minimum for SW50 against steel ball. This is due to the highest fracture toughness for the SiC ceramics with a large amount of WC. The impact of better toughness property is reflected in less grain pull-out or fracture during sliding. Wear volume against SiC ball is increased from 2.3×10^{-2} mm^3 to 5.2×10^{-2} mm^3 with the decrease in WC content from 50 to 10 wt%. While a small change in wear volume is observed against steel ball, an order of magnitude less wear volume is obtained against WC-Co ball.

9.3.2 Mechanisms of Material Removal

Figure 9.3 illustrates representative SiC-WC composites' surfaces after sliding at 20 N load against different counterbodies. Since the addition of WC led to an increase in fracture toughness of SiC ceramics, the grain pull-out and fracture from the disk surfaces are decreased when slid against SiC ball. The removal of WC particles from the disk surfaces due to SiC ball penetration causes a large amount of material loss in case of SiC ceramics with a large amount of WC (Figure 9.3a and b). When worn

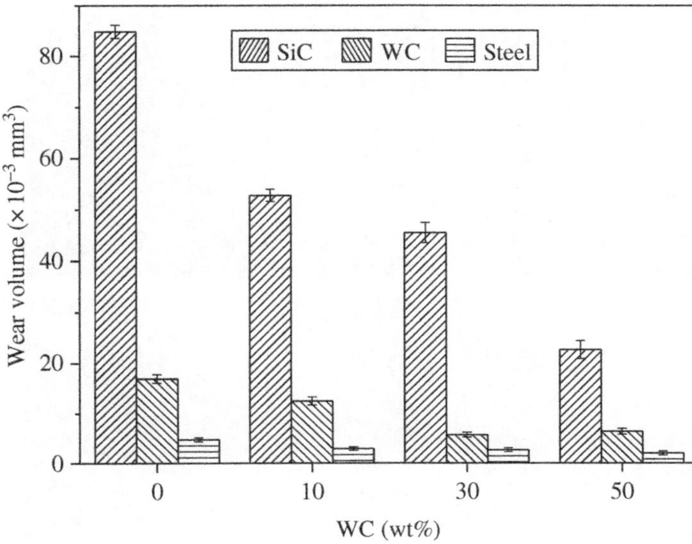

Figure 9.2 Wear volume data for SiC-WC composites against different counterbody balls: SiC, WC-Co, and steel as a function of WC content at 20 N load. *Source:* Adapted from Sharma et al. [5, 6].

against WC-Co, SW10 worn surface shows pulled-out WC particles or debris and noteworthy abrasion, while SW50 worn surface is significantly covered with a thick layer (Figure 9.3c and d). After sliding against steel ball, the SW10 surface appears rough with notable abrasion and fractured surface, while the SW50 surface shows reduced abrasion and fracture (Figure 9.3e and f).

The mode of material removal is further understood by studying the counterbody ball surfaces (Figure 9.4) after sliding at 20 N against SiC-WC composites. The worn surfaces of SiC balls show mainly abrasion and pits, while that of WC-Co and steel balls show polishing and oxide layer. The oxide layer formation is dominant in the case of steel balls worn against composite containing a large amount of WC. The reduced abrasion and fracture observed on the worn surface of SW50 (see Figure 9.3f) also suggests that the contact was completely covered by layer. Owing to less fracture and penetration with increased WC content in SiC ceramics, the scar diameter decreased for any counterbody. The scar diameter decreased from 1.04 to 0.86 mm for SiC, from 1.15 to 0.96 mm for WC-Co, and from 1.24 to 1.07 mm for steel counterbody with increased WC from 0 to 50 wt%.

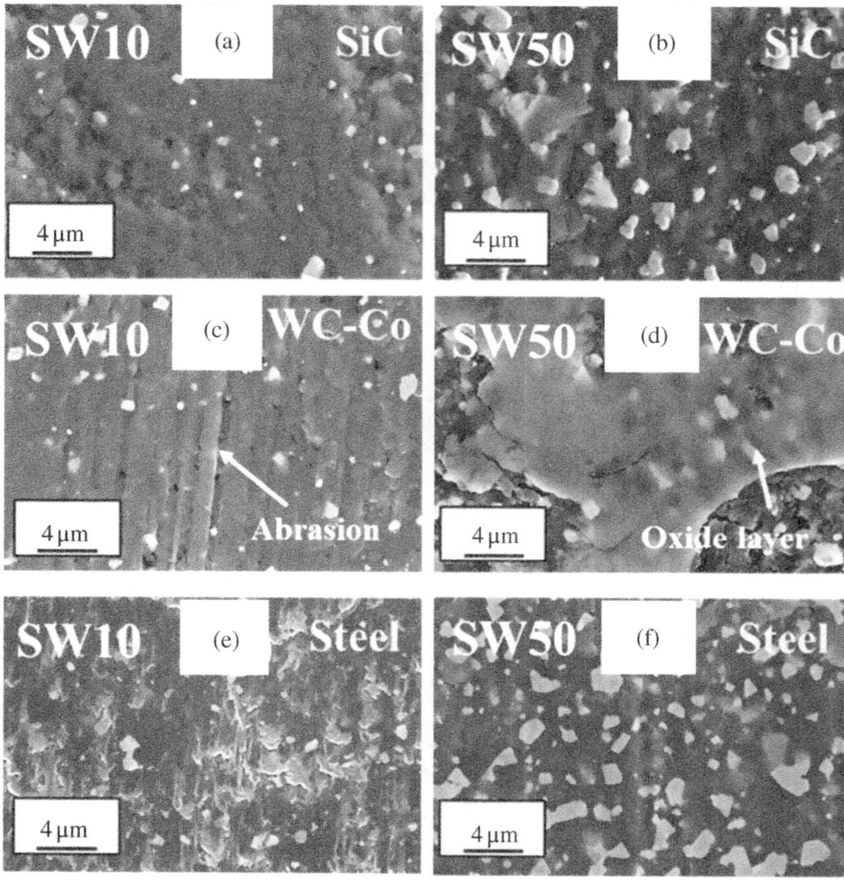

Figure 9.3 Typical worn surface images of (a) monolithic SiC against SiC ball, (b) SiC-50 wt% composite against SiC ball, (c) monolithic SiC against WC-Co ball, (d) SiC-50 wt% composite against WC-Co ball, (e) monolithic SiC against steel ball, and (f) SiC-50 wt% composite against steel ball. Other sliding conditions: 20 N load and room temperature. *Source:* Adapted from Sharma et al. [5, 6].

Further, Raman spectroscopy analysis of debris collected after sliding SW50 composite against different counterbodies (Figure 9.5) confirms the formation of silicon oxide [20], tungsten oxide [21], and cobalt tungsten oxide [22] during sliding of SiC-WC composites against SiC or WC-Co. Raman analysis also supports the formation of iron oxide [23] and iron tungsten oxide [24] during sliding of the composites against steel.

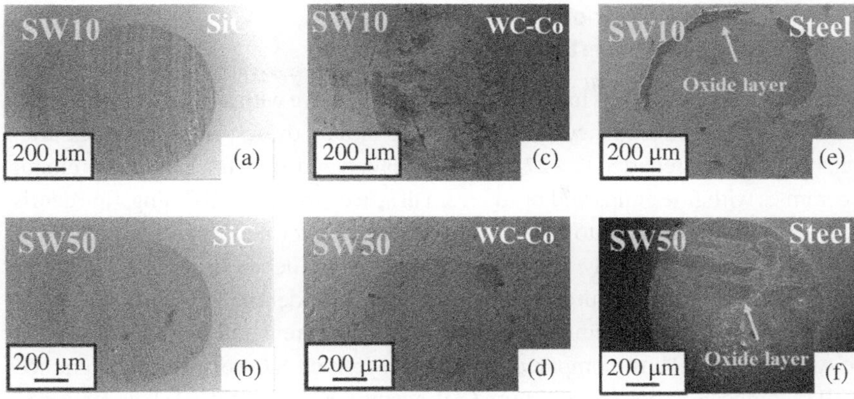

Figure 9.4 Typical worn surfaces of counterbody: (a) SiC ball against SiC-10 wt% composite, (b) SiC ball against SiC-50 wt% composite, (c) WC-Co ball against SiC-10 wt% composite, (d) WC-Co ball against SiC-50 wt% composite, (e) steel ball against SiC-10 wt% composite, and (f) steel ball against SiC-50 wt% composite. Other sliding conditions: 20 N load and room temperature. *Source:* Adapted from Sharma et al. [5, 6].

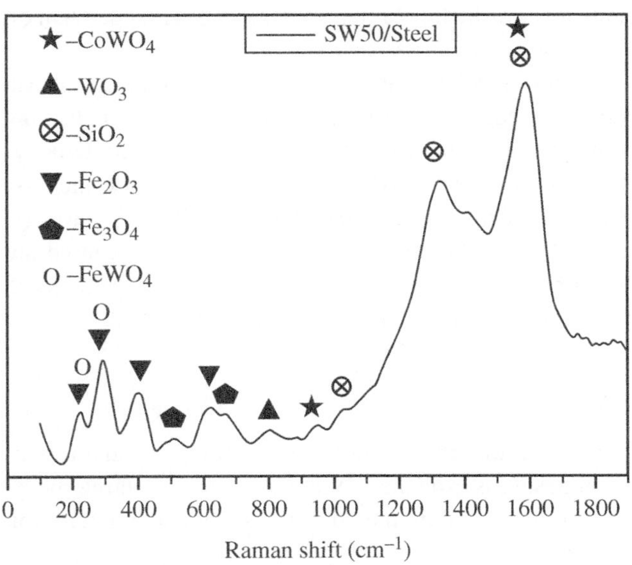

Figure 9.5 Raman spectroscopy analysis of wear debris collected after sliding SiC-50 wt% WC (SW50) composite against steel ball. Other sliding conditions: 20 N load and room temperature [5]. *Source:* Reproduced with permission of Springer Nature.

9.3.3 Friction and Wear of SiC-WC Composites: System-Dependent Properties

The above-described results indicate that reduced fracture with a large amount of WC resulted in increased tendency for the formation of hard oxides, and thereby less friction. It is generally reported [25–27] that deformation and grain fracture are dominant in ceramics with fine grains and result in small size debris. During sliding, fine debris particles are compressed into layers on surfaces and deformed, resulting in reduced friction or wear. Accordingly, fine grains resulted in fine debris size for SiC-WC composites with the large amount of WC. Also, small size debris particles are likely to connect to each other in humid conditions to form adherent and protective layers on the worn surface. Thus, compared with coarse-grained SiC ceramics, fine-grained SiC-WC composites showed minimum COF against SiC ball or WC-Co ball. However, high COF observed for SW50 against steel ball can be attributed to the formation of hard iron tungsten oxide ($FeWO_4$) as per the following reactions:

$$2Fe + O_2 \rightarrow 2FeO \tag{9.1}$$

$$FeO + WO_3 \rightarrow FeWO_4 \tag{9.2}$$

The above-mentioned reactions are thermodynamically feasible ($\Delta G < 0$) in the 273–973 K range. Therefore, it is worthwhile to state that friction is independent of the hardness of the counterbody.

In the present study, coarse-grained SiC ceramics exhibited maximum wear against any counterbody. Higher fracture toughness of SiC-WC composites led to a reduction in grain pull-out and fracture, causing a reduction in wear against SiC counterbody. The material loss from fine grain composites is decreased with less fracture and protective tribolayer formation during sliding against steel or WC-Co ball (see Figure 9.4). For coarse grain composites, the stability of tribolayer is reduced and caused an increase in wear by cracking or delamination [27]. Compared with wear against steel or WC-Co ball, higher wear is obtained for composites slid against SiC ball.

9.3.4 Wear Mechanisms

In this section, the influence of counterbody and WC content on dominant wear mechanisms of SiC-WC composites is explained. With the change in counterbody from ceramics to metal, transition is the dominant mechanism of material removal from fracture to adhesive wear for the SIC-WC composites. The large tendency for fracture of SiC ceramics (SW0) is found against SiC ball as compared against steel or WC-Co ball. Surface polishing is dominant in sliding against steel ball. The debris particles generated in sliding resulted in high wear because of three-body abrasion for SiC ceramics. Material from the disk surface is removed due to abrasion and cracking during sliding of hard SiC counterbody.

> SiC-50 wt% WC composite with higher fracture toughness restricts crack propagation and exhibits less wear.

On the other hand, the formation of tribolayer influenced the friction and wear properties of SiC-WC composites against steel or WC-Co ball. The strongly adherent tribolayer avoided further wear by limiting penetration depth. The counterbody ball of less hardness penetrated less into the composite surface and resulted in less wear. The severity of grain pull-out, fracture, and abrasion are reduced with the formation of protective tribolayer at the contact as the hardness of counterbody decreases from 28 GPa (for SiC ball) to 7 GPa (for steel ball) (see Figure 9.4). The dominant mechanisms of material removal for monolithic SiC and SiC-WC composites against different counterbodies are schematically illustrated in Figure 9.6.

Further, it is widely reported that wear behavior of brittle materials is affected by mechanical properties like fracture toughness, hardness, and elastic modulus. For example, lower specific wear rate for the ceramic with higher hardness is reported by Tatarko et al. [28]. However, neither fracture toughness nor hardness exhibited direct relation with the specific wear rate of Si_3N_4 in a study of Miyazaki et al. [29] In the present investigation, the wear of investigated SiC-WC composites is dependent on fracture toughness than hardness. SW50 composites with maximum fracture toughness (6.7 MPa·m$^{1/2}$) and moderate hardness (24 GPa) showed minimum wear against any counterbody.

> For SiC-WC composites, mechanical fracture is observed against ceramic counterbody, whereas tribochemistry is the dominant material removal mechanism for cermet and metal counterbody.

Finally, results obtained in the sliding of SiC-WC composites concluded that pull-out and fracture can be reduced with an increase in WC content.

9.4 Reciprocated Sliding Wear Behavior of SiC-WC Composites

In this section, salient results obtained in reciprocated sliding wear study of SiC-WC composites are discussed in terms of test temperature (room temperature and 500 °C) and composition (0–50 wt% WC content).

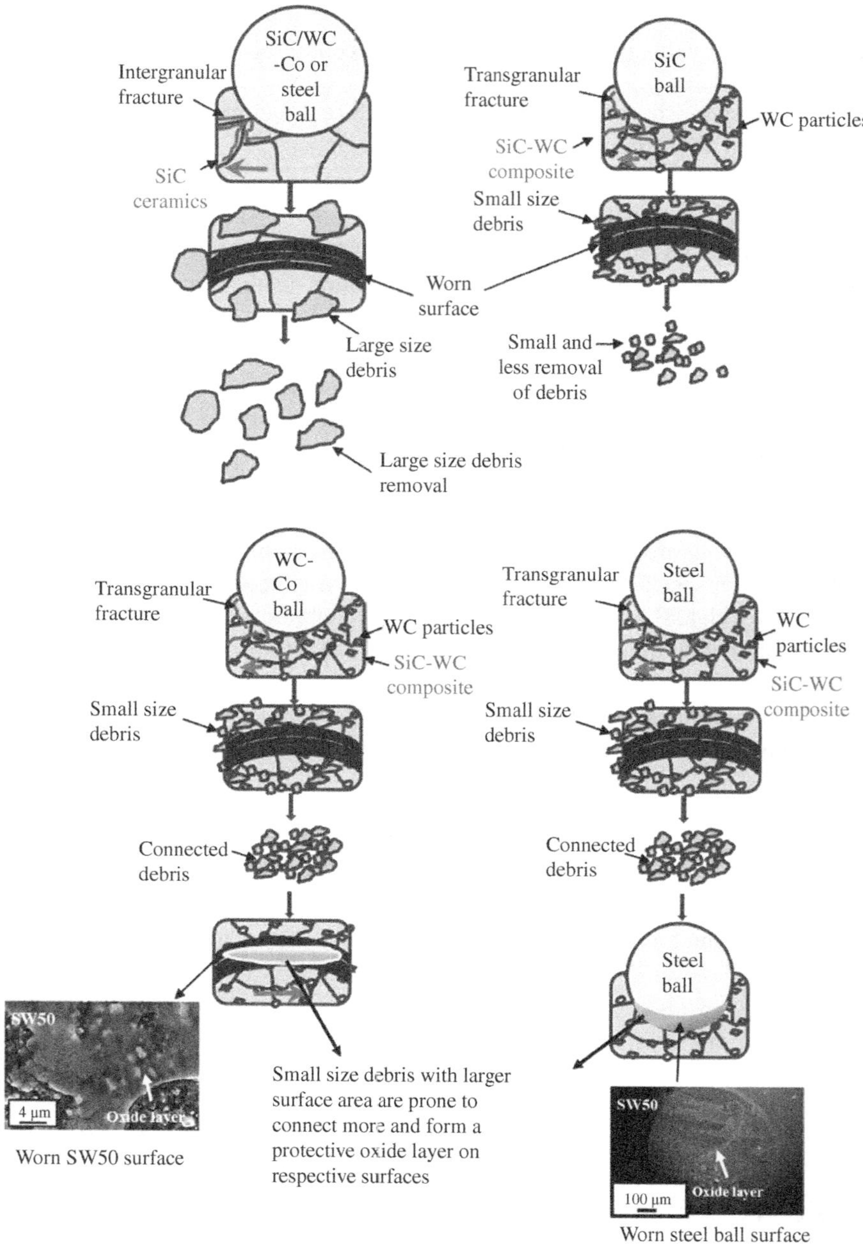

Figure 9.6 A schematic representation of major mechanisms of material removal for monolithic SiC ceramics and SiC-WC composites when slid against different counterbodies [5]. *Source:* Reproduced with permission of Springer Nature.

9.4.1 Frictional and Wear Behavior

It is observed that the steady-state COF for SiC ceramics decreased from 0.48 to 0.35 with an increase in WC amount from 0 to 50 wt% in room-temperature reciprocated sliding test. At elevated temperature (500 °C), friction between the SiC ball and composite surface is 0.7 during initial running-in-period and continuously increased beyond 1.0. This can be attributed to the incessant generation and fracture of debris during sliding.

Wear volume data of SiC ceramics and SiC-WC composites after sliding in reciprocating sliding conditions at ambient (RT) temperature and at 500 °C are shown in Figure 9.7. In ambient conditions, wear volume is reduced from 2.7×10^{-3} mm^3 to 2.8×10^{-4} mm^3 for SiC-WC composites with large WC content when tested at 19 N load. The propagation of fatigue-induced cracks is barred by a large amount of WC particles to provide minimum wear for SW50 composites. During sliding at 500 °C, monolithic SiC ceramics do not exhibit a notable difference in wear volume (see Figure 9.7). This is due to the less degradation in strength at elevated temperatures for SiC ceramics [30]. The wear volume is reduced from 2.9×10^{-3} mm^3 for SiC ceramics to 2.0×10^{-3} mm^3 for SW10 composite but remains unaltered for SW30 composite. Conversely, wear volume increased almost 3.5 times, i.e. from 2.1×10^{-3} mm^3 to 7.6×10^{-3} mm^3 with an increase in WC amount from 30 to 50 wt%. This notable difference in wear in case of SW50 is due to the change in major material removal mechanisms.

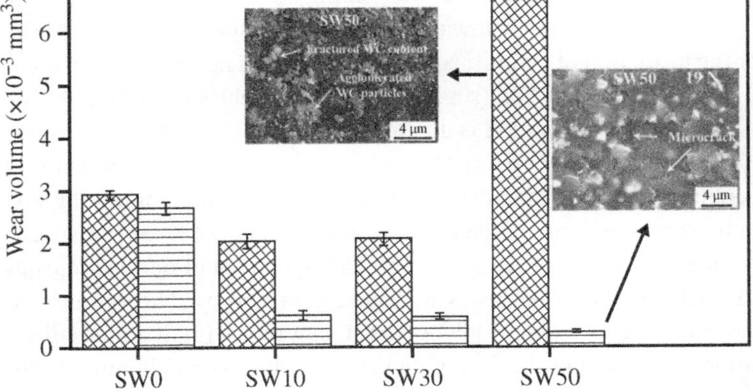

Figure 9.7 Wear volume of SiC-WC composites at 19 N load at RT and at 500 °C. Insets show the worn surfaces to reveal the differences in wear mechanism for SW50 composites at RT and at 500°C [7]. *Source:* Reproduced with permission of Elsevier.

9.4.2 Critical Analysis of Wear Mechanisms

In general, SEM images of SiC-WC composites after sliding at 19 N load and room temperature show shear fracture and compaction of SiC grains, pull-out of WC from the top surface. The reduced extent of fracture resulted in less wear for the composite with large WC content. Typically worn surface of SW50 composite, possessing the maximum fracture toughness of 6.7 MPa $m^{1/2}$, shows well-bonded WC particles and minimum wear among all composites (see Figure 9.7). Earlier reports on sliding wear of cemented carbide also indicated well-bonded WC grains to the matrix, less pull-out, and fracture [31–33].

In case of reciprocating sliding wear at 500 °C, SiC ceramics show rough surface with significant fracture of softened SiC grains. The fracture reduced with an increase in WC content to 30 wt%. However, SW50 composite surface shows severe fracture and increased pull-out of WC particle agglomerates or clusters (see inset in Figure 9.7). Stresses induced due to thermal expansion mismatch are reported to be larger at higher temperatures [34]. It is possible that stress concentrations at increased defect sites of WC particle agglomerates in SW50 composites are further increased in high-temperature sliding conditions and result in a large amount of material removal. Therefore, reinforcement of 30 wt% WC is said to be optimum for SiC ceramics in high-temperature reciprocated sliding wear conditions.

9.4.2.1 Wear Debris Analysis

Representative SEM images of wear debris collected after reciprocated sliding wear of SiC-WC composites at room temperature and at 500 °C are shown in Figure 9.8. In general, wear debris particles collected after sliding of SW0 ceramics at room temperature appear to be loose, small in size, and irregular in shape (see Figure 9.8a), while those collected after sliding of SiC-WC composites are large in size with a clear tendency towards cylindrical shapes. The size, and the tendency for the formation of cylindrical shape debris particles, is higher in composite with a large amount of WC (see Figure 9.8b for debris from SW50). The cylindrical-shaped debris formation is due to two major facts: (i) fine size of SiC grains in SiC-WC composites with large WC content and (ii) (moderately) humid (~55% RH) environment. The small size debris from the surface containing fine grains is likely to connect and achieve the cylindrical shape [35, 36]. Also, the presence of noticeable peaks of oxygen in the EDS spectrum of debris indicates that debris is rich in silica (see insets in Figure 9.8a and b). The cylindrical-shaped debris is reported to form in silicon-based materials with hydrated silicon oxide formation, as a result of the following reactions in moderately humid sliding contacts [35, 37–39]:

$$SiC + 2O_2 \rightarrow SiO_2 + CO_2 \tag{9.3}$$

Figure 9.8 SEM images of debris particles collected after reciprocated sliding: (a) monolithic SiC at room temperature, (b) SiC-50 wt% WC at room temperature, (c) monolithic SiC at 500 °C, and (d) SiC-50 wt% WC at 500 °C. EDS analysis for (a), (b), and (d) are shown as insets in respective images. Sliding load: 19 N [7]. *Source:* Reproduced with permission of Elsevier.

$$SiC + 2H_2O \rightarrow SiO_2 + CH_4 \tag{9.4}$$

$$SiO_2 + 2H_2O \rightarrow Si(OH)_4 \tag{9.5}$$

The cylindrical debris formation is usually related to reduced friction or wear in sliding [37–40]. Based on the debris analysis and worn surface analysis, it can be said that both fracture and tribochemistry influence the wear properties of the investigated SiC-WC composites in room-temperature reciprocated sliding conditions.

Debris particles collected after sliding at 500 °C (see Figure 9.8c and d) appear dry, loose, and irregularly sized. While dry debris is believed to be the result of higher friction and wear at high-temperature sliding, small size debris from composites with a large amount of WC supports fracture of fine grains. In addition, EDS analysis (inset in Figure 9.8d) of collected debris particles after sliding at 500 °C does not reveal any change, as compared with that collected after sliding at room temperature. Therefore, it can be concluded that cracking-induced fracture is dominant in reciprocated sliding wear of the SiC-WC composites at 500 °C.

9.4.2.2 Effect of Temperature

Results obtained in the present study indicate that test temperature considerably influences friction and wear properties of SiC-WC composites in reciprocated sliding wear conditions.

In case of high-temperature sliding, the COF increased above 1.0 in running-in-period due to the hard and dry debris particles generated at the contact. Though the size is small, debris particles are not hydrated to connect with each other, as otherwise observed in room-temperature sliding.

> The softening of matrix resulted in higher wear in high-temperature conditions than in room-temperature conditions.

The material loss decreased with an increase in the amount of WC to 30 wt%. However, WC particle agglomerates induce severe stresses and result in an increased material loss at high temperature for the composites containing WC content beyond 30 wt%.

9.4.2.3 Effect of Test Configuration on Wear Behavior

Since all other test parameters like initial maximum Hertzian stresses, mating materials, environmental conditions, etc., are maintained, the results obtained from unidirectional sliding and reciprocatory sliding tests can be safely compared.

> The dominant wear mechanism is alike for monolithic SiC ceramics in both test conditions, but a change in dominant wear mechanism from mechanical fracture to tribochemistry is observed for the composites when wear conditions changed from unidirectional to reciprocated sliding.

The comparison of wear rates obtained for SiC-WC composites against SiC ball in unidirectional sliding and reciprocating sliding conditions at ambient temperature is shown in Figure 9.9. Wear rate varied from 15.0×10^{-6} $mm^3/N{\cdot}m$ to 4.5×10^{-6} $mm^3/N{\cdot}m$ in unidirectional sliding and from 14.8×10^{-6} $mm^3/N{\cdot}m$ to 1.5×10^{-6} $mm^3/N{\cdot}m$ in reciprocating sliding conditions with WC content. Wear rate remained almost the same for monolithic SiC ceramics in both the wear conditions. For SiC-50 wt% WC composites, wear rate in reciprocating sliding conditions is reduced to almost 30% than in unidirectional sliding conditions. This particular observation essentially indicates that the extent of wear damage of SiC-WC composites is influenced by the wear test configuration. The SiC-50 wt% WC composite is beneficial in reducing the wear damage to a maximum extent when used in reciprocating sliding test conditions. This can be attributed to the domination of retained tribochemistry-induced wear debris in reciprocating sliding contacts compared with unidirectional sliding conditions.

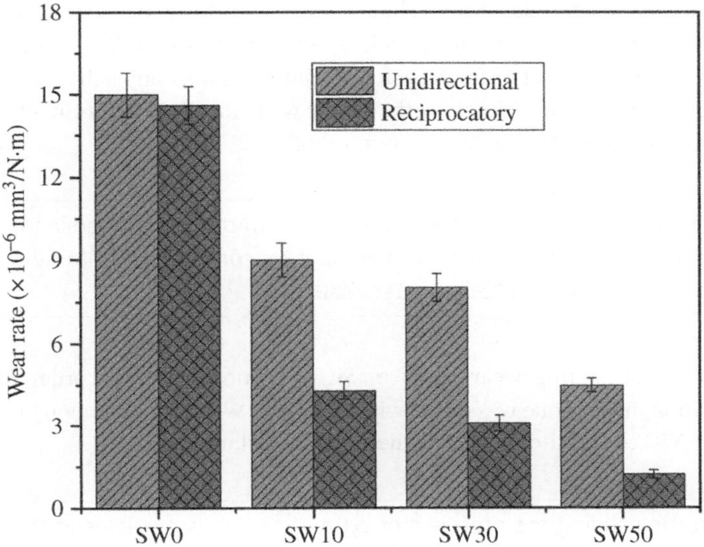

Figure 9.9 Wear rate data for SiC-WC composites against SiC ball in unidirectional sliding and reciprocating sliding conditions at ambient temperature.

In the case of composites, the finer microstructure leads to finer debris particles (as explained earlier). The change in the extent of wear is understood based on long contact time of debris in reciprocated sliding conditions and the responsible mechanism of material removal. Fine wear debris connects easily with each other and forms cylindrical debris in reciprocating sliding conditions, compared with unidirectional sliding conditions. The cylindrical debris particles are rolled and resulted in reduced friction and wear for the composites. The characteristic difference in wear debris obtained after wear in unidirectional sliding and reciprocatory sliding conditions is attributed to the difference in wear induced damage.

Based on the results obtained in reciprocated sliding wear tests, it can be said that SW50 composites showed superior wear resistance at room temperature and SW30 composites at 500 °C. The fracture and tribochemistry are major mechanisms of material removal in reciprocating sliding at room temperature, while fracture controlled material removal at 500 °C.

9.5 Closure

The discussion on the influence of counterbody on sliding wear of SiC-(0–50 wt%) WC composites implies that the friction decreased with WC content against SiC or WC-Co ball, while it increased against steel ball. Generation of hard iron tungsten oxide ($FeWO_4$) debris in ceramics during sliding against steel is believed to cause high friction. The increased fracture toughness of SiC ceramics with WC content caused a reduction in the extent of fracture during sliding against any ball and led to reduced wear. A significant change in the major wear mechanism of SiC-WC composites is observed with the change in counterbody.

Against SiC ball, SiC-WC composites showed mechanical fracture as the dominant wear mechanism, while tribochemistry is dominant for composites containing a large amount of WC against WC-Co or steel ball.

In the reciprocating sliding wear study of SiC-WC composites, an order of magnitude variation in wear rate was found with change in WC content or temperature. SiC-30 wt% WC composite showed highest wear resistance at 500 °C.

Under reciprocating sliding, the presence of cylindrical debris particles at the contact led to reduced wear and friction for the SiC-composites at room temperature, while loose and dry debris particles generated at 500 °C resulted in high wear and friction.

For more in-depth understanding on the sliding wear behavior of SiC-WC composites, readers can refer to author's published reports [5–7].

References

1 Padture, N.P. (1994). In situ-toughened silicon carbide. *J. Am. Ceram. Soc.* 77: 519–523.

2 Kim, Y.W., Chun, Y.S., Nishimura, T. et al. (2007). High-temperature strength of silicon carbide ceramics sintered with rare-earth oxide and aluminium nitride. *Acta Mater.* 55: 727–736.

3 Lafon-Placette, S., Delbe, K., Denape, J., and Ferrato, M. (2015). Tribological characterization of silicon carbide and carbon materials. *J. Eur. Ceram. Soc.* 35: 1147–1159.

4 Kovalcikova, A., Kurek, P., Balko, J. et al. (2014). Effect of the counterpart material on wear characteristics of silicon carbide ceramics. *Int. J. Refract. Met. Hard Mater.* 44: 12–18.

5 Sharma, S.K., Manoj Kumar, B.V., and Kim, Y.W. (2019). Tribology of WC reinforced SiC ceramics: influence of counterbody. *Friction* 7: 129.

6 Sharma, S.K., Manoj Kumar, B.V., and Kim, Y.W. (2015). Effect of WC addition on sliding wear behavior of SiC ceramics. *Ceram. Int.* 41 (3): 3427–3437.

7 Sharma, S.K., Venkata Manoj Kumar, B., Zugelj, B.B. et al. (2017). Room and high temperature reciprocated sliding wear behavior of SiC-WC composites *Ceram. Int.* 43 (18): 16827–16834.

8 Elfallagh, F. and Inkson, B.J. (2009). 3D analysis of crack morphologies in silicate glass using FIB tomography. *J. Eur. Ceram. Soc.* 29: 47–52.

9 Kanematsu, W. (2004). Subsurface damage in scratch testing of silicon nitride. *Wear* 256: 100–107.

10 Desa, O. and Bahadur, S. (1999). Material removal and subsurface damage studies in dry and lubricated single-point scratch tests on alumina and silicon nitride. *Wear* 225: 1264–1275.

11 Zarudi, I., Zhang, L., and Mai, Y.-W. (1996). Subsurface damage in alumina induced by single-point scratching. *J. Mater. Sci.* 31: 905–914.

12 Wu, H., Inkson, B.J., and Roberts, S.G. (2001). Subsurface deformation of machined Al_2O_3 and Al_2O_3/5 vol% SiC nanocomposite. *J. Microsc.* 201: 212–220.

13 Wu, H.Z., Roberts, S.G., Mobus, G., and Inkson, B.J. (2003). Subsurface damage analysis by TEM and 3D FIB crack mapping in alumina and alumina/5 vol.% SiC nanocomposites. *Acta Mater.* 51: 149–163.

14 Hockey, B.J. and Lawn, B.R. (1975). Electron microscopy of microcracking about indentations in aluminium oxide and silicon carbide. *J. Mater. Sci.* 10: 1275–1284.

15 Sharma, S.K., Venkata Manoj, B., Kumar, K.-Y. et al. (2014). Erosion behavior of SiC–WC composites. *Ceram. Int.* 40: 6829–6839.

16 Lopez, O.B., Ortiz, A.L., Guiberteau, F., and Padture, N.P. (2005). Effect of microstructure on sliding-wear properties of liquid-phase-sintered α-SiC. *J. Am. Ceram. Soc.* 88 (8): 2159–2163.

17 Gates, R.S. and Hsu, S.M. (2004). Tribochemistry between water and Si_3N_4 and SiC: induction time analysis. *Tribol. Lett.* 17 (3): 399–407.

18 Hsu, S.M. and Shen, M.C. (1996). Ceramic wear maps. *Wear* 200: 154–175.

19 Kato, K. and Adachi, K. (2002). Wear of advanced ceramics. *Wear* 253 (11–12): 1097–1104.

20 Cortes, A., Celedon, C., and Zarate, R. (2015). CVD synthesis of graphene from acetylene catalyzed by a reduced CuO thin film deposited on SiO_2 substrates. *J. Chil. Chem. Soc.* 60: 2911–2913.

21 Luo, J.Y., Chen, X.X., Li, W.D. et al. (2013). Variable-temperature Raman spectroscopic study of the hydrogen sensing mechanism in Pt-WO_3 nanowire film. *Appl. Phys. Lett.* 102: 104–113.

22 Anspoks, A., Kalinko, A., Timoshenko, J., and Kuzmin, A. (2014). Local structure relaxation in nanosized tungstates. *Solid State Commun.* 183: 22–26.

23 Zhang, X., Niu, Y., Meng, X. et al. (2013). Structural evolution and characteristics of the phase transformations between α-Fe_2O_3, Fe_3O_4 and γ-Fe_2O_3 nanoparticles under reducing and oxidizing atmospheres. *CrystEngComm* 15: 8166–8172.

24 Qian, J., Peng, Z., Wu, D., and Fu, X. (2014). $FeWO_4$/FeS core/shell nanorods fabricated by thermal evaporation. *Mater. Lett.* 122: 86–89.

25 Rice, R.W. (1985). Micromechanics of microstructural aspects of ceramic wear. In: *Proceedings of the 9th Annual Conference on Composites and Advanced Ceramic Materials*, Ceramic Engineering and Science Proceedings, vol. 6, 940–958. New York: Wiley.

26 Gahr, K.H.Z. (1989). Sliding wear of ceramic-ceramic, ceramic-steel and steel-steel pairs in lubricated and unlubricated contact. *Wear* 133: 1–22.

27 Gahr, K.H.Z., Bundschuh, W., and Zimmerlin, B. (1993). Effect of grain size on friction and sliding wear of oxide ceramics. *Wear* 162: 269–279.

28 Tatarko, P., Kasiarova, M., Dusza, J. et al. (2010). Wear resistance of hot-pressed Si_3N_4/SiC micro/nanocomposites sintered with rare-earth oxide additives. *Wear* 269 (11): 867–874.

29 Miyazaki, H., Hyuga, H., Yoshizawa, Y.I. et al. (2009). Correlation of wear behavior and indentation fracture resistance in silicon nitride ceramics hot-pressed with alumina and yttria. *J. Eur. Ceram. Soc.* 29 (8): 1535–1542.

30 Keppeler, M., Reichert, H.G., Broadley, J.M. et al. (1998). High temperature mechanical behaviour of liquid phase sintered silicon carbide. *J. Eur. Ceram. Soc.* 18: 521–526.

31 Pirso, J., Letunovits, S., and Viljus, M. (2004). Friction and wear behaviour of cemented carbides. *Wear* 257: 257–265.

32 Bonny, K., Baets, P.D., Perez, Y. et al. (2010). Friction and wear characteristics of WC–Co cemented carbides in dry reciprocating sliding contact. *Wear* 268: 1504–1517.

33 Muthuraja, A. and Senthilvelan, S. (2015). Adhesive wear performance of tungsten carbide based solid lubricant material. *Int. J. Refract. Met Hard Mater.* 52: 235–244.

34 Watts, J., Hilmas, G., Fahrenholtz, W.G. et al. (2011). Measurement of thermal residual stresses in ZrB_2–SiC composites. *J. Eur. Ceram. Soc.* 31 (9): 1811–1820.

35 Zhao, X., Liu, J., Zhu, B. et al. (1997). Tribological characteristics of Si_3N_4 ceramic sliding on stainless steel. *Wear* 206 (1): 76–82.

36 Danyluk, S., McNallan, M., and Park, D.S. (1994). Friction and wear of silicon nitride exposed to moisture at high temperatures. In: *Friction and Wear of Ceramics* (ed. S. Jahanmir), 61–77. New York: Marcel Dekker.

37 Tomizawa, H. and Fisher, T.E. (1985). Friction and wear of silicon nitride at 150°C to 800°C. *ASLE Trans.* 29 (4): 481–488.

38 Zanoria, E., Danyluk, S., and McNallan, M. (1991). Effects of load and temperature on the formation of rolls by reciprocal sliding of silicon-on-silicon at 34%

humidity. In: *Proceedings of the 18th Leeds–Lyon Symposium on Tribology* (ed. D. Dowson), 501–506. Amsterdam: Elsevier.

39 Dong, X. and Jahanmir, S. (1993). Wear transition diagram for silicon nitride. *Wear* 165: 169–180.

40 Kumar, B.V.M., Basu, B., Murthy, V.S.R., and Gupta, M. (2005). The role of tribochemistry on fretting wear of Mg-SiC particulate composites. *Compos. A: Appl. Sci. Manuf.* 36: 13–23.

10

Sliding Wear of Zirconia-Toughened Alumina

One of the major challenges in limiting the wider clinical acceptability of bioceramics is the poor reliability in terms of fracture resistance. The tailoring of composition and microstructure is needed to address such an issue. Another performance-limiting property of bioceramics at articulating joints is the wear resistance. While illustrating twin requirements, this chapter reports the result of the published study from the author's own research group on ZrO_2-toughened Al_2O_3 (ZTA). The load-dependent friction and wear rate of ZTA at a sliding contact with ZrO_2 or steel counterbody are explained. In an effort to explain the cracking-induced wear of ZTA, theoretical calculations are discussed to reveal the role of material properties.

10.1 Introduction

The last few decades have witnessed the gradual increase in the use of engineering ceramics for biomedical applications. Significant research efforts have been invested in investigating the application potential of several oxide ceramics, including Al_2O_3, ZrO_2, etc. [1–8]. Zirconia-toughened alumina (ZTA) with a low amount of ZrO_2 (<20 wt%) dispersed in Al_2O_3 is one such important oxide ceramic.

> The size and volume fraction of tetragonal ZrO_2 is important, as zirconia induces stress induced transformation, which enhances the fracture toughness.

Apart from biomedical applications, these ceramics are used as cutting tool inserts and bearing components. All such applications require high wear resistance coupled with a good combination of fracture resistance properties [4–9].

In the above perspective, this chapter quantitatively analyzes the wear resistance property and establishes a correlation between microstructure and

Friction and Wear of Ceramics: Principles and Case Studies, First Edition.
Bikramjit Basu, Mitjan Kalin, and B. Venkata Manoj Kumar.

mechanical/tribological properties. The attempts to use theoretical models will be highlighted. As part of the tribological study, the sliding wear tests were further carried out on $95Al_2O_3$-$5ZrO_2$-800 ppm against ZrO_2 and steel in dry unlubricated conditions to evaluate friction and wear properties. In a series of papers from the authors' group, the processing and mechanical properties of ZTA ceramics as well as ZTA-based femoral head and acetabular socket development are described [10–12].

10.2 Mechanical Properties of ZTA

The powder mixtures of alumina, zirconia, and sintering additive magnesium nitrate were sintered in air at 1600 °C for six hours. The density and ceramic composition of 95A-5Z-400, 85A-15Z-400, 95A-5Z-800, and 85A-15Z-800 with sinter density and grain sizes are summarized in Table 10.1.

In conventionally sintered ZTA, the equiaxed grains of ZrO_2 and Al_2O_3 are the microstructural features (not shown). Based on the experimental protocols described in an earlier chapter of this book for the mechanical property measurements, a series of properties of ZTA were measured and reported in detail in published papers, as mentioned above. Although the sintered microstructure predominantly contains the tetragonal ZrO_2, it is perceived that the phase transformation of tetragonal ZrO_2 to monoclinic ZrO_2 can take place on fracture surfaces. Therefore, the XRD technique was employed to study the phase assemblage on both polished and fracture surfaces. While correlating mechanical properties with microstructure, X-ray diffraction patterns were critically analyzed to determine the amount of monoclinic zirconia (X_m), based on the intensity of the characteristic X-ray peaks of t-ZrO_2 and m-ZrO_2 phases. Subsequently, % transformability is defined as the difference in m-ZrO_2 content (V_m) measured on the fracture and polished surfaces.

Table 10.1 The composition, sample designations, and grain size of ZTA composites, sintered at 1600 °C for six hours in air [10].

Composition	Al_2O_3 (wt%)	ZrO_2 (wt%)	MgO (ppm)	Sinter density (% ρ_{th})	Grain size (μm)	
					Al_2O_3	ZrO_2
95A-5Z-400	95	5	400	98.2±0.3	3.3	1.2
85A-15Z-400	85	15	400	96.7±0.2	2.8	1.3
95A-5Z-800	95	5	800	98.3±0.4	3.2	0.9
85A-15Z-800	85	15	800	97.9±0.1	2.9	1.1

Source: Reproduced with permission of SAGE Publications.

Table 10.2 Comparison of mechanical properties of ZTA composites in the present work with previous studies. Toughness is measured using indentation cracking method unless otherwise mentioned [10].

ZTA-composition (ZrO_2 addition to Al_2O_3)	Sintering conditions	Fracture toughness ($MPa{\cdot}m^{1/2}$)		Vickers hardness (GPa)	Tensile strength (MPa)	Compressive strength (MPa)	Reference
5.9 wt% ZrO_2	1550 °C, 4 h	4.8		16.6	—	—	10
21.9 wt% t-ZrO_2	1600 °C, 30 h	5		15	—	—	11
15 wt% ZrO_2	1500 °C, 2 h	8.5		—	—	—	12
20 wt% ZrO_2	1500 °C, 2 h	5.1		16.1	—	484.3	12
10 wt% ZrO_2	1600 °C, 1 h	4.9		16.8	—	—	1
7.5 wt% t-ZrO_2	1600 °C; 2 h	7		16.3	—	—	2
20 wt% t-ZrO_2		11.5		16.4			
7.5 wt% YSZ		7.9		16.4			
Pure alumina	1600 °C; 6 h	3.2	—	18.4 ± 1.1	134 ± 12.0	587.0 ± 21.0	Present work
5 wt% ZrO_2 + (400 ppm MgO)		4.4	3.5 ± 0.24[a]	19.1 ± 0.7	195.1 ± 8.7	877.9 ± 28.7	
15 wt% ZrO_2 + (400 ppm MgO)		5.7	3.3 ± 0.30[a]	18.4 ± 0.6	195.3 ± 12	927.2 ± 31.3	
5 wt% ZrO_2 + (800 ppm MgO)		4.5	4.3 ± 0.15[a]	19.3 ± 0.5	213.3 ± 10.3	1101.1 ± 25.6	
15 wt% ZrO_2 + (800 ppm MgO)		5.9	3.1 ± 0.41[a]	18.5 ± 0.4	182.2 ± 11.8	786.1 ± 21.4	

[a] Measured using single edge V-notch beam technique.

Source: Reproduced with permission of SAGE Publications.

The compression strength property is important in the context of tribological performance. As shown in Table 10.2, 95A-5Z-800 exhibits the highest compressive strength of around 1100 MPa. This is better than an earlier study, which reported the compression strength to be around 500 MPa for similar ceramic or pure alumina (587 MPa) [13].

Similar ceramic, i.e. 95A-5Z-800 composite, exhibited superior tensile failure strength of 213 MPa, 150% higher compared to pure alumina (134 MPa). In contrast to the hardness of earlier developed ZTA (16–17 GPa), much higher hardness of around 19 GPa could be achieved even with 15 wt% ZrO_2 addition to Al_2O_3. This result is interesting as with more ZrO_2 addition to Al_2O_3, and one would expect a lower hardness. A better hardness is also indicative of fine grain sizes in the sintered microstructure. The fracture toughness of ZTA composites, determined by the single edge V-notch beam (SEVNB) method, is the highest for the 95A-5Z-800 ceramic (4.3 MPa·m$^{1/2}$).

Concerning microstructure–property correlation, the toughness in ZTA composites is influenced by the extent of transformation from t-ZrO_2 to m-ZrO_2. The volume fraction of m-ZrO_2 was calculated to determine the transformability of t-ZrO_2 during crack propagation. The results, shown in Figure 10.1, demonstrate

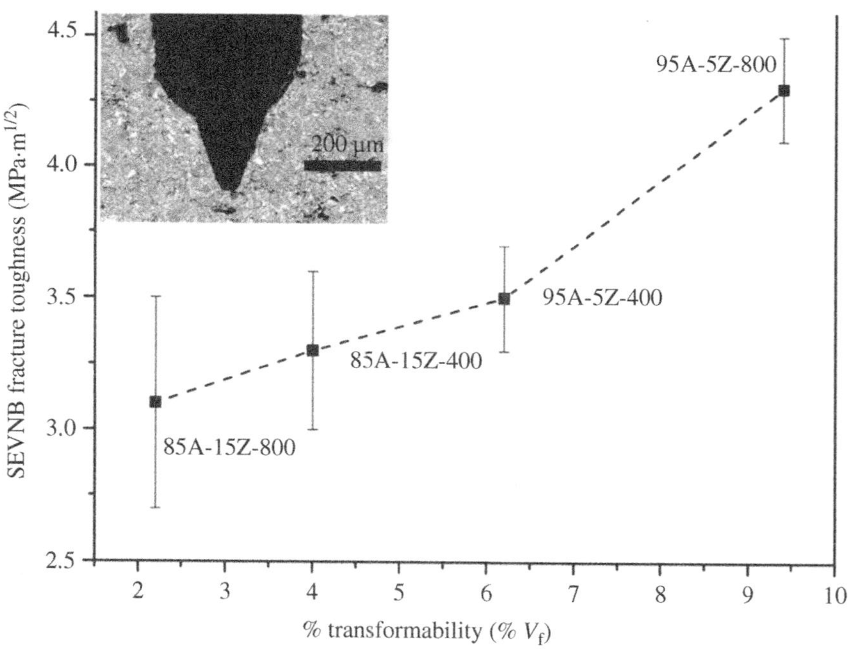

Figure 10.1 SEVNB fracture toughness of ZTA composites with respect to % transformability (V_f) of t-ZrO_2 after fracture. The inset shows the SEM image of the V-notch in one of the experimental samples for SEVNB test. SEVNB, single edge V-notch beam; ZTA, zirconia-toughened alumina [10]. *Source:* Reproduced with permission of SAGE Publications.

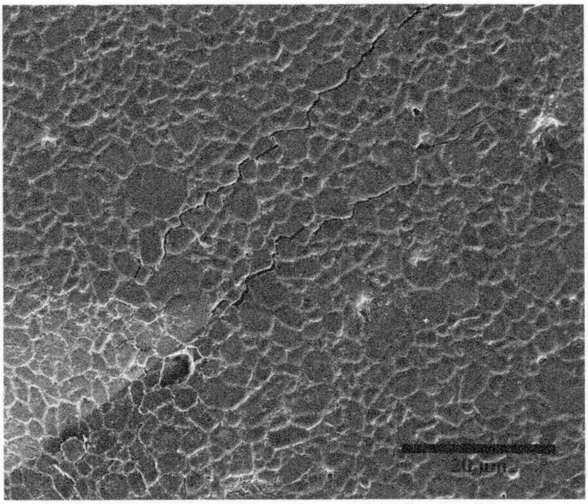

Figure 10.2 The crack deflection behavior of the indentation-induced cracks around the edge of a Vickers indent on ZTA composite [10]. *Source:* Reproduced with permission of SAGE Publications.

a direct correlation between the transformability of t-ZrO_2 and fracture toughness. In addition to transformation toughening, the additional contribution from crack deflection, as evident in Figure 10.2, also contributes to achieve moderately high fracture toughness.

10.3 Sliding Wear Properties of ZTA

A ball-on-disc-type wear tester was used to determine the tribological properties of ZTA composites against 10 mm dia steel or ZrO_2 balls. A relative sliding speed of 30 rpm (sliding velocity 6 mm/s) was maintained while varying the normal loads of 5, 10, and 15 N to simulate the most severe unlubricated wear conditions. The total test duration was 120 minutes for all the tribocouple pairs. The total sliding distance for each tribocouple was 45 m.

Figure 10.3 illustrates the frictional behavior of ZTA disc against zirconia ball at varying normal loads. In particular, the steady-state coefficient of friction (COF) of 0.41, 0.39, and 0.42 was attained at various normal loads of 5, 10, and 15 N, respectively. These COF values are relatively less than the ceramic-on-ceramic contact. The difference in steady-state COF between ZTA and counterbodies can be related to the interfacial wear mechanism, as discussed later.

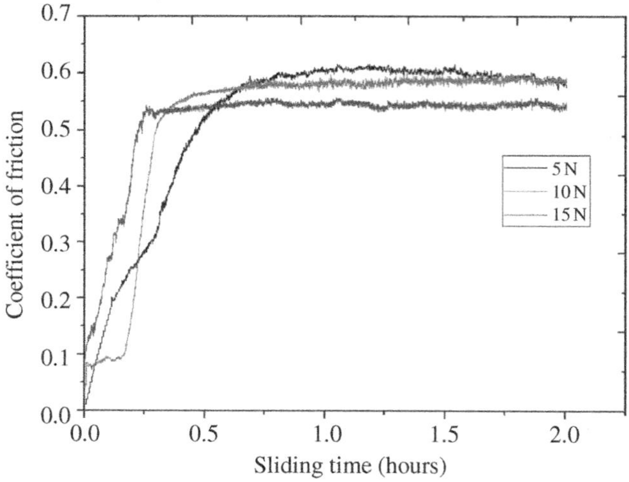

Figure 10.3 Frictional behavior of ZTA disc against cubic ZrO_2 ball with varying load. Sliding experiment was conducted for the duration of two hours at 30 rpm. ZTA, zirconia-toughened alumina [10]. *Source:* Reproduced with permission of SAGE Publications.

The wear rate of ZTA flat against ZrO_2 and steel ball is measured, which varies around 10^{-9} $mm^3/N{\cdot}m$, as shown in Figure 10.4. At the lowest load regime (5 N), the specific wear rate is 0.1×10^{-9} $mm^3/N{\cdot}m$ and -0.35×10^{-9} $mm^3/N{\cdot}m$ against ZrO_2 and steel ball, respectively. A similar negative wear rate of alumina-coated YZTP and alumina versus WC and ZTA with Cr (C10Z10) versus steel ball tribo-couple has been reported by Scheppokat et al. [14].

A linear increase in wear with respect to a load of ZTA against ZrO_2 with load is the signature of abrasive wear. In contrast, negligible or negative wear against steel suggests adhesive wear mechanisms mediated by the transfer of oxidized iron transfer film on ZTA and subsequent interaction with the transfer film [15, 16].

While assessing the wear-induced damage, some representative surface profiles taken at different locations (marked as A, B, C, and D) are shown in Figure 10.5. The width of the worn region or maximum depth of worn surface at those chosen locations varies within a narrow window, indicating homogeneity in wear-induced damage of ZTA against ZrO_2. In contrast, the representative worn surface profile, taken from any random location on worn ZTA after sliding against steel, clearly indicates the adhesion of transfer layer and as such, the transfer layer forms as a collection of wear fragments from steel counterbody in different sizes, as revealed from the spikes on the surface profile. The details of the worn surface on ZTA after sliding against ZrO_2 are displayed in Figure 10.6.

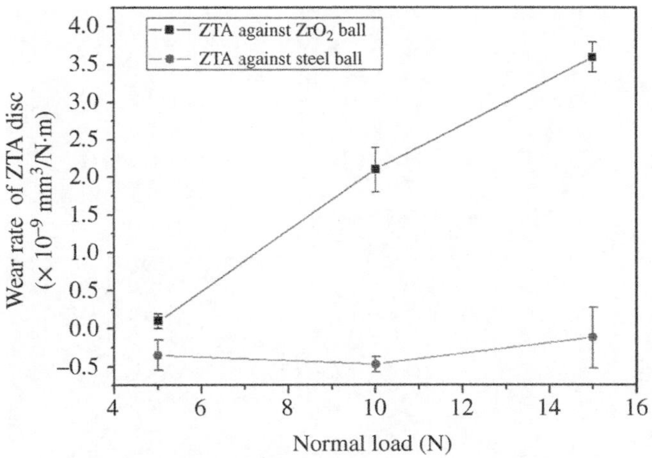

Figure 10.4 Specific wear rate of ZTA disc when subjected to unidirectional sliding against ZrO_2 and steel ball normal load. Sliding experiment was conducted for the duration of two hours at 30 rpm. ZTA, zirconia-toughened alumina [10]. *Source:* Reproduced with permission of SAGE Publications.

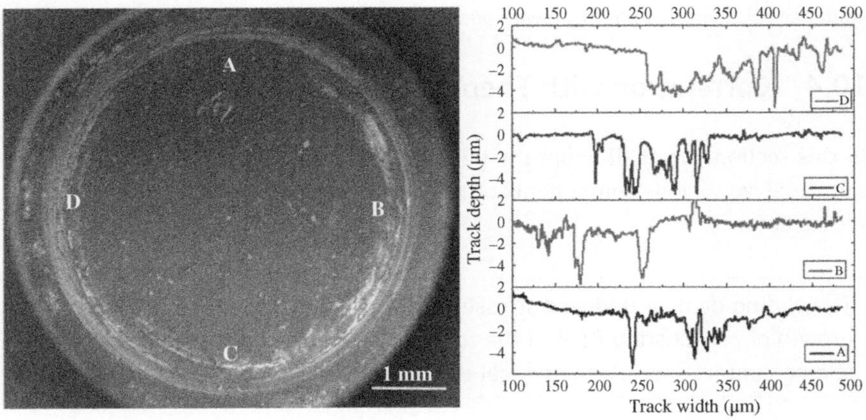

Figure 10.5 SEM image of the entire wear track on ZTA after sliding against cubic ZrO_2 ball (center). The different surface profiles were taken at various locations (A, B, C, and D) to understand the depth of wear removal phenomenon. The sliding conditions were normal load 15 N, sliding speed 30 rpm, and sliding duration two hours. ZTA, zirconia-toughened alumina [10]. *Source:* Reproduced with permission of SAGE Publications.

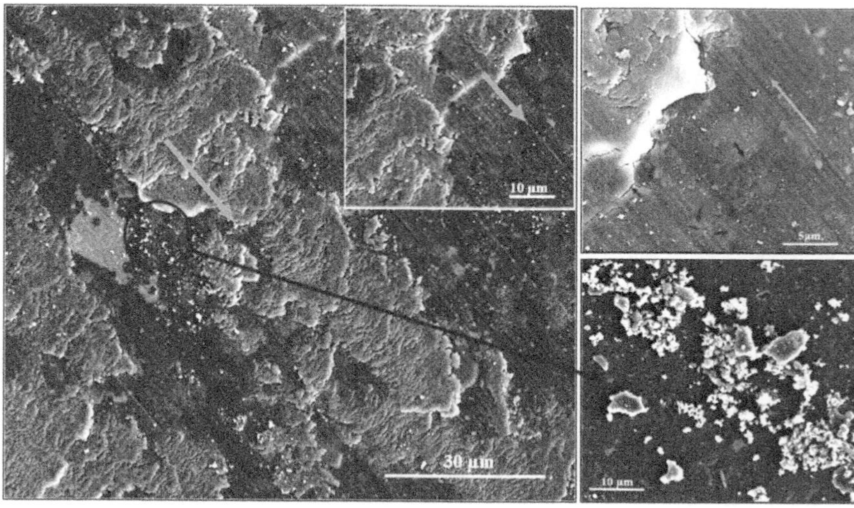

Figure 10.6 SEM images of the debris fragment layer adhered on the worn surface of ZTA sample against ZrO_2 ball and the micrograph of tribolayer at higher magnification is shown in the inset; the cracks on the worn surface are due to tensile stress at the trailing edge of the contact surface of sliding pairs and selective debris accumulated on the worn surface. The sliding conditions were normal load 15 N, sliding speed 30 rpm, and sliding duration two hours. The arrow indicates the sliding direction. ZTA, zirconia-toughened alumina [10]. *Source:* Reproduced with permission of SAGE Publications.

10.4 Correlation with Theoretical Analysis

In this section, we shall adopt the published theoretical models [17,20] to critically analyze various components of the overall stress scenario at the ZTA/ZrO_2 sliding contact.

> The sliding damage-induced stresses would trigger crack nucleation or crack growth of pre-existing flaws if the tensile stress field at the trailing edge of a sliding contact exceeds the critical damage stress.

Cho et al. [18] reported the modulation of the damage stress in a multiphase microstructure, leading to the transition from mild to severe wear.

Such analysis can rationalize the observation of cracking on the worn surfaces, as shown, for example, in Figure 10.6. The maximum tensile stress at the trailing edge of the contact (σ_{max}) can be expressed as

$$\sigma_{max} = \frac{3F}{2\pi a^2}\left[\frac{1-2\vartheta}{3} + \frac{(4+\vartheta)\pi\mu}{8}\right] \tag{10.1}$$

Table 10.3 Summary of Hertzian contact parameters, maximum tensile stress at trailing edge, and damage stress for ZTA ceramic composites for sliding against ZrO_2 ball [10].

Normal load (N)	Contact radius, a (μm)	Hertzian contact pressure (MPa)	σ_{max} (MPa)	σ_s (MPa)
5	38	389	2930.8	
10	48	490	3532.1	280
15	55	561	4094.9	

Source: Reproduced with permission of SAGE Publications.

where F is the normal load, a is the Hertzian contact radius, ϑ is Poisson's ratio, and μ is the COF. The values of σ_{max} are summarized in Table 10.3. The critical tensile stress (σ_s) induced for microcrack propagation can be expressed by the following equation:

$$\sigma_s = \sigma_I^* \left(\frac{d^*}{d}\right)^{1/2} - \sum_{i=0}^{n} \sigma_{Ii} \tag{10.2}$$

The second term on the right-hand side of Eq. (10.2) is the summation of the internal stresses, and the first term is the normalized frictional stresses at a grain size (d) with respect to critical grain size (d*). The second term of Eq. (10.2) essentially comprises related internal stress (σ_{I1}), thermal residual stress (σ_{I2}), and phase transformation-induced compressive stresses (σ_{I3}).

The internal stresses, σ_{I1}, can be computed by the following equation:

$$\sigma_{I1} = \frac{1}{2}\Delta\alpha E_M \left(T_e - T_{Room}\right) \tag{10.3}$$

where the critical temperature T_e is the characteristic temperature above which all stresses can be relaxed. $\Delta\alpha$ is the change in the coefficient of thermal expansion between the a- and c-axes. For alumina, the maximum thermal expansion anisotropy stress can be obtained as 299 MPa.

The thermal stresses between the alumina and zirconia phases (σ_{I2}) at the grain boundary can be estimated using the following equation:

$$\sigma_{I2} = \frac{\Delta\alpha\Delta T}{\frac{1+\vartheta_m}{2E_m}} + \frac{1-2\vartheta_p}{E_p} \tag{10.4}$$

where the subscripts m and p refer to matrix alumina and particle zirconia, respectively; $\Delta\alpha$ is the differential thermal expansion coefficient $(\alpha_p - \alpha_m)$; $\Delta T = 1575\,°C$ is the temperature range of cooling; ϑ and E are Poisson's ratio and elastic modulus, respectively. For ZTA, $\alpha_m = 9\times10^{-6}\ K^{-1}$ and $\alpha_p = 10\times10^{-6}\ K^{-1}$, respectively, and $E_m = 380\,GPa$ and $E_p = 210\,GPa$, $\vartheta_m = 0.3$ and $\vartheta_p = 0.3$. For such a set of values, we compute $\sigma_{I2} = 920\,MPa$.

Finally, the transformation from tetragonal to monoclinic ZrO_2 phase transformation-induced compressive stresses (σ_{I3}) can be computed as follows:

$$\sigma_{I3} = -E_p \frac{\Delta V}{V} \frac{V_v}{3(1-\vartheta_p)} \tag{10.5}$$

where E and ν are the elastic modulus and Poisson's ratio of ZrO_2, $V\nu$ is the volume fraction of ZrO_2, and $\Delta V/V$ is the fractional volume change due to ZrO_2 phase transformation. If we use typical values for ZrO_2 phase transformation, i.e. $\Delta V/V = 0.007$, $E = 210\,GPa$, $V\nu = 0.03$, and $\nu = 0.3$, then it can be estimated that $\sigma_{I3} = -23.2\,MPa$, assuming the absence of stress relaxation. Considering the presence of only 5% ZrO_2, the value of $\sigma_I^*\left(d^*/d\right)^{1/2}$ can be estimated to be 1475.8 MPa.

After incorporating various stress values, we obtain $\sigma_s = 280\,MPa$. Looking at data in Table 10.3, it would appear that σ_{max} surpasses σ_s at all the sliding conditions. It is clear that the σ_{max} has a higher magnitude than σ_s to trigger cracking-induced abrasive damage at sliding contact. It is important to note that the grain size and residual stress need to be minimized.

> Wang et al. [19] and Hamilton [20] reported that when tensile stress exceeds the critical damage stress, the cracking-induced delamination leads to the wear of sliding contact.

10.5 Closure

In several published papers, the wear rate of ceramics is reported to be $10^{-6}\ mm^3/N{\cdot}m$. The results summarized in this chapter demonstrate that a much better wear resistance (wear rate $\sim10^{-9}\ mm^3/N{\cdot}m$) can be achieved by Al_2O_3-5% ZrO_2 ceramics within a narrow variation of ZrO_2 addition. It has been shown that it is possible to obtain a better combination of compression strength, tensile strength, hardness, and fracture toughness in ZTA ceramic developed in author's research group than earlier published results. Also, the sliding wear resistance of ZTA against ZrO_2 shows the dominance of abrasive wear. Importantly, the published theory based on residual stresses is adopted to rationalize the abrasion-induced cracking in ZTA ceramics. Overall, the relationship between microstructure and sliding wear of ZTA ceramics is explained in the present chapter. However, author's published report [10] can be referred for further understanding.

References

1 Basu, B. (2017). *Biomaterials Science and Tissue Engineering: Principles and Methods*. Cambridge University Press.
2 Kurtz, S.M., Kocagöz, S., Arnholt, C. et al. (2014). Advances in zirconia toughened alumina biomaterials for total joint replacement. *J. Mech. Behav. Biomed. Mater.* 31: 107–116.
3 Basu, B. and Ghosh, S. (2017). *Biomaterials for Musculoskeletal Regeneration: Applications; Concepts*. Springer Nature.
4 Basu, B., Vleugels, J., and Van Der Biest, O. (2004). ZrO_2–Al_2O_3 composites with tailored toughness. *J. Alloys Comp.* 372: 278–284.
5 Mills, H. and Blackburn, S. (2000). Zirconia toughened aluminas by hydrothermal processing. *J. Eur. Ceram. Soc.* 20: 1085–1090.
6 Casellas, D., Nagl, M., Llanes, L., and Anglada, M. (2003). Fracture toughness of alumina and ZTA ceramics: microstructural coarsening effects. *J. Mater. Process. Technol.* 143: 148–152.
7 Rao, P., Iwasa, M., Tanaka, T. et al. (2003). Preparation and mechanical properties of Al_2O_3–15wt.% ZrO_2 composites. *Scr. Mater.* 48: 437–441.
8 Zheng, Y., Li, H., Zhou, T. et al. (2013). Microstructure and mechanical properties of Al_2O_3/ZrO_2 eutectic ceramic composites prepared by explosion synthesis. *J. Alloys Comp.* 551: 475–480.
9 Sarkar, D., Adak, S., and Mitra, N. (2007). Preparation and characterization of an Al_2O_3–ZrO_2 nanocomposite, part I: powder synthesis and transformation behavior during fracture. *Compos. A Appl. Sci. Manuf.* 38: 124–131.
10 Ravikumar, K., Sarkar, D., and Basu, B. (2018). ZrO_2-toughened Al_2O_3 composites with better fracture and wear resistance properties. *J. Biomater. Appl.* 32 (9): 1174–1186.
11 Sarkar, D., Mandal, S., Reddy, B.S. et al. (2017). ZrO_2-toughened-Al_2O_3 based near-net shaped femoral head: unique fabrication approach, 3D microstructure, burst strength and muscle cell response. *Mater. Sci. Eng. C* 77: 1216–1227.
12 Sarkar, D., Reddy, B.S., and Basu, B. (2018). Implimenting statistical modelling approach towards development of ultrafine grained bioceramics: case of ZrO_2 toughened Al_2O_3. *J. Am. Ceram. Soc.* 101 (3): 1333–1343.
13 Nevarez-Rascon, A., Aguilar-Elguezabal, A., Orrantia, E., and Bocanegra-Bernal, M. (2011). Compressive strength, hardness and fracture toughness of Al_2O_3

whiskers reinforced ZTA and ATZ nanocomposites: Weibull analysis. *Int. J. Refract. Met. Hard Mater.* 29: 333–340.
14 Scheppokat, S., Hannink, R., Janssen, R. et al. (2005). Sliding wear of Cr–Al_2O_3–ZrO_2 and Mo–Al_2O_3–ZrO_2 composites. *J. Eur. Ceram. Soc.* 25: 837–845.
15 Pearson, S., Shipway, P., Abere, J., and Hewitt, R. (2013). The effect of temperature on wear and friction of a high strength steel in fretting. *Wear* 303: 622–631.
16 Buckley, D.H. (1981). *Surface Effects in Adhesion, Friction, Wear, and Lubrication.* Elsevier.
17 Basu, B. (2005). Toughening of yttria-stabilised tetragonal zirconia ceramics. *Int. Mater. Rev.* 50: 239–256.
18 Cho, S.J., Hockey, B.J., Lawn, B.R., and Beniison, S.J. (1989). Grain size and R-curve effects in abrasive wear of alumina. *J. Am. Ceram. Soc.* 72: 1249–1252.
19 Wang, Y., He, C., Hockey, B. et al. (1995). Wear transitions in monolithic alumina and zirconia-alumina composites. *Wear* 181: 156–164.
20 Hamilton, G.M. (1983). Explicit equations for the stresses beneath a sliding spherical contact. *Proc. Inst. Mech. Eng. C J. Mech. Eng. Sci.* 197: 53–59.

11

Abrasive Wear of Detonation Sprayed WC-12Co Coatings

This chapter provides a brief description of the key results on the abrasive wear of detonation sprayed WC-Co coatings using a planned set of dry sand rubber wheel abrasion tests against three abrasives, i.e. SiO_2, Al_2O_3, and SiC. A linear increase in abrasive wear with the hardness of abrasive was recorded in all the experiments. Apart from presenting the published quantitative results of the abrasive wear resistance, the discussion primarily focuses on the surface and subsurface material removal mechanisms. An attempt has been made to discuss the difference in wear mechanisms with decarburization and coating deposition conditions.

11.1 Introduction

In determining the tribological performance, the abrasive wear behavior of thermal sprayed WC-Co coatings has been widely investigated [1–12]. It has been repeatedly noted that the abrasion of liquid-phase sintered bulk WC-Co of identical composition [8] is always superior to thermal sprayed WC-Co coatings. This particularity in performance primarily occurs due to considerable decarburization experienced by the powder particles during the spray process [1, 6]. The volume fraction of WC in the coatings (as compared with that in the powder) is reduced by decarburization that permits the development of undesirable phases like W_2C, W, which is responsible for the dissolution of carbon and tungsten in the cobalt binder phase, thus making it hard and brittle [6, 9]. Further, the mechanically weaker regions in the coatings [6, 8] are determined to be the inter-splat boundaries, characteristic of thermal spray coatings.

It is well known that the comprehensiveness of decarburization of powder particles of WC-Co, comprising the coating, shall affect its abrasion behavior; yet the trend of this phenomenon is not completely understood. For example, Schwetzke and Kreye [9] employed several coating techniques in order to obtain thermal

Friction and Wear of Ceramics: Principles and Case Studies, First Edition.
Bikramjit Basu, Mitjan Kalin, and B. Venkata Manoj Kumar.

sprayed coatings using WC-17Co. The results indicated that decarburization hardly affected the abrasive wear rate of WC-17Co when carbon loss was below 60%. A comparative study on the rubber wheel abrasion resistance on vacuum plasma sprayed (VPS) and air plasma sprayed (APS) WC-17Co coatings was executed by Chen and Hutchings [1] and concluded that the performance of VPS WC-17Co coatings to be superior. This variation in wear behavior was substantiated on the basis that, in VPS coatings the decarburization was negligible whereas, for APS coatings the reverse was true, which was due to the presence of W_2C and W phases with a concomitant decrease in the volume fraction of WC.

In general, the abrasion of the coating increases with increase in hardness of the abrasive. Bozzi and de Mello [5] proposed that the mechanism of abrasion in WC-Co coatings transforms from micropolishing to microplowing and cutting and eventually to interlamellar brittle fracture SiO_2, Al_2O_3, and SiC abrasives, respectively. To this end, the results defined in the present chapter would explain the influence of abrasives on wear behavior of WC-Co coatings with respect to WC-Co bulk. The surface and subsurface damage mechanisms will also be uncovered in this chapter.

11.2 Coatings and Abrasive Wear

Commercially available WC-12 wt% Co powder feedstock (Sulzer Metco, USA, particle size: 10–41 μm) was deposited at three levels of oxygen to fuel ratio (referred to as OF ratio): 1.0, 1.5, and 2.0, using a detonation spray gun to obtain a coating of ~350 μm thickness on a mild steel substrate. Details of deposition are explained elsewhere [13].

The important physical and mechanical characteristics of coatings deposited at three OF ratios, bulk WC-12Co, and mild steel substrate are shown in Table 11.1. It is observed that an increase in the OF ratio partially increases the density of the coating due to increased decarburization and lower porosity and improved intersplat bonding [13]. Average thickness of all the three coatings was 350 μm as can be noted from Table 11.1. Further, the decrease in the coating was due to the increase in OF ratio that was probably due to an increase in the flattening of powder particles during the coating process, which resulted due to an increase in the temperature of powder particle with increasing OF ratio [13]. The coating porosity decreased with the increase in OF ratio from 1.16 to higher values due to the similar reasons as explained above. A comparison of the cross-section of the detonation sprayed coatings obtained at three different OF ratios with that of bulk WC-12Co is shown in Figure 11.1. The microstructure of the OF-1.16 coatings and bulk WC-12Co appear quite similar even though WC cuboids are smaller in size in the coating due to finer cuboid size of powder feedstock. Further, the number density of cuboids in OF-2.0 coatings (Figure 11.1d) is lower than that of OF-1.50

Table 11.1 Properties of various materials investigated in the present study [13].

Material designation/OF (oxy-fuel) ratio	Coating thickness (μm)	Coating roughness, R_a (μm)	Density (g/cm^3)	Porosity (%)	Representative hardness (GPa)	E (GPa)
OF-1.16	347±30	5.0±0.4	13.75	0.65±0.13	9.20	300
OF-1.50	345±25	4.1±0.4	14.64	0.30±0.10	11.15	300
OF-2.0	352±32	3.5±0.3	14.95	0.35±0.12	11.00	290
WC-12Co	Bulk	1.2±0.05	14.30	≃0	12.85[a]	565
Mild steel	Bulk	0.5±0.03	7.84	≃0	2.00[a]	210

[a] Bulk hardness value. *Source:* Reproduced with permission of Elsevier.

coating (Figure 11.1b). The bright regions/zones increased with OF ratio and predominant in OF-2.0 coating (Figure 11.1d).

The nature of the abrasive used is of prime importance in the abrasion test. In the present study, the synergetic part of type of abrasives and degree of decarburization of the WC-12Co coatings on the abrasion resistance are estimated. Hence, three

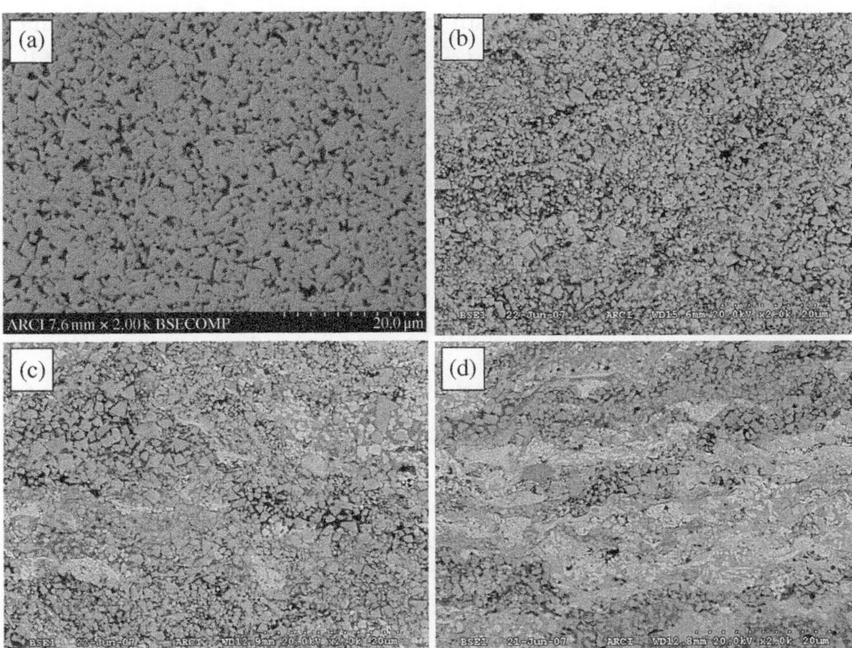

Figure 11.1 Typical cross-sectional SEM images of WC-Co coatings, deposited at an oxy-fuel ratio of 1.16 (b), 1.50 (c), and 2.0 (d). The sintered bulk WC-12Co is also presented in (a) [13]. *Source:* Reproduced with permission of Elsevier.

abrasive particles, namely SiC, SiO_2, and Al_2O_3, with different hardness values were used whose properties are provided in Section 11.3. The morphological features of the coatings, i.e. the abraded surface and the degree of subsurface damage beneath the abraded surfaces, were determined using the scanning electron microscope. Further, the combustion method was used in evaluating the carbon content and hence decarburization of the WC-Co coatings, which were achieved at different OF ratios.

During abrasion, coating surface was intended perpendicular by the abrasive's indent. Thus, it becomes pertinent that the hardness evaluation is executed on the coated surface post polishing, in order to maintain the perpendicular direction of indentation to the coating surface. In the previous literature [13], higher hardness values of WC-Co coatings are at lower indentation loads or at lower indentation depths. Further, with an increase in indentation load, an initial continuous decrease in the hardness occurs, which thereafter remains almost unchanged over a large range of loads, and finally decreases again at very large loads due to the substrate effect. The constant hardness value is referred to as representative hardness of coating. Representative hardness of the investigated WC-Co coatings was determined in the same manner.

It can be observed from Table 11.1 that the hardness is the least for OF-1.16 coatings, which further increases with increase in the OF ratio. However, the hardness measurements of the WC-12Co coatings (9.2–11.15 GPa) are significantly lower than the bulk WC-Co of identical composition (12.85 GPa). Correspondingly, the modulus of elasticity of WC-12Co coatings (290–300 GPa) is considerably lower than the bulk WC-Co of identical composition (565 GPa). The lower hardness and modulus of elasticity of coating in comparison to bulk material is comprehended to the lamellar structure of thermal sprayed coatings, associated with porosity, resulting in weakly bonded lamellae [6, 8]. As mentioned in earlier discussion, due to the undeveloped cracks at the indentation corners even at 10 kg load, the indentation fracture toughness of the coating surface could not be determined.

11.3 Abrasive Wear Results

Figure 11.2 displays the abrasive wear rate in steady-state condition as a function of ratio of hardness of abrasive to hardness of surface (H_p/H) used for all the investigated materials. The prime observations obvious from Figure 11.2 are as follows:

1) Among the investigated, OF-1.50 and OF-1.16 coatings demonstrated comparable abrasive wear rates for any abrasive. On the other hand, a remarkable higher abrasion rate was exhibited by OF-2.0 coatings.

2) The abrasive wear rate of WC-12Co coatings (ranging from 4×10^{-3} mm^3/rev to 16×10^{-3} mm^3/rev) is considerably superior to that of uncoated mild steel (ranging from 210×10^{-3} mm^3/rev to 430×10^{-3} mm^3/rev). Thus, a substantial protection could be provided by the WC-12Co coatings against abrasive wear.
3) Moreover, abrasive wear rates endured by WC-12Co coatings are almost one order of magnitude higher compared with bulk WC-12Co hard material.
4) The highest and the lowest wear rate for WC-12Co coatings and bulk WC-12Co are obtained against SiC and SiO_2 abrasives, respectively. An intermediate wear rate to SiC and SiO_2 is obtained as a result of Al_2O_3 abrasive.
5) An increase in the abrasive wear rate is observed with an increase in H_p/H value for all materials. However, the increasing rate of abrasive wear with an increase in H_p/H is dependent on OF ratio. Wear rate is almost double when OF-1.16 or OF-1.50 coatings are abraded by SiC abrasive compared with SiO_2 abrasive. In contrast, almost four times wear rate is obtained when abrasive is changed from SiO_2 to SiC for bulk WC-12Co or OF-2.0 coatings.
6) The increasing values of H_p/H with continuously increasing abrasion rate without any kind of indication of saturation is probably due to the class of three-body abrasion in the present experiment, where the test material is in contact with the freely rotating abrasive.

> The increase in abrasion rate necessarily suggests increased load transferring ability of abrasive particles of higher hardness.

11.4 Surface and Subsurface Damage Mechanisms

The determination of the abrasive wear mechanisms of the abraded surfaces of the bulk WC-12Co and WC-12Co coatings (with three OF ratios) was analyzed. The material removal in bulk WC-12Co initially includes the removal of the soft binder phase initially, which is further followed by the WC cuboids cracking (not shown as figure). The mode of material removal mechanism remains unaltered irrespective of the abrasive used for abrasion testing. In case of OF-1.16 and OF-1.50 coatings, the material is initially removed by removal of soft Co binder phase, as observed in bulk WC-12Co. But, weak bonding of WC cuboids with Co binder phase results in pull-out of WC cuboids before WC cuboid cracking (not shown as figure) during repeated stresses in abrasive impact. Thus, higher abrasion rates are observed in OF-1.16 and OF-1.50 coatings compared with bulk WC-12Co. The surface of OF-2.0 coating after

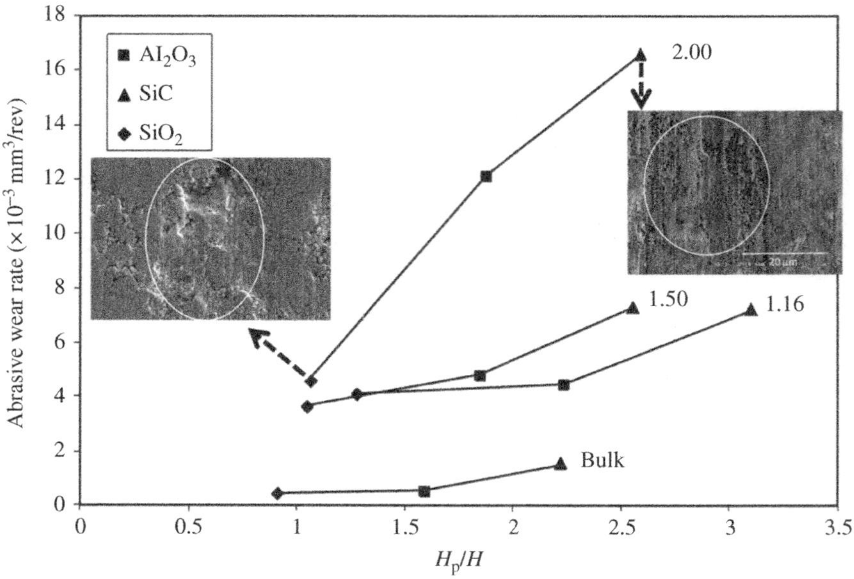

Figure 11.2 The variation of abrasive wear rate of bulk and coated WC-12Co as a function of ratio of abrasive hardness (H_p) and abraded material hardness (H). Numbers for each plot indicate OF ratio used in coating. Insets show surfaces of coating after abrasion by SiO_2 and SiC particles [13]. *Source:* Reproduced with permission of Elsevier.

abrasion (see insets in Figure 11.2) shows severe damage in terms of large patches. It is to note here that the coating with large OF ratio is severely decarburized by dissolving carbon and tungsten in cobalt [4, 8, 9, 13]. The cracks along the boundaries of inter-splats in the coating containing brittle binder phase readily propagate even at low levels of stresses and cause the highest abrasive wear rate. The extent of wear increases with increase in hardness of abrasive for the coating or bulk material. The deeper and wider cracks under the worn surface of OF-2.0 coating are observed after abrasion by harder SiC particles compared with Al_2O_3 particles (see Figure 11.3).

The cross-sectional examination of abraded coatings reveals the development and propagation of cracks along inter-splat boundaries. Subsurface cracking is dominant in OF-2.0 coating. The change in abrasive wear rate with the width of subsurface cracked zone is shown in Figure 11.4. The width to which cracks are present underneath the worn surface is linearly proportional to abrasion rate. This suggests that subsurface cracking is responsible for the observed variation in severity of abrasive wear. Insets in Figure 11.4

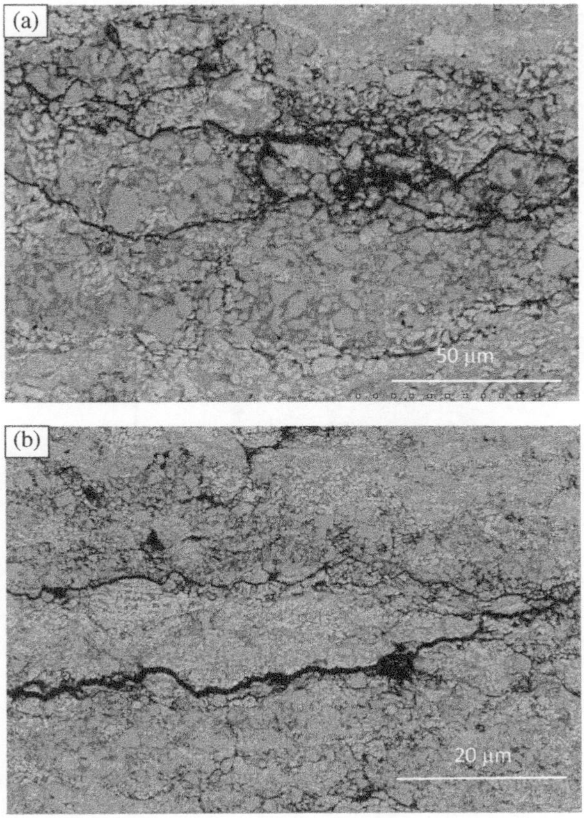

Figure 11.3 Crack paths below the worn OF-2.0 coatings after abrasion by (a) Al_2O_3 abrasive and (b) SiC abrasive [13]. *Source:* Reproduced with permission of Elsevier.

demonstrate larger subsurface crack zone width in case of SiC abrasive compared with SiO_2 abrasive for OF-2.0 coating.

11.5 Closure

The discussion in this chapter unambiguously establishes a linear correlation between abrasive wear of detonation sprayed WC-12Co coatings and abrasive hardness. The material removal mechanism is dominated by WC cuboid cracking, WC cuboid pull-out, and inter-splat cracking-induced delamination. Importantly, the abrasive wear also quantitatively scales well with the subsurface cracking zone width for the worn coatings. More detailed discussion on abrasive wear results presented in the present chapter can be found in our published paper [13].

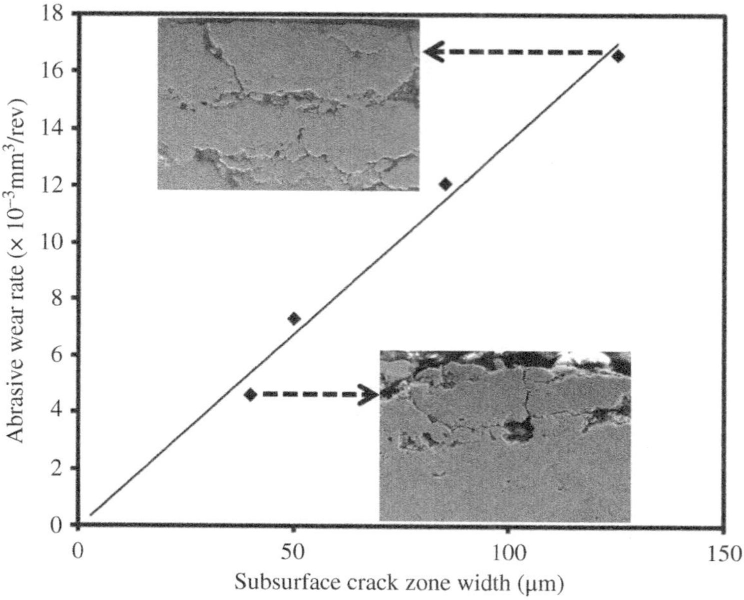

Figure 11.4 Abrasive wear rate versus width of subsurface crack zone in OF-2.0 coating. Insets show typical subsurfaces of the coating after abrasion by SiO_2 and SiC abrasive [13]. *Source:* Reproduced with permission of Elsevier.

References

1 Chen, H. and Hutchings, I.M. (1998). Abrasive wear resistance of plasma sprayed tungsten carbide-cobalt coatings. *Surf. Coat. Technol.* 107: 106–114.

2 Usmani, S., Sampath, S., Houck, D.L., and Lee, D. (1997). Effect of carbide grain size on the sliding and abrasive wear behavior of thermal sprayed WC-Co coatings. *Tribol. Trans.* 40: 470–478.

3 Li, C.J., Ohmori, A., and Tani, K. (1999). Effect of WC particle size on the abrasive wear of thermally sprayed WC-Co coatings. *Mater. Manuf. Process.* 14: 175–184.

4 Ramnath, V. and Jayaraman, N. (1989). Characterization and wear performance of plasma sprayed WC-Co coatings. *Mater. Sci. Technol.* 5: 382–388.

5 Bozzi, A.C. and de Mello, J.D.B. (1999). Wear resistance and wear mechanism of WC-12%Co thermal sprayed coatings in three-body abrasion. *Wear* 233–235: 575–587.

6 Stewart, D.A., Shipway, P.H., and McCartney, D.G. (1999). Abrasive wear behaviour of conventional and nanocomposite HVOF-sprayed WC-Co coatings. *Wear* 225–229: 789–798.

7 Guilemany, J.M., Dosta, S., and Miguel, J.R. (2006). The enhancement of the properties of WC-Co HVOF coatings through the use of nanostructure and microstructure feedstock powders. *Surf. Coat. Technol.* 201: 1180–1190.

8 Wayne, S.F. and Sampath, S. (1992). Structure/property relationship in sintered and thermal sprayed WC-Co. *J. Therm. Spray Technol.* 1: 307–316.

9 Schwetzke, R. and Kreye, H. (1999). Microstructure and properties of tungsten carbide coatings sprayed with various high-velocity oxygen fuel spray systems. *J. Therm. Spray Technol.* 8: 433–439.

10 Vuoristo, P., Niemi, K., Makela, A., and Mantyla, T. (1993). Spray parameter effects on structure and wear properties of detonation fun sprayed WC+17%Co coatings. In: *Proceedings: National Thermal Spray Conference (7–11 June 1993)*, 173–178. ASM International.

11 Stewart, D.A., Shipway, P.H., and McCartney, D.G. (1998). Influence of heat treatment on the abrasive wear behaviour of HVOF sprayed WC-Co coatings. *Surf. Coat. Technol.* 105: 13–24.

12 Liao, H., Normand, B., and Coddet, C. (2000). Influence of coating microstructure on the abrasive wear resistance of WC/Co cermet coatings. *Surf. Coat. Technol.* 124: 235–242.

13 Babu, P.S., Basu, B., Sundararajan, G. (2010). Abrasive wear behavior of detonation sprayed WC–12Co coatings: Influence of decarburization and abrasive characteristics. *Wear* 268: 1387–1399.

12

Solid–Lubricant Interaction and Friction at Lubricated Contacts

The frictional behavior at any lubricated tribological interface largely depends on the wetting and surface energy of mating surfaces. This chapter presents a summary of key research results to describe the relation between surface–lubricant interaction and friction. In particular, the published results obtained with various types of materials (steels, DLC coatings, ceramics, and polymers) in elastohydrodynamic (EHD) contact with oils and water will be discussed. It will be demonstrated that a new wetting parameter, namely the spreading parameter, is to be employed in place of contact angle to understand the wetting characteristics of lubricants. Furthermore, we present that the polar component of the surface energy plays a key role for reducing the coefficient of friction in the solid–liquid slip mechanism.

12.1 Introduction

Friction mechanisms are typically discussed in terms of adhesion, deformation of asperities, and cutting or plowing action of harder material over the softer mating material. In boundary or even mixed lubrication, these mechanisms will remain active, but become more complex, as the boundary films interfere to change the contributions to shear stress, while significantly dominating the friction. However, in full-film lubrication (like hydrodynamic or EHD), the viscous properties of lubricants influence the friction force. This is to some extent true also for mixed lubrication, in the part of the liquid film friction contribution.

However, in recent years, another phenomenon, namely the solid–liquid interactions, has been recognized to be relevant for the full-film lubrication friction in macroscale contacts. This is based on many published studies at the nanoscale and even from biomimetics. In these studies, it was shown that the wetting behavior,

Friction and Wear of Ceramics: Principles and Case Studies, First Edition.
Bikramjit Basu, Mitjan Kalin, and B. Venkata Manoj Kumar.

which governs the interactions and mechanisms between the very top surface layers and liquids, gives these surfaces, especially of many species in nature, completely new and special functionalities. These include very high adhesion, superhydrophobic property, reduced friction, self-cleaning, etc. [1–9]. However, these studies used nonengineering materials and so the wetting phenomenon for common lubricants on engineering surfaces and correlation between wetting and surface energy for engineering contacts are still poorly understood. For example, lubricating oils typically spread over the surfaces and change the contact angle, which is the most common parameter for wetting evaluation. This makes oil wetting evaluation more difficult to understand and to evaluate in terms of oleophobicity or oleophilicity compared with common wetting situations with water, where a droplet is immediately formed in a static form.

Another problem in recognizing the importance of solid–liquid interactions for friction is the lack of direct evidence itself, especially in engineering macro contacts. Namely, often these studies were only theoretical [10, 11] or at very small scale, i.e. nano level [12–14] and only a few involved macro contacts [13–17]. As a result, it is difficult to compare them directly and make solid general conclusions; however, it is very obvious that the surface energy and the wetting in these studies affected the friction significantly. Furthermore, they often reported that the mechanism that causes the wetting and/or surface energy to influence friction was the solid–liquid boundary slip. This therefore suggests the boundary interface slip as the actual cause affecting the friction performance [13–15, 17].

The concept of boundary slip is shown in Figure 12.1. Moreover, recently it was suggested that because of the slip, the surfaces contacting with oils and/or water in highly loaded macro contacts can reduce the friction as well [14, 15, 17]. However, it still remains unclear what are the effective surface mechanisms and

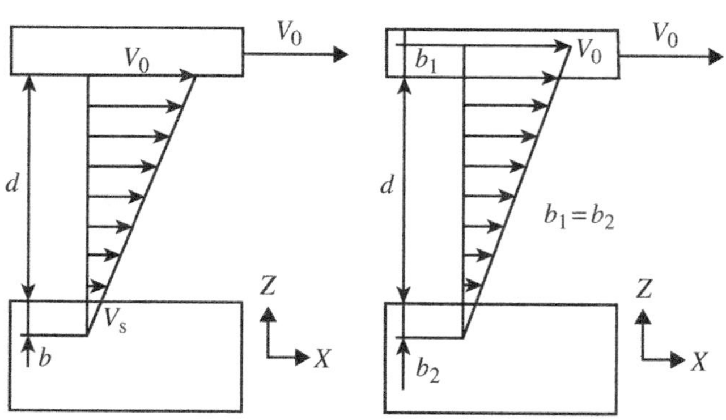

Figure 12.1 A concept of boundary slip. Slip on just one surface or both surfaces is shown. *b* equals to slip length.

what are the controlling parameters for slip to occur. This suggests that it is therefore not possible to controllably predict or tailor the friction behavior. Especially, in some cases poor wetting or high contact angle [13, 15, 16] was suggested, while in some cases, low surface energy [16] was claimed as a parameter to affect slip, but obviously these cannot be equally important for the slip and interface behavior. Namely, the surface energy depends only on the surfaces, while wetting and contact angles are results of the surfaces and liquids.

Nevertheless, the solid–liquid slip has been already proved to occur in EHD contacts [12–14], but there is still lack in understanding the relation between wetting and slip between lubricating oils and common engineering surfaces. In this chapter, several key results are presented that show main issues in understanding the role of wetting, surface energy, and solid–liquid interactions on friction of EHD contacts. The discussion will evolve around wetting of engineering surfaces with lubricating oils, its evaluation and correlation between surface energy and wetting, how wetting effects EHD friction, and how wetting effects on solid–liquid slip.

12.2 Materials and Sliding Wear Experiments

Steel, DLC coatings, ceramics, and polymers, thus very different types of materials were employed for wetting study. For further tribological tests and slip observation, only the most promising materials based on the wetting results were selected. DIN 100Cr6/AISI 52100 steel was used as a reference material. A great part of evidence is used on DLC coatings that can have very particular surface and bulk properties. Accordingly, several DLC coatings, namely five, were employed to obtain various wetting and surface energies. These include tetrahedral amorphous DLC (ta-C) that is non-hydrogenated and amorphous DLC (a-C:H), which is hydrogenated. These two are both non-doped. Further, we used doped DLC coatings containing silicon (Si-DLC), nitrogen (N-DLC), and fluorine (F-DLC); all being hydrogenated. In addition, four types of structural ceramics were used, namely two classes of oxide ceramics and two classes of non-oxide ceramics were selected. Disc samples were prepared from zirconia (ZrO_2, i.e. TZ-3Y), high-purity alumina (Al_2O_3), silicon carbide (SiC), and silicon nitride (Si_3N_4). As the last type of material, various unreinforced tribology-relevant polymers were used. These were polyamide (PA6), polyoxymethylene (POM), polyether ether ketone (PEEK), and polytetrafluoroethylene (PTFE). The latter has very low surface energy as a very nonpolar polymer material.

The MTM tribotester (PCS Instruments, London, UK) was used for tribo-tests. Steel/steel and DLC/DLC contacts were employed, and they were lubricated using the nonpolar base PAO 9 oil. For tribological study, doped N-DLC, F-DLC, Si-DLC, and non-doped a-C:H coatings were employed. Ambient temperature was selected for the tests. Load was set to 0.52 GPa (5 N) as a Hertzian pressure. Slide-to-roll ratio was always at 50%. Entrainment velocity was 0.8–1.6 m/s, where the lambda values were between 10 and 22, see Eq. (12.1) [15], so that conditions correspond to full-film EHD regime.

$$\lambda = \frac{h_0}{\sqrt{R_{qD}^2 + R_{qB}^2}} \tag{12.1}$$

The minimum film thickness h_0 was determined according to Hamrock–Dowson equation [16].

12.3 Wetting and Spreading Properties

Three different types of oils in terms of viscosity and polarity were used to evaluate the wetting, without any additives. Demineralized water was employed as a reference and sometimes used as a lubricant. Water has high polarity, so demineralized water can serve as a complementary fluid for nonpolar oils.

The surface energy (γ) of materials was determined using the Owens–Wendt–Rabel–Kaelble (OWRK) model (Eq. 12.2) [17]. In order to calculate the surface energy, the contact angles (θ) with proper model fluids that have fully defined surface tension need to be measured. These are demineralized water, which is polar fluid, and diodomethane that is nonpolar.

$$\gamma_L \cdot (1 + \cos\theta) = 2 \cdot \left(\sqrt{\gamma_S^D \cdot \gamma_L^D} + \sqrt{\gamma_S^P \cdot \gamma_L^P}\right) \tag{12.2}$$

Pendant-drop method was selected to measure the polar, dispersive, and total surface tension of the oils. The dispersive component was obtained via the contact angles on PTFE, and then applying the Wu and Brzozowski [18] method from Eq. (12.3). The dispersive component of the surface energy (γ^D) of the PTFE was determined as 23.53 mJ/m^2. The polar component (γ^P) of the surface tension was then determined using the Fowkes method from Eq. (12.4) [19].

$$\gamma_L \cdot \frac{(1 + \cos\theta)}{4} = \frac{(\gamma_S^P \cdot \gamma_L^P)}{(\gamma_S^P + \gamma_L^P)} + \frac{(\gamma_S^D \cdot \gamma_L^D)}{(\gamma_S^D + \gamma_L^D)} \tag{12.3}$$

$$\gamma = \gamma^P + \gamma^D \tag{12.4}$$

Wetting of applied engineering materials in this work with the oils was measured via the contact angle, which is vastly and most typically used method for wetting characterization. However, when measuring contact angles with oils, they are changing over the time, so in this case it requires about 15 seconds for oil drop to come to a final and stable condition.

In this view, 16 seconds were allowed for drop to wet the surface before the final contact angle was obtained. This procedure was used for all the measurements, even though the contact angle would reach a final value earlier, or even immediately. Figure 12.2a and b presents the characteristic measurements for both, the spreading-wetting and the adhesion-wetting. Each measurement was repeated at least five times; the average and the standard deviation are reported.

In addition to the contact angle, wettability using the spreading parameter (S) [20] was evaluated. The spreading parameter corresponds to the difference between the work of the solid–liquid interface (W_a) adhesion and the cohesion work in the liquid (W_c). A common form is shown in Eq. (12.5).

$$S = W_a - W_c = \gamma_S - (\gamma_{SL} + \gamma_L) = \gamma_L \cdot (\cos\theta - 1) \tag{12.5}$$

Based on the OWRK model [17], we have derived a new equation that allows to calculate the properties of oil spreading on the surface for spreading-wetting (Figure 12.2a) and adhesion-wetting (Figure 12.2b). This is not applicable otherwise from Eq. (12.5) directly. We denote our new spreading parameter as SP (Eq. 12.6) to make a difference versus the common denotation "S," such as in Eq. (12.5).

$$\text{SP} = 2\left(\sqrt{\gamma_S^D \gamma_L^D} + \sqrt{\gamma_S^P \gamma_L^P} - \gamma_L\right) \tag{12.6}$$

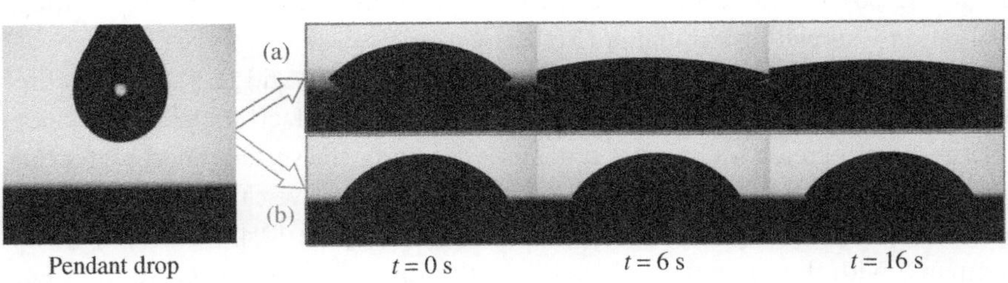

Figure 12.2 Pendant drop and sessile drops for (a) spreading-wetting experiments on steel with PAO 9 oil and (b) adhesion-wetting experiments on F-DLC with PAO 9 oil. Tests were measured during 16 seconds.

12.4 Surface Energies of Different Classes of Materials

A wide range of materials was selected for surface energy and wetting evaluation to find most promising materials in terms of inducing solid–liquid slip. This is used to implement a new mechanism of elastohydrodynamic lubrication (EHL) friction reduction.

All DLC coatings (with the exception of ta-C coating) have lower surface energy than steel (see Figure 12.3a). The ta-C coating is characteristic for containing almost no hydrogen, which passivates the dangling bonds at the DLC surface, thus reducing its reactivity [22–26]. All other DLC coatings contain hydrogen. In this view, all non-hydrogenated coatings have a higher adsorption ability and reactivity with polar groups and molecules [27]. On the other hand, ceramics have very high surface energy; especially the polar surface energy. In fact, these were the highest among all the materials, even higher of the steel (Figure 12.3b). Si_3N_4 has the most pronounced and high polar surface energy. Moreover, oxide ceramics, namely the Al_2O_3 and ZrO_2, showed lower surface energy than the non-oxide Si_3N_4 and SiC. Polymers, apart from the PTFE, also have high surface energy, but in contrary to ceramics, their polar surface energy was low (similar to some DLC coatings and steel, and lower than ceramics) (Figure 12.3c). The most characteristic features were measured for the PTFE and the F-DLC. This was related to their very low surface energy: as is well known, PTFE has a very low surface energy; however, in this work, we obtained even lower value for F-DLC. And what is more, PTFE and F-DLC have extremely low or close to negligible polar surface energy (Figure 12.3a and c).

12.5 Wetting Evaluation of Engineering Surfaces

Contrary to water (Figure 12.2a), oil on most of used surfaces spreads over it resulting in changing contact angle, stabilizing after about 15 seconds (Figure 12.2b). Due to this fact, we introduce an additional parameter for wetting evaluation – spreading parameter (Eq. 12.6).

The contact angle and spreading parameter (Figures 12.4 and 12.5) highlight the relevant differences that these parameters possess. Contact angles with water (Figure 12.4) increase in the same order as it decreases the surface energy – for the same type of materials (Figure 12.3). On the other hand, we cannot see any clear relation of contact angles measured with the oils (Figure 12.4). However, when compared with the same type of materials, the spreading parameter SP from Eq. (12.6) shows a nice, clear, and same relation as the surface energy (Figure 12.4) with all the materials, and this is the case with the water and with the oils. Moreover,

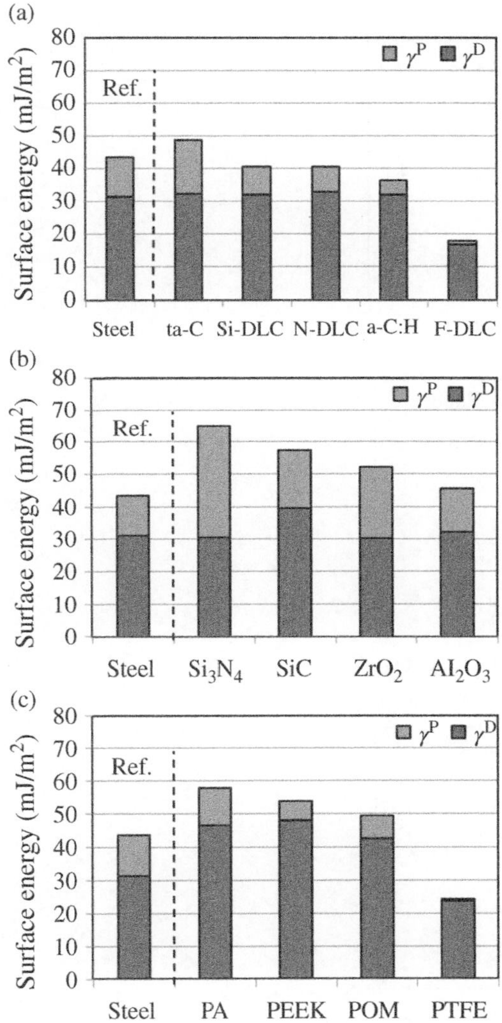

Figure 12.3 Surface energies of studied materials: (a) DLC coatings, (b) ceramics, and (c) polymers. Steel is always added for reference [21]. *Source:* Reproduced with permission of Springer Nature.

this is irrespective of the positive or negative values, and there is no exception for this rule (Figure 12.5).

This can be clearly seen if we plot contact angle (Figure 12.6a and b) and spreading parameter (Figure 12.6c and d) versus the surface energy for all classes of materials. It is clear from Figure 12.6a that water, only liquid that does not spread

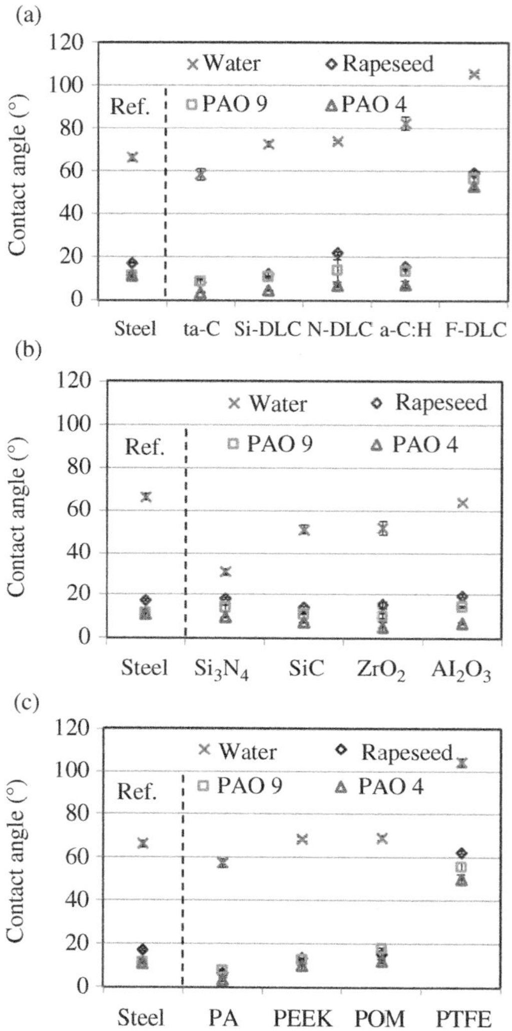

Figure 12.4 Steady-state contact angles of water and lubricating oils on: (a) DLC coatings, (b) ceramics, and (c) polymers. Steel is always added for reference.

over the studied surfaces (Figure 12.2a), has clear correlation between contact angle and surface energy, but on the other hand, that kind of correlation cannot be established for other three oils (Figure 12.6b). On the contrary, there is always a clear correlation between spreading parameter and surface energy for water (Figure 12.6c) and all oils (Figure 12.6d). This means that for proper wetting evaluation of oils over engineering surfaces, the spreading parameter should be used instead of contact angle.

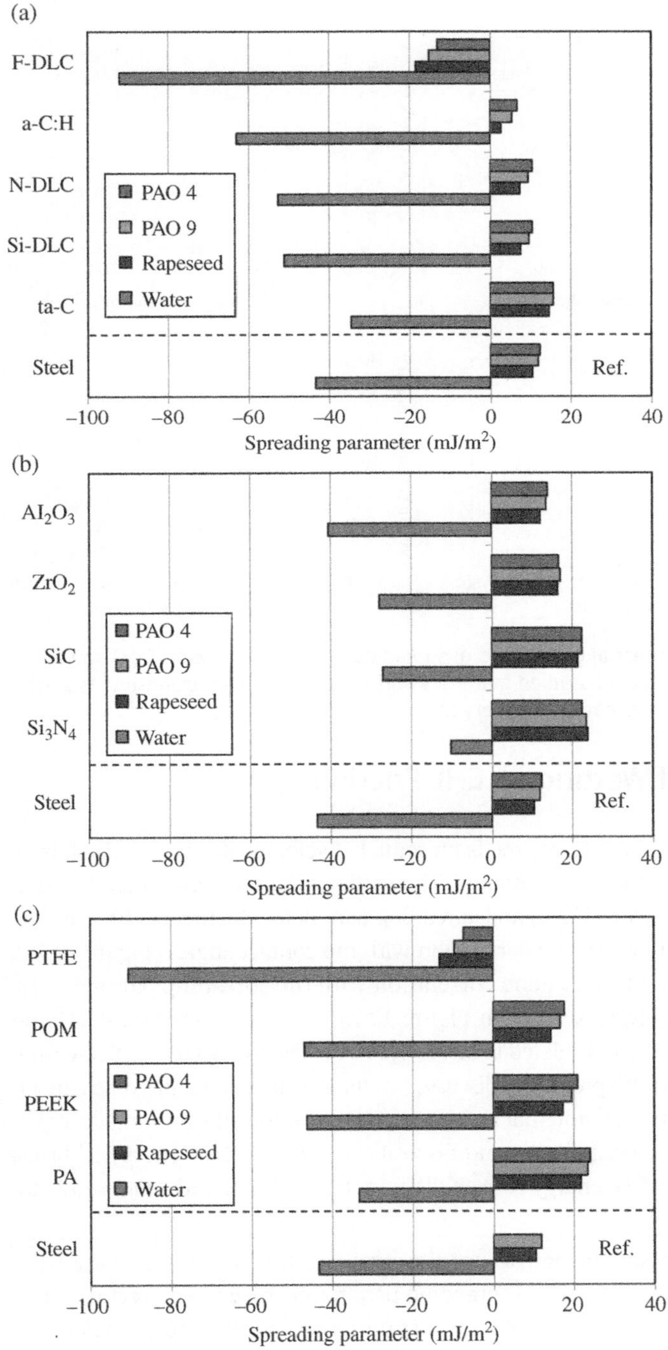

Figure 12.5 Spreading parameter values of water and lubricating oils on: (a) DLC coatings, (b) ceramics, and (c) polymers. Steel is always added for reference.

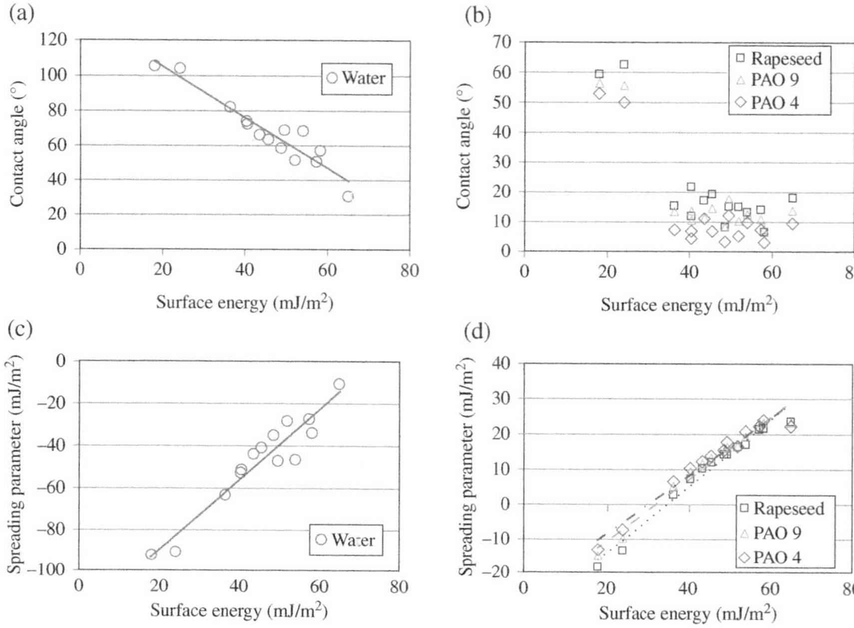

Figure 12.6 (a, b) Contact angles and (c, d) spreading parameter of water (a, c) and all three oils (b, d) versus the all studied materials' surface energy. Linear trend-line is added for liquid where clear correlation is observed.

12.6 Effect of Wetting on EHL Friction

Coefficient of friction in EHL regime is presented at velocity of 1.2 m/s, which is an intermediate value in these experiments. The friction values are presented together with contact angle (Figure 12.7a) and spreading parameter (Figure 12.7b). The friction results show very low or no correlation with the contact angles (Figure 12.7a), and on the contrary, the very good correlation with the spreading parameter SP (Figure 12.7b). Namely, as seen from Figure 12.7a, contact angles for a-C:H and N-DLC, and of steel, are not related to any trend in the friction behavior. These findings are in agreement with previously discussed wetting evaluation for oils on engineering surfaces, where it was found that there is no relation among the oil contact angles, irrespective of their type, polarity, and viscosity, and the surface energy in broad range of materials surface energy or even different types of materials (Figure 12.6a).

> The contact angle should not be used to evaluate the wetting of materials with lubricating oils, instead, the spreading parameters give the correct results. They are valid both in adhesional-wetting and spreading-wetting behavior.

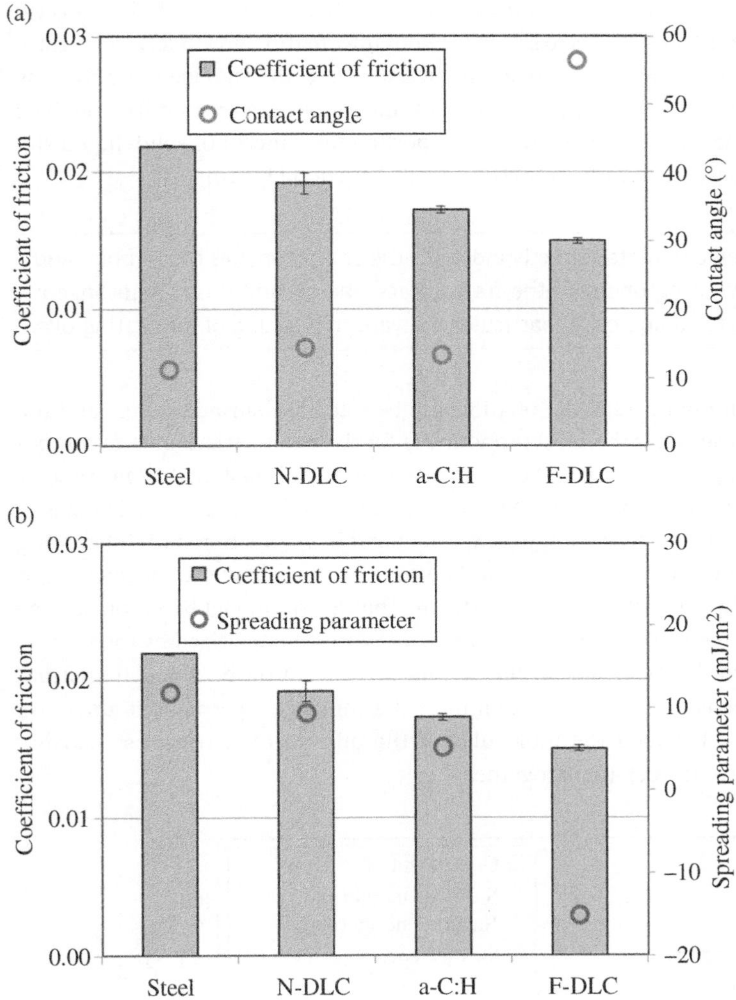

Figure 12.7 Coefficient of friction versus (a) contact angle and (b) spreading parameter when steel, N-DLC, a-C:H, and F-DLC coatings are used. Tests were performed at 1.2 m/s entrainment speed.

12.7 Correlation Between Spreading Parameter and Friction

From the above results it can be concluded that the slower will be the oil spread on the specific engineering material surface, the smaller will be the spreading parameter SP, and this will eventually result in the lower coefficient of friction

(Figure 12.7b). Such small spreading parameter values correspond to low surface energy (Figure 12.6d). That is to say, it provides information how likely the surface will interact with the oil. Weaker interactions that are characteristic for low spreading parameter SP suggest more slip and consequently low friction. The results presented above are, therefore, supporting the concept of solid–liquid slip [28] and its consequence on the EHL friction reduction [13–16].

> The spreading parameter directly indicates the characteristics of the polar and the dispersive component of the free surface energy, and this is relevant for solid surfaces and liquids; in particular, relevant for wetting of lubricating oils.

As shown in Figure 12.8, the relationship between the surface energy and the coefficient of friction is the most pronounced for the polar surface energy component. More specific, EHL friction decreases with materials having the lowest polar surface energy. This shows that the most emphasis to reduce the EHL friction is to be placed on polar surface energy component and its interaction with lubricating oils. In agreement to this, the friction behavior would follow the mechanism of solid–liquid slip, described in "Slip-inducing interaction model based on surface forces" [21]. That is to say, the polar interactions are characterized by the electric dipole between the molecules of the oil and the solid surface, which are permanent. This means that when they dominate the contacts, a permanent attraction will be formed between the molecules of the oil and the surface, so causing permanent force to keep them together.

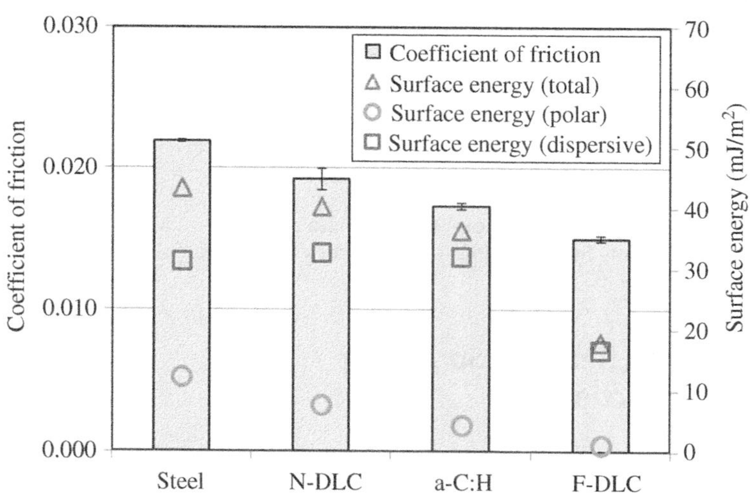

Figure 12.8 Coefficient of friction versus total, polar, and dispersive surface energy for steel, N-DLC, a-C:H, and F-DLC coatings. Tests were performed at 1.2 m/s entrainment speed.

In contrast, nonpolar molecules have intermittent electric dipole. They form instantaneous dipoles. Such dipoles polarize adjacent neutral molecule, which after becoming an oriented dipole, create an induced dipole. Accordingly, the molecules orient to form temporary interactions in the form of dispersive forces. The attractive forces, those that require external force to separate these interactions, are therefore overall much weaker than the polar ones. Previous reports showed that the polar forces are up to 10 times larger than the dispersion forces, and this clearly affects the adhesion mechanisms [12].

In this view, steel that has the highest polar component of surface energy (Figure 12.8) results in the strongest attractive interactions with the lubricating oils and therefore has the highest full-film EHL friction. In contrast, DLC coatings, for example, which have low polar component of surface energy, provide weak interactions with the lubricating oil. Accordingly, the oil may slip more easily at the interface, so causing less resistance and showing lower friction (Figure 12.8).

12.8 Closure

Compared with steel, DLC coatings (except non-hydrogenated ta-C) have lower surface energy, especially polar component. The F-DLC has very low surface energy, similar to PTFE, but it is notable that its polar component of surface energy is almost negligible. On the other hand, ceramics and polymers (except PTFE) have higher surface energy, but for ceramics due to higher polar and for polymers due to higher dispersive component.

When contact angle is used to measure wetting of the water, linear correlation was established with the surface energy of several types of materials (steel, ceramics, DLC, and polymer). However, there is no correlation with the lubricating oils. On the contrary, a clear and very good linear correlation is found for the spreading parameter SP and the surface energy. This was true for lubricating oils with different viscosities and polarities, and with water. Accordingly, a much more reliable predictive tool to establish wetting performance with oils is the spreading parameter SP and is clearly favored over contact angle. The latter is not appropriate for the spreading-wetting, while spreading parameter SP is suitable for both, spreading-wetting and adhesional-wetting.

By employing DLC surface coating in the tribological contacts, the wetting with oils becomes poorer, namely the spreading parameter SP, as well as the surface energy: in particular its polar component; are reduced, and thus, up to 30% (or more) reduction in EHL friction was achieved compared to conventional steel/steel contacts.

A "slip-inducing interaction model based on surface forces" explains the solid–liquid slip by encountering the strength of dispersive interactions, which are much smaller than polar ones and so provide poor wetting behavior, which enhances slip at the interface, and so the lower friction.

Taken together, this chapter establishes spreading parameter as an important factor to influence the friction of EHD-lubricated contacts.

References

1 Bhushan, B. and Jung, Y.C. (2011). Natural and biomimetic artificial surfaces for superhydrophobicity, self-cleaning, low adhesion and drag reduction. *Prog. Mater. Sci.* 56: 1–108. https://doi.org/10.1016/j.pmatsci.2010.04.003.

2 Bhushan, B. (2009). Biomimetics: lessons from nature – an overview. *Philos. Trans. R. Soc. A* 367: 1445–1486. https://doi.org/10.1098/rsta.2009.0011.

3 Guo, Z., Liu, W., and Su, B.-L. (2011). Superhydrophobic surfaces: from natural to biomimetic to functional. *J. Colloid Interface Sci.* 353: 335–355. https://doi.org/10.1016/j.jcis.2010.08.047.

4 Chen, B., Wu, P.D., and Gao, H. (2008). Hierarchical modelling of attachment and detachment mechanisms of gecko toe adhesion. *Proc. R. Soc. A.* 464: 1639–1652. https://doi.org/10.1098/rspa.2007.0350.

5 Jung, Y.C. and Bhushan, B. (2010). Biomimetic structures for fluid drag reduction in laminar and turbulent flows. *J. Phys. Condens. Matter* 22: 035104. https://doi.org/10.1088/0953-8984/22/3/035104.

6 Autumn, K. and Gravish, N. (2008). Gecko adhesion: evolutionary nanotechnology. *Philos. Trans. R. Soc. A* 366: 1575–1590. https://doi.org/10.1098/rsta.2007.2173.

7 Feng, L., Li, S., Li, Y. et al. (2002). Super-hydrophobic surfaces: from natural to artificial. *Adv. Mater.* 14: 1857–1860. https://doi.org/10.1002/adma.200290020.

8 Gorb, S.N. (2008). Biological attachment devices: exploring nature's diversity for biomimetics. *Philos. Trans. R. Soc. A* 366: 1557–1574. https://doi.org/10.1098/rsta.2007.2172.

9 Azimi, G., Dhiman, R., Kwon, H.-M. et al. (2013). Hydrophobicity of rare-earth oxide ceramics. *Nat. Mater.* 12: 315–320. https://doi.org/10.1038/nmat3545.

10 Vinogradova, O.I. (1999). Slippage of water over hydrophobic surfaces. *Int. J. Miner. Process.* 56: 31–60. https://doi.org/10.1016/S0301-7516(98)00041-6.

11 Vinogradova, O.I. (1995). Drainage of thin liquid film confined between hydrophobic surfaces. *Langmuir* 11: 2213–2220. https://doi.org/10.1021/la00006a059.

12 Lugscheider, E. and Bobzin, K. (2001). The influence on surface free energy of PVD-coatings. *Surf. Coat. Technol.* 142–144: 755–760. https://doi.org/10.1016/S0257-8972(01)01315-9.

13 Ponjavic, A., Chennaoui, M., and Wong, J.S.S. (2013). Through-thickness velocity profile measurements in an elastohydrodynamic contact. *Tribol. Lett.* 50: 261–277. https://doi.org/10.1007/s11249-013-0118-x.

14 Wong, P.L., Li, X.M., and Guo, G. (2013). Evidence of lubricant slip on steel surface in EHL contact. *Tribol. Int.* 61: 116–119. https://doi.org/10.1016/j.triboint.2012.12.009.

15 Stachowiak, G.W. and Batchelor, A.W. (2005). *Engineering Tribology*, 3e. Oxford: Elsevier/Butterworth-Heinemann.

16 Hamrock, B.J. and Dowson, D. (1977). Isothermal elastohydrodynamic lubrication of point contacts. Part III. Fully flooded results. *J. Lubric. Tech.* 99: 264–276. https://doi.org/10.1115/1.3453074.

17 Owens, D.K. and Wendt, R.C. (1969). Estimation of the surface free energy of polymers. *J. Appl. Polym. Sci.* 13: 1741–1747. https://doi.org/10.1002/app.1969.070130815.

18 Wu, S. and Brzozowski, K.J. (1971). Surface free energy and polarity of organic pigments. *J. Colloid Interface Sci.* 37: 686–690. https://doi.org/10.1016/0021-9797(71)90347-X.

19 Fowkes, F.M. (1964). Attractive forces at interfaces. *Ind. Eng. Chem.* 56: 40–52. https://doi.org/10.1021/ie50660a008.

20 Holmberg, K. (2002). *Handbook of Applied Surface and Colloid Chemistry*, vol. 1. Chichester, UK: Wiley.

21 Kalin, M. and Polajnar, M. (2013). The effect of wetting and surface energy on the friction and slip in oil-lubricated contacts. *Tribol. Lett.* 52: 185–194. https://doi.org/10.1007/s11249-013-0194-y.

22 Fontaine, J., Donnet, C., Grill, A., and LeMogne, T. (2001). Tribochemistry between hydrogen and diamond-like carbon films. *Surf. Coat. Technol.* 146–147: 286–291. https://doi.org/10.1016/S0257-8972(01)01398-6.

23 Edremir, A. (2001). The role of hydrogen in tribological properties of diamond-like carbon films. *Surf. Coat. Technol.* 146–147: 292–297. https://doi.org/10.1016/S0257-8972(01)01417-7.

24 Erdemir, A. (2004). Genesis of superlow friction and wear in diamondlike carbon films. *Tribol. Int.* 37: 1005–1012. https://doi.org/10.1016/j.triboint.2004.07.018.

25 Li, H., Xu, T., Wang, C. et al. (2007). Tribochemical effects on the friction and wear behaviors of a-C:H and a-C films in different environment. *Tribol. Int.* 40: 132–138. https://doi.org/10.1016/j.triboint.2006.03.007.

26 Erdemir, A. and Donnet, C. (2006). Tribology of diamond-like carbon films: recent progress and future prospects. *J. Phys. D Appl. Phys.* 39: R311–R327. https://doi.org/10.1088/0022-3727/39/18/R01.

27 Kalin, M., Simič, R., Hirayama, T. et al. (2014). Neutron reflectometry of alcohol adsorption on various DLC coatings. *J. Appl. Surf. Sci.* 288: 405–410. https://doi.org/10.1016/j.apsusc.2013.10.047.

28 Kalin, M., Igor, V., and Jože, V. (2009). The Stribeck curve and lubrication design for non-fully wetted surfaces. *Wear* 267 (5): 1232–1240.

Section IV

Case Study: Erosive Wear of Ceramics

13

Erosive Wear of SiC-WC Composites

In this chapter, salient observations in solid particle erosive wear of SiC-WC composites against SiC particles at ambient temperature and high temperature (800 °C) are discussed. Importantly, the effect of WC content in SiC matrix and angle of impingement of erodent particles on material loss of SiC-WC composites are analyzed with a particular emphasis on changes in material degradation mechanisms. The detailed analysis of the experimental results obtained from room-temperature and high-temperature erosion behavior of SiC-WC composites can be found elsewhere [1, 2].

13.1 Introduction

Silicon carbide (SiC) ceramic materials are used in various wear-resistant applications due to their superior properties like high-temperature strength, high hardness, and superior corrosion and wear resistance [3–5]. As described in Chapter 7, erosive wear is the most important material damage in various engineering and tribological applications like gas burner parts, turbine parts, fluidized bed combustion systems, cutting tools, etc. [6].

In order to assess the true potential of SiC-based ceramics for high-temperature wear applications, an understanding on high-temperature erosion behavior is needed. The wear resistance of SiC ceramics can be improved via composite approach, i.e. by incorporating suitable reinforcement in SiC matrix [7–9]. As per the observation of Colclough and Yeomans [7], the erosion loss for SiC-TiB_2 composite and monolithic SiC ceramics at room temperature was lower compared with erosion at 1000 °C. However, improved fracture toughness resulted in superior erosion resistance for SiC-TiB_2 compared with monolithic SiC at any given temperature. Wang and Levy [8] observed inverse relation of hardness and density to the erosion loss of SiC-SiC_f composite at both room temperature and elevated

Friction and Wear of Ceramics: Principles and Case Studies, First Edition.
Bikramjit Basu, Mitjan Kalin, and B. Venkata Manoj Kumar.

temperature (850 °C). However, almost an order of magnitude less erosive wear rate was observed at elevated temperature due to increased deformation, decreased chipping, and cracking of fibers and matrix. In another study [9], the erosive wear rate of SiC-Si_3N_4 composite increased to maximum with an increase in temperature to 800 °C due to stripping of aggregates of composite and its brittle fracture. However, further increase in temperature to 1400 °C resulted in decreased erosive wear due to increased plastic deformation and protection from oxidation. In a study by Ham et al. [10], erosive wear rates were found higher for $CaAl_2SiO_5$/SiC glass ceramic-SiC_f composite at 200 °C compared with room temperature. But their study did not show any change in erosive wear rate of the composite with further increase in test temperature to 726 °C.

In the backdrop of the above, this chapter describes the published results on the erosive wear of SiC-WC composites to assess the performance for their use in applications like turbine blades, exhaust systems for turbopump engines, etc. SiC-WC composites were eroded by stream of SiC particles at ambient temperature and high temperature (800 °C). In particular, the influences of the impingement angle of SiC particles and WC content (0, 10, 30, and 50 wt%) in the composite on the erosive wear are investigated. The dominant mechanisms of material removal are also highlighted.

13.2 Materials and Erosion Tests

SiC-WC composites were prepared by hot pressing at 1800 °C for one hour under a pressure of 40 MPa in argon. The SiC, SiC-10 wt%WC, SiC-30 wt%WC, and SiC-50 wt%WC composites are, respectively, designated as SW0, SW10, SW30, and SW50. Details of preparation, and microstructural and mechanical characteristics may be found elsewhere [1]. An air jet erosion facility was used for erosion tests at various impingement angles (30°, 60°, and 90°) and test temperatures, ambient (room) temperature [1] and high temperature (800 °C) [2]. SiC or Al_2O_3 erodent particles of 40–70 μm in size were mixed with air and passed through a 3 mm diameter WC nozzle to erode the SiC-WC composite at 4 mm stand-off distance and 3 g/min flow rate. The erosion was continued until a steady state in weight loss was obtained. The steady-state average volume loss was divided by the amount of erodent to obtain the steady-state erosive wear rate.

13.3 Influence of Type of Erodent on Erosive Wear Behavior

The erosive wear rates for SiC ceramics against normal incidence of two types of erodent, Al_2O_3 and SiC, are presented in Figure 13.1. One important observation from the present study is that the erosive wear rate is almost 50–130 times

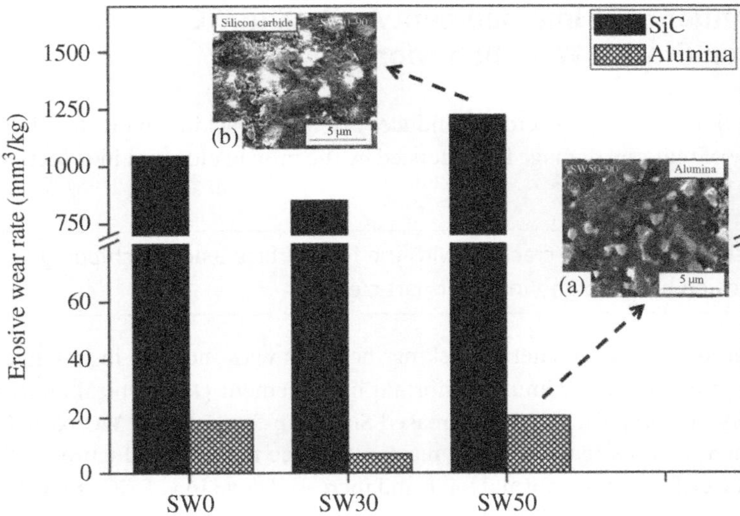

Figure 13.1 The steady-state erosive wear rate of the investigated SiC-WC composites. Insets show SEM images of SW50 composite after erosion at 90° using (a) alumina (Al_2O_3) and (b) silicon carbide (SiC) erodent. The designation of materials: SW0, SiC-0% WC composites; SW30, SiC-30% WC composites; and SW50, SiC-50% WC composites [1]. *Source:* Reproduced with permission of Elsevier.

lower when SiC ceramics were impinged by Al_2O_3 particles compared with SiC particles.

Regarding wear mechanisms, WC particle pull-out and SiC grain fracture in SiC-WC composites are observed when impinged by any erodent (see insets in Figure 13.1). But, the severity of material loss was decreased against Al_2O_3 erodent in terms of relatively mild indentation and smaller extent of chipping in the composite. Earlier reports also highlighted the difference in erosive wear of ceramics by different particles [11–13].

Furthermore, plastic flow in erodent particles or eroded material is possible under severe impact stresses [14, 15].

> The hardness ratio of erodent (H_p) to eroded material (H_e) is related to the erosive wear behavior in case of brittle solids.

In the present investigation, the H_p/H_e ratio decreased for the given SiC-WC composite, when eroded by Al_2O_3 erodent. Thus, reduced damage of the composites in case of Al_2O_3 erodent can be attributed to the lower H_p/H_e ratio.

13.4 Influence of Impingement Angle and WC Content on Erosive Wear Behavior

It is known that solid particle erosion induces extensive cracking in brittle solids. The severity of material damage is influenced by the propagation and intersection of cracks [11].

> The intersection of lateral cracking with the free surface leads to chipping in case of erosion by sharp and irregular particles.

The material removal by lateral cracking increases with increase in impingement angle, reaching a maximum at normal impingement [16]. Typical eroded surface of SW30 composite shows increased SiC grain fracture and WC particle pull-out when the impingement angle reaches 90° (see the inset in Figure 13.2). This is because of increased normal force and increased cracking of SiC grains for SiC-WC composites at normal incidence.

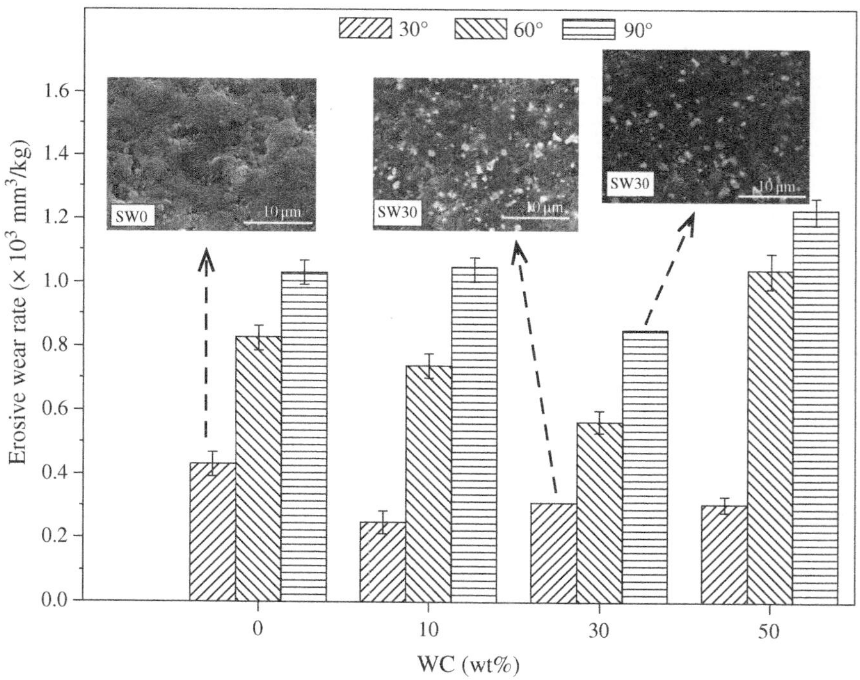

Figure 13.2 Erosive wear rate of the investigated SiC-WC composites at ambient temperature. Typical SEM images of eroded surfaces are shown as insets. The designation of materials: SW0, SiC-0% WC composites; SW30, SiC-30% WC composites; and SW50, SiC-50% WC composites [1]. *Source:* Reproduced with permission of Elsevier.

Also, SW30 composite exhibited lowest erosive wear rate with reduced fracture at a given impingement angle. The homogeneous distribution of WC particles is responsible for reduced SiC grain fracture. However, easy pull-out of a large amount of heavy WC particle and subsequent easy fracture of SiC grain are attributed to the higher erosive wear rate for SW50 composite. The microstructural and mechanical characteristics can be used to understand erosive wear behavior of SiC-WC composites.

13.5 Correlating Erosive Wear Behavior with Microstructural Characteristics

Image analysis software was used to determine the grain size of SiC, volume fraction, and interparticle distance of reinforcement (WC) particles. Correlating with erosion results, the higher erosive wear resistance is observed for SiC-30 wt% WC composite with 640 nm grain size compared with SiC ceramics with 835 nm grain size. But, further reduction in grain size to 580 nm in SW50 composites resulted in least erosive wear resistance. The erosive wear resistance increased with decrease in interparticle distance of WC particles and increase in volume fraction of WC up to 30 wt%. Further decrease in interparticle distance and increase in WC content did not result in improvement of erosive wear resistance. The inferior erosive wear resistance for SW50 composites is due to the removal of coarse WC particles in large amounts.

13.6 Correlating Erosive Wear Behavior with Mechanical Properties

The hardness and fracture toughness significantly affect erosive wear loss of SiC-WC composites. Regarding fracture toughness, SiC ceramics with the least fracture toughness (~5.8 MPa·$m^{1/2}$) exhibited lower erosive wear resistance, while SW30 composite with intermediate fracture toughness (~6.5 MPa·$m^{1/2}$) showed maximum erosive wear resistance. However, further increase in fracture toughness (to ~6.7 MPa·$m^{1/2}$) in SiC-50 wt% WC composite resulted in least erosive wear resistance at high impingement angle.

For brittle ceramics, contribution from "short-crack" toughness toward wear resistance is significant compared with "long-crack" toughness.

Correlating the hardness with erosive wear, SW30 composite possessing maximum hardness (~26 GPa) showed the highest erosive wear resistance. On the other hand, SW50 composite possessing a lower hardness (~24 GPa) exhibited least erosive wear resistance at 60° and 90° impingement angle.

Based on the results obtained in the present study, SW30 composite with improved mechanical properties and superior erosive wear resistance is preferred for components subjected to erosion conditions.

13.7 Erosive Wear Behavior at High Temperature

The erosive wear rate of SiC-WC composites as a function of impingement angle and WC content at 800 °C is presented in Figure 13.3. A wide range of erosive wear rates of 10–10^3 mm^3/kg with varying erosion conditions is reported for SiC ceramics [8, 9, 17, 18]. The erosive wear rate obtained in the present study is found to vary from ~2×10^2 mm^3/kg to ~8×10^2 mm^3/kg. The erosive wear rate for SiC-WC composites increased with increase in impingement angle due to increased fracture. With respect to WC content, the highest erosive wear resistance is found in SW30 composite at any impingement angle. Further increase in WC content to 50 wt% resulted in less erosive wear resistance at normal impingement angle.

SiC ceramics after erosion at 800 °C at higher angle reveal grain pull-out and indentation-induced fracture (see insets in Figure 13.3), while compact surface deformation is present at lower impingement angles (not shown in the figure).

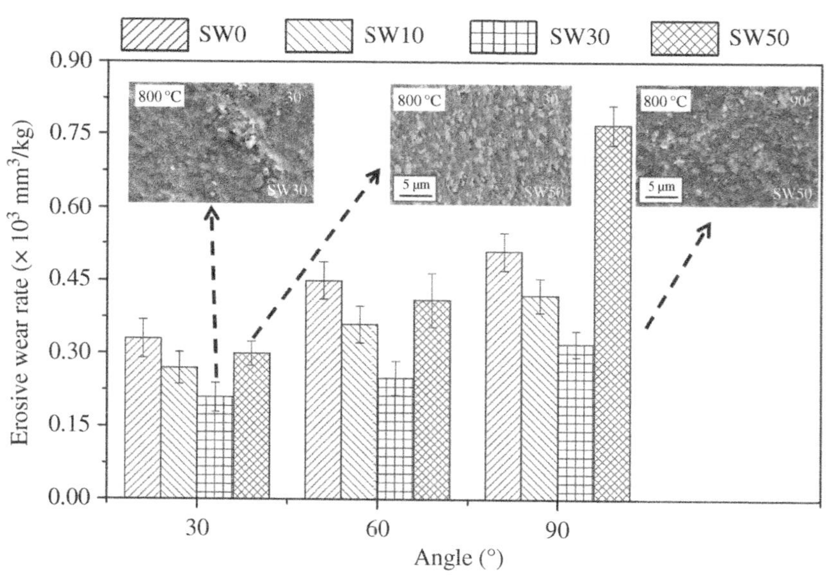

Figure 13.3 Erosive wear rate of the investigated SiC-WC composites at 800 °C. Eroded surfaces of selected composites are shown as insets. The designation of materials: SW0, SiC-0% WC composites; SW10, SiC-10% WC composites; SW30, SiC-30% WC composites; and SW50, SiC-50% WC composites [2]. *Source:* Reproduced with permission of Elsevier.

> During erosion of brittle SiC ceramics, multi-indentations of sharp SiC erodent particles cause elastic–plastic deformation-mediated microfracture and lateral crack formation below the plastic zone of subsurface.

During high-temperature erosion of SiC-Si_3N_4 ceramics, material loss is also attributed to the extrusion of Si_3N_4 phase and fracture of SiC grains [9]. During erosion at elevated temperatures, the formation of glassy silica phase prevents the fracture [9, 19–21].

Representative worn surfaces of selected SiC-WC composites after erosion at 800 °C are shown as insets in Figure 13.3. The WC particles are pulled-out from eroded surfaces at given impingement angle. The fracture is reduced for SiC-WC composites compared with SiC ceramics. Also, mild WC particle pull-out and SiC grain fracture are observed at any impingement angle in case of SW30 composite. However, SW50 composite eroded surface shows easy pull-out of heavy WC agglomerates and fracture (see insets in Figure 13.3). This is in line with the erosive wear rate data provided in Figure 13.3.

The presence of Si, W, C, and O is observed in EDS analysis of SW50 composite after erosion at normal impingement (not shown in the figure). The presence of

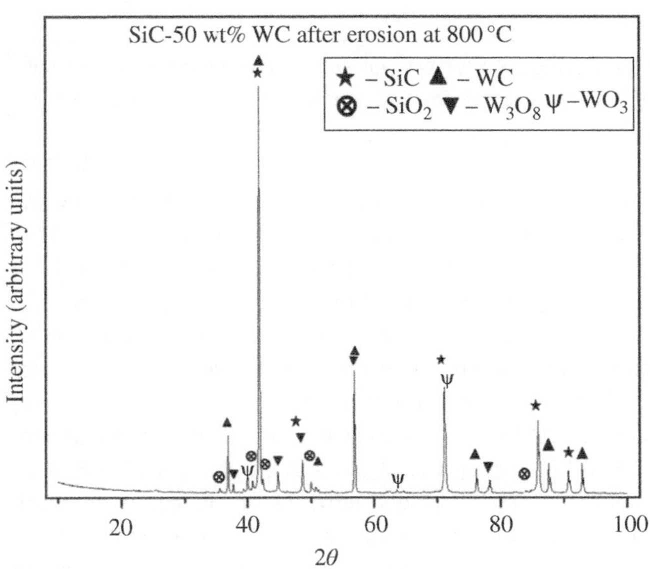

Figure 13.4 Phase analysis of SW50 composite after high-temperature (800 °C) erosion [2]. *Source:* Reproduced with permission of Elsevier.

silica and/or tungsten oxide was recorded after erosion at 800 °C and this can be explained by the following set of equations:

$$SiC\ (s) + 2O_2\ (g) \rightarrow SiO_2\ (s) + CO_2\ (g) \quad (13.1)$$

$$2WC\ (s) + 5O_2\ (g) \rightarrow 2WO_3\ (s) + 2CO_2\ (g) \quad (13.2)$$

$$3WC\ (s) + 7O_2\ (g) \rightarrow W_3O_8\ (s) + 3CO_2\ (g) \quad (13.3)$$

Further, SiO_2, W_3O_8, and WO_3 formation is also supported by the XRD analysis of eroded SW50 composite (see Figure 13.4). The reduced material loss in SW30 composite is attributed to the homogeneous distribution of WC particles when compared with heavy WC agglomerates in SW50 composite. Among the investigated composites, the highest hardness contributes to the lowest wear for the SW30 composite.

13.8 Closure

Summarizing the discussion on erosion at ambient temperature, it can be said that SiC-WC composites showed WC particles pull-out and SiC grains fracture in erosion by SiC or Al_2O_3 particles.

> The time needed for achieving steady-state erosion and the overall severity of material damage of SiC-WC composites is considerably influenced by the hardness of the impinging particles.

The material loss in the given erosion conditions was also affected by the amount of WC in the composite and impingement angle of the erodent. Also, SiC-30% WC composites exhibited higher hardness and superior erosive wear resistance because of the homogeneous distribution of WC particles in SiC matrix.

Concerning high-temperature erosive wear of SiC-WC composites at 800 °C, the wear rate increased with the increased fracture at higher impingement angles of erodent, but decreased with increase in amount of WC up to 30 wt%. In particular, SiC-30% WC composites exhibited least erosive wear at lower impingement angle (30°), while SiC-50% WC composites showed the highest erosive wear at higher impingement angle (90°). The eroded surfaces of SiC ceramics dominantly showed grain fracture, while that of SiC-WC composites exhibited WC particle pull-out from SiC matrix and subsequent SiC grain fracture. The extent of fracture is reduced in SiC-WC composites with maximum WC content of 30 wt% at lower impingement angles.

References

1 Sharma, S.K., Kumar, B.V.M., Lim, K.Y. et al. (2014). Erosion behavior of SiC-WC composites. *Ceram. Int.* 40: 6829–6839.

2 Sharma, K., Kumar, B.V.M., and Kim, Y.W. (2017). Effect of impingement angle and WC content on high temperature erosion behavior of SiC-WC composites. *Int. J. Refract. Met. Hard Mater.* 68: 166–171.

3 Roewer, G., Herzog, U., Trommer, K. et al. (2002). Silicon carbide – a survey of synthetic approaches, properties and applications. *Struct. Bond.* 101: 59–135.

4 Manoj Kumar, B.V., Kim, Y.W., Lim, D.S., and Seo, W.S. (2011). Influence of small amount of sintering additives on unlubricated sliding wear properties of SiC ceramics. *Ceram. Int.* 37 (8): 3599–3608.

5 Padture, N.P. (1994). In situ-toughened silicon carbide. *J. Am. Ceram. Soc.* 77: 519–523.

6 Kosel, T.H. (1992). Solid particle erosion. In: *Friction, Lubrication and Wear Technology, ASM Handbook*, vol. 18 (ed. P.J. Blau), 199–213. Materials Park, OH: ASM International.

7 Colclough, A.F. and Yeomans, J.A. (1997). Hard particle erosion of silicon carbide and silicon carbide-titanium diboride from room temperature to 1000 °C. *Wear* 209: 229–236.

8 Wang, B. and Levy, A.V. (1990). Erosion behavior of SiC fiber-SiC matrix composites. *Wear* 138: 125–136.

9 Li, X., Ding, H., Huang, Z. et al. (2014). Solid particle erosion-wear behavior of SiC–Si_3N_4 composite ceramic at elevated temperature. *Ceram. Int.* 40: 16201–16207.

10 Ham, A.L., Yeomans, J.A., and Watts, J.F. (1997). Elevated temperature solid particle erosion of silicon carbide continuous fibre reinforced calcium aluminosilicate glass-ceramic matrix composite. *Wear* 203: 387–392.

11 Shipway, P.H. and Hutchings, I.M. (1996). The role of properties in the erosion of brittle materials. *Wear* 193: 105–113.

12 Akimune, Y., Katano, Y., and Matoba, M. (1989). Spherical impact damage and strength degradation in silicon nitrides for automobile turbocharger rotors. *J. Am. Ceram. Soc.* 72: 1422–1428.

13 Gulden, M.E. (198l). Correlation of experimental erosion data with elastic-plastic impact models. *J. Am. Ceram. Soc.* 64: C59–C60.

14 Hutchings, I.M. (1992). Ductile-brittle transition and wear maps for the erosion and abrasion of brittle materials. *J. Phys. D App. Phys.* 25: A212–A221.

15 Srinivasan, S. and Scattergood, R.O. (1988). Effect of erodent hardness on erosion of brittle materials. *Wear* J28: 139–152.

16 Hutchings, I.M. (1992). *Tribology: Friction and Wear of Engineering Materials.* London: Edward Arnold.

17 Routbort, J.L. and Scattergood, R.O. (1980). Anomalous solid-particle erosion rate of hot pressed silicon carbide. *J. Am. Ceram. Soc.* 63: 593–595.

18 Routbort, J.L. and Matzke, H. (1983). On the correlation between erosion and fracture parameters in SiC ceramics. *J. Mater. Sci.* 18: 1491–1496.

19 John, C.S. (1975). The brittle-to-ductile transition in pre-cleaved silicon single crystals. *Philos. Mag.* 32: 1193–1212.

20 Mendiratta, M.G., Wimmer, J., and Bransky, I. (1977). Dynamic K_{IC} and dynamic flexural strength in HS-130 Si_3N_4. *J. Mater. Sci.* 12: 212–214.

21 Sharma, S.K., Selokar, A.W., Kumar, B.V.M., and Venkateswaran, T. (2017). High temperature erosion behavior of spark plasma sintered ZrB_2-SiC composites. *Ceram. Int.* 43: 8982–8988.

14

Thermo-Erosive Behavior of ZrB_2-SiC Composites

The thermo-erosive stability of spark plasma sintered ZrB_2-SiC-based multiphase ceramic composites using a combination of experimental work and computational analysis is critically analyzed in this chapter. High-temperature erosion tests were conducted at 800 °C in thermo-erosive environment for 1200 seconds. The careful selection of test parameters allows us to conduct the experiment with the net energy deposition per unit area of 50.5 MJ/m^2. Based on the recently published data from the author's research group [1], the thermo-erosive mechanisms are qualitatively discussed together with the results of finite element (FE) analysis.

14.1 Introduction

The leading edges of supersonic aircrafts experience high temperatures close to 800 °C due to intense aerodynamic heating along with high stagnation pressure [2]. These aircrafts fly at supersonic speed in the range of Mach 2–3 at altitudes of 10–14 km. Most conventional materials fail under such thermomechanical loads and this necessitates the use of ultra-high temperature ceramics (UHTCs). UHTCs have high melting temperatures in excess of 3000 °C and possess good thermal and mechanical properties up to high temperatures [2]. Zirconium diboride (ZrB_2)-based UHTCs are one of the most studied materials for high-speed aircrafts [2–4].

> ZrB_2-based ceramics are suitable for high-temperature applications as they possess high melting point (3245 °C), high thermal conductivity (~60 W/m·K), high hardness (>18 GPa), good chemical inertness, and wear resistance.

Friction and Wear of Ceramics: Principles and Case Studies, First Edition.
Bikramjit Basu, Mitjan Kalin, and B. Venkata Manoj Kumar.

ZrB_2-based composites are developed with various reinforcements like SiC, ZrC, ZrO_2, carbon nanotube, Ti, $TiSi_2$, TaB_2, and $TaSi_2$ [4–13]. Recently, ZrB_2-SiC-based composites were investigated and suitable guidelines for design of thermal protection systems for space vehicle are proposed [12]. In these applications, the components are likely to face wear due to continuous impingement of atmospheric particulates. Accordingly, an understanding on the erosive wear behavior is necessary to assess the potential of UHTC composites for thermal protection systems. High-temperature erosive wear behavior of non-oxide ceramics is not extensively investigated. Among the limited studies, at least two orders of magnitude high erosive wear rate is reported when SiC-WC were eroded against SiC than that eroded by Al_2O_3 particles [14]. Similar observations were found when silica, soda-lime and borosilicate glass, zirconia, alumina, silicon carbide, and boron carbide were eroded by four different erodent materials: silicon, silica, alumina, and silicon carbide. Higher erosive wear rates of 200–600 μg/g were recorded when ceramics were eroded by SiC erodent (hardness 33 GPa), while lower erosive wear rates of 0.1–50 μg/g were obtained when eroded by SiO_2 erodent (hardness 13 GPa) [15]. It was reported in another study that the surface morphology of eroded SiC ceramics varied with size of the erodent [16], and the mechanism of material removal was dependent not only on mechanical properties but also on microstructure and residual stresses [17–21]. A critical literature search could not however reveal considerable study on high-temperature erosive wear behavior of ZrB_2-based ceramics.

The oxidation resistance of ZrB_2-SiC-Ti ceramic during arc jet exposure has also been reported recently [22]. It is to note that metallic airframe in supersonic vehicles are subjected to temperatures of around 1100 K and stresses above 32 MPa [1, 2]. Besides, supersonic space vehicle structural components like strut and leading edges experience erosive environment for extended durations at high temperatures [23, 24].

In a recent study, reduced erosive wear of ZrB_2-SiC composites with an increase in SiC content is reported, and this was attributed to the formation of protective layer of silica and borosilicate glass on the eroded surface [5]. However, it was also observed that the erosive wear properties were not directly related to mechanical properties of bulk ceramic composites [5, 6, 8–10]. As high-temperature erosion is driven not only by temperature but also by material yielding, stress evolution in the exposed region becomes important to assess the material erosion. While determining stress levels in high-temperature eroded regions is not possible by experimental techniques, computational analysis can provide necessary insights in estimating the stress or temperature distribution in eroded regions. On a similar note, the molecular dynamics simulation was necessary to understand the anisotropic mechanical amorphization in diamond polishing [25]. In other sliding study, simulation was used to observe the change in high-temperature frictional behavior of NaCl near melting temperature [26]. Accordingly, a novel approach of a combination of experimental erosion testing and computational analysis is necessary to assess the true potential of ceramic composites for critical components in thermal protection systems for supersonic space vehicles. In the above perspective, ZrB_2-SiC-based multiphase ceramic composites were subjected to high-velocity impingement of SiC particles at 800 °C and distributions of stress and temperature in the exposed erosion region were estimated by finite element (FE) analysis to understand the underlying reasons for high-temperature erosive wear behavior. Thermodynamic feasibility of possible reactions in high-temperature erosion region is also discussed.

14.2 High-Temperature Erosion Tests and Computational Modeling

Two compositions, i.e. ZrB_2-18 wt% SiC and ZrB_2-18 wt%–SiC-10 wt% Ti, respectively, designated as ZST0 and ZST10, were processed. The milled powders of each composition were sintered by using spark plasma sintering (SPS) apparatus (Dr. Sinter, Model 515S, SPS Syntax Inc., Japan) via multistage sintering (MS-SPS) route with final holding at 1600 °C for 10 minutes [12].

Erosion tests were performed with SiC erodent particles (40–70 μm size) at the particle velocity of 47 m/s with a flow rate of 3 g/min. The sample was placed in a holder and heated to 1073 K for 40 minutes, and erosion tests were conducted at an impingement angle of 90°. The average weight loss in steady state was converted to average volume loss and steady-state erosive wear rate E in mm^3/kg was determined as per Eq. (14.1).

$$E = \frac{V}{M} \tag{14.1}$$

where V is the volume of the material removed in steady state in mm^3 and M is the mass of the erodent particles used in kg.

The erosion testing experiments were carried out by matching the energy deposition experienced by a supersonic aircraft flying at 900 m/s at 10 km altitude for 30 minutes. The particle mass flow rate as experienced by a supersonic vehicle was estimated based on the impingement speed and concentration of particulate matter in 129.6 μg/kg of air atmosphere at 10 km altitude [27]. The kinetic energy (KE) deposited per unit area of ceramic surface of aerospace vehicle was calculated at Mach 2–3 speeds, as per the following equations:

$$\mathrm{KE} = \frac{1}{2}\frac{\dot{m}}{A} v^2 \cdot t_{\mathrm{flight}} \tag{14.2a}$$

$$\dot{m} = \rho \cdot A \cdot v \cdot C \tag{14.2b}$$

where KE is the kinetic energy deposited per unit area, v is the velocity and $\dot{m}$ is the mass flow rate of impinging erodent particles, t_{flight} is the flight time, A is the cross-sectional area of exposed supersonic aircraft [2, 23], and C is the maximum of annual dust concentration (129.6 μg/kg of air) [25]. Based on the above calculations, KE deposition is estimated to be 38.4 MJ/m^2 for aircraft flying at 900 m/s at 10 km altitude. In accordance with this calculation, erosion experiments were designed using more conservative energy deposition value of 50.5 MJ/m^2 (130% of in-flight values) as calculated using Eqs. (14.3a) and (14.3b). Accordingly, various experimental parameters, viz. particle velocity and particle feed rate, were set (Table 14.1).

The sample surface was exposed to equivalent heat deposition (Eq. 14.4) and a uniform force equivalent of momentum lost by erodent per unit time (F) was applied on the exposed portion of flat face of ceramic sample in thermo-structural analysis, which was calculated using Eq. (14.5).

$$\text{KE} = \frac{1}{2}\dot{m} * \left(v_{\text{erodent}}\right)^2 * t_{\text{exposure}} \tag{14.3a}$$

$$\dot{m} = \rho_{\text{erodent}} * A_{\text{sample}} * v_{\text{erodent}} \tag{14.3b}$$

$$\dot{q}_e = \frac{1}{2} * \rho_{\text{erodent}} * A_{\text{sample}} * \left(v_{\text{erodent}}\right)^3 * t_{\text{exposure}} \tag{14.4}$$

$$F = \dot{m} * v_{\text{erodent}} \tag{14.5}$$

Table 14.1 Summary of parameters used in erosion experiments [1].

Parameter	Value
Erodent particle size	>25 μm
Equivalent force	0.025 N
Particle velocity	47 ± 1 m/s
Particle feed rate	3 ± 0.2 g/min
Impact angle	90°
Time of testing	1200 s
Operating temperature	1073 ± 100 K
Pressure	101.325 kPa
Energy deposited	50.5 MJ/m^2

Source: Reproduced with permission of Elsevier.

Considering the temperature limitation with the available experimental setup, erosion tests were conducted at 1073 K. Normal impact was selected to simulate erosion behavior of ZrB_2-SiC-based composites, as the erosive wear rate of the brittle materials is the maximum at normal impact [28].

14.3 Computational Modeling of Thermo-Erosive Behavior

FE modeling was carried out to simulate the thermo-structural behavior of ZrB_2-SiC ceramic composite (ZST0 and ZST10 composition) during high-temperature erosion testing using the ANSYS Mechanical 15 software. High-quality structured grid with second-order elements for a better accuracy was generated using the ANSYS to discretize the solid domain (refer Figure 14.1). The displacement constraints were present due to contact of the sample with the filler along the circumference and contact of sample with the holder at the back wall. The mode geometry and computational domain are briefly shown in Figure 14.1. A two-step methodology was adopted for FE analysis. In the first step, transient thermal

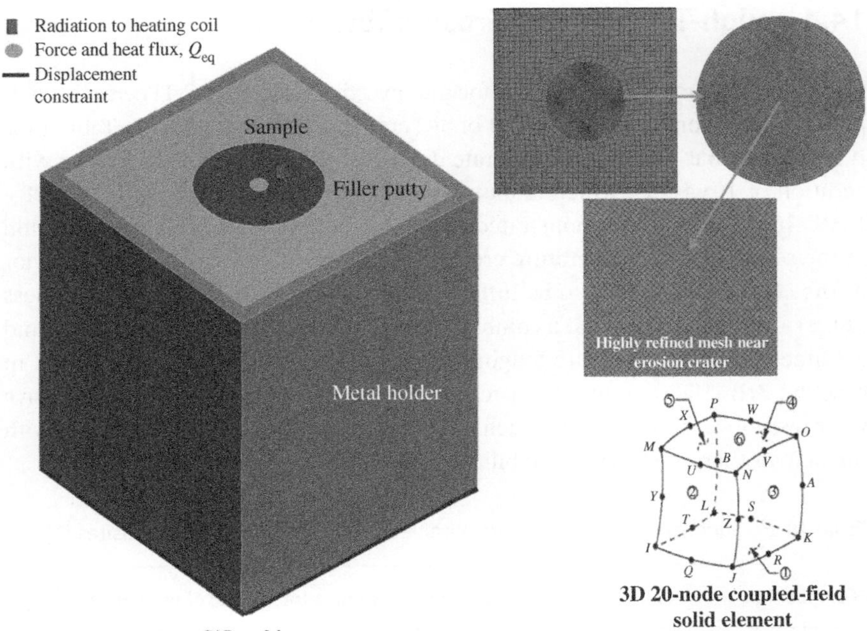

Figure 14.1 CAD model of the ZrB_2-SiC-Ti sample in the sample holder during test, refined mesh and 20-noded hex element used for discretizing the solid domain [1]. *Source:* Reproduced with permission of Elsevier.

analysis (TTA) was performed to estimate temperature distribution in the sample. Then, the nodal temperature distribution obtained from TTA was mapped to the structural domain.

1) The surface of sample is exposed to equivalent heat deposition $\dot{q}_e$ and a uniform force equivalent of momentum lost by erodent per unit time (F) on the exposed portion of flat face of composite.
2) The conduction is assumed as the major heat transfer mode.
3) The displacement of rear face of sample holder is constrained in all the directions. Thermal expansion is arrested in all the directions along back wall of tested ceramic sample. This subjects the ceramic disk to an intense stress environment in a high-temperature erosive environment above 1000 K.

The properties including thermal conductivity, specific heat, density, elastic modulus, and Poisson's ratio were calculated for ZrB_2-18SiC composite using the rule of mixture assuming that there were no sintering reaction products for this composite. For calculating these material properties, temperature-dependent property data available from literature [2, 29, 30] were used. On the other hand, experimentally measured material properties were used for ZST10 composite [12].

14.4 High-Temperature Erosion Test Results

The high-temperature erosion loss for the investigated ZrB_2-SiC-Ti ceramic composites due to normal impingement of SiC erodent are summarized in Table 14.2. It is evident that the erosive wear rate decreases from 35.9 to 33.5 mm^3/kg with addition of 10 wt% Ti, which amounts to 6.6% decrease in erosion. Thus, ZrB_2-18SiC-10Ti composite demonstrated to provide better resistance against erosion in the selected high-temperature erosion conditions. The erosive wear behavior of the ceramics is known to be influenced by hardness and fracture toughness [20, 31–33]. It is to note that a combination of 1.7 times increase in hardness and 2.5 times increase in fracture toughness obtained with the addition of 10% Ti in baseline ZrB_2-18SiC composition resulted in only 6.6% improvement in erosive wear resistance. The reason for such observation is explained in later section with the help of thermo-structural stability analysis.

Table 14.2 Summary of erosion results obtained for the investigated composites [1].

Composition	Mass loss per kg of erodent (g)	Volume loss per kg of erodent (mm^3/kg)
ZST0	180.0 ± 7.6	35.9 ± 1.5
ZST10	170.0 ± 0.5	33.5 ± 0.1

Source: Reproduced with permission of Elsevier.

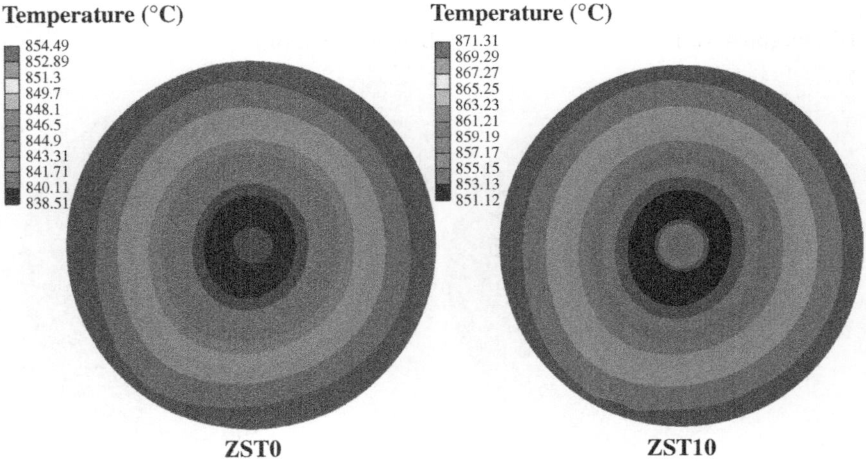

Figure 14.2 Transient thermal analysis: Simulated temperature contours (ZST0 and ZST10) at the end of 1200 seconds of high-temperature erosion test. Maximum temperatures of 840 and 852 °C are recorded for ZST0 and ZST10, respectively [1]. *Source:* Reproduced with permission of Elsevier.

14.5 Transient Thermal Studies Using FE Analysis

In the first step, TTA was carried out to calculate the temperature experienced by exposed ZST0 and ZST10 samples during high-temperature erosion testing. It is imperative to capture the temperature field evolution during high-temperature erosive testing, because of the development of thermo-structural stresses in the crater region. The temperature contours of the exposed surfaces of ZST0 and ZST10 are plotted in Figure 14.2. The spatial temperature distribution shows that the sample experiences a temperature in the interval of 840–841 °C for ZST0 and 855–862 °C for ZST10 in the vicinity of the crater region, which amounts to a thermal gradient of approximately 300 K/m in case of ZST0 and 4500 K/m in case of ZST10.

14.6 Coupled Thermo-Structural Analysis

The displacement field of high-temperature erosion tested ceramic disk of ZST0 composition is estimated by performing thermo-structural analysis (Figure 14.3). The sample experiences maximum displacement of 0.38 and 0.40 mm near the sample circumference for ZST0 and ZST10, respectively. But only a small reduction in displacement field is recorded as we move radially towards the center. The exposed surface of sample experiences spatially varying stress field near the

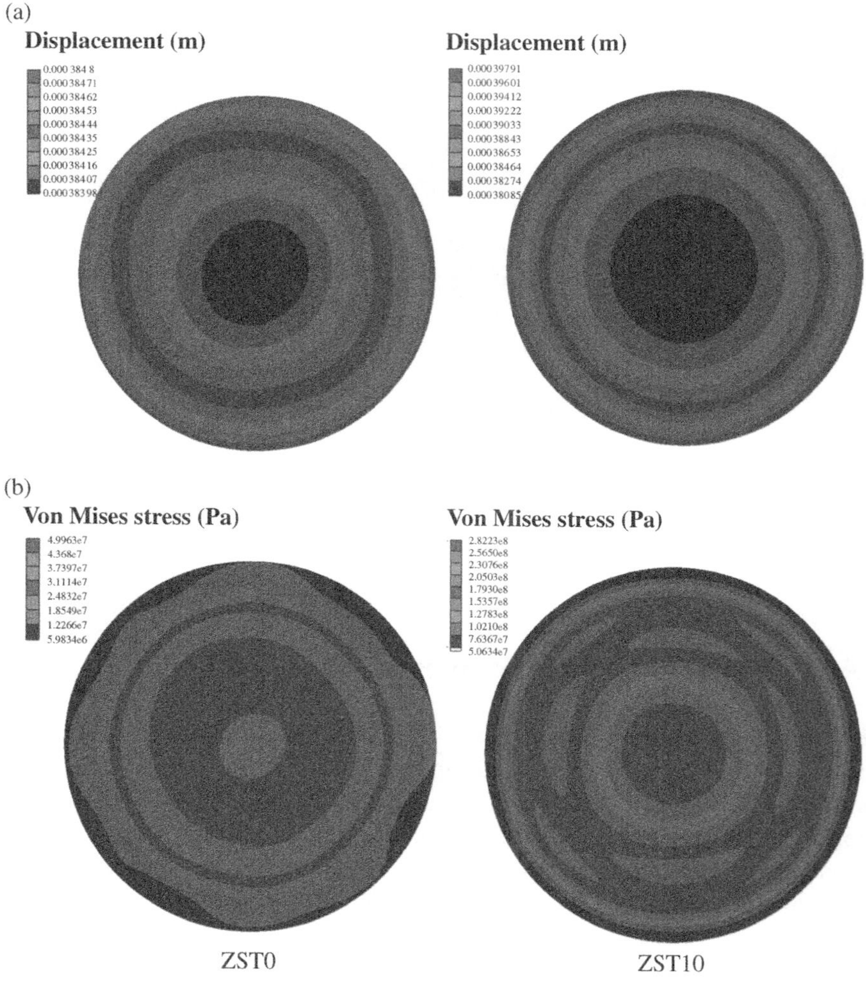

Figure 14.3 Coupled thermal-structural analysis: Contours (ZST0 and ZST10) of (a) deformation and (b) von Mises stress contours during high-temperature erosion test [1]. *Source:* Reproduced with permission of Elsevier.

erosion crater. In case of ZST0, stress at the center of crater is 38.6 MPa, whereas stress as high as 49 MPa is recorded at the crater periphery. This can be explained by the fact that the temperature in the crater region of ZrB_2-18SiC is almost uniform within the crater periphery. However, there is a clear thermal gradient visible in the temperature contours of ZrB_2-18SiC-10Ti sample around crater periphery. The temperature beyond crater periphery is lower than the temperature within. Thus, the region near the crater periphery experiences displacement

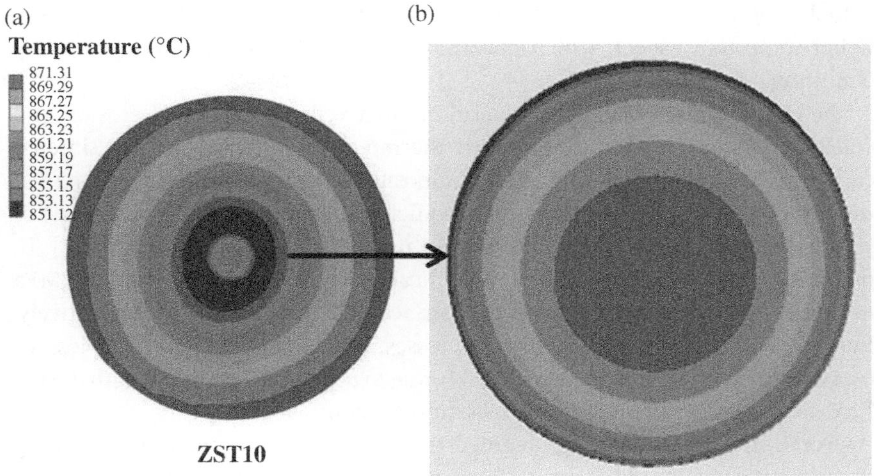

Figure 14.4 Temperature distribution in (a) sample and (b) temperature distribution in the crater region (as shown by the arrow) for eroded ZrB_2-18SiC-10Ti composite [1]. *Source:* Reproduced with permission of Elsevier.

constraint due to material beyond periphery, which does not expand thermally at the same rate. Hence, higher stresses are recorded at crater periphery than crater center. On the other hand, maximum stress of 282.2 MPa is estimated at the crater center and stress falls to 191.8 MPa near crater periphery, in case of ZrB_2-18SiC-10Ti. This can be explained by the fact that temperature at the crater center is higher than the temperature near the crater periphery, as shown in Figure 14.4. It is to note that such contour plot (Figure 14.4b) has been generated by selecting the nodes in the crater region of the sample only. The stress falls to 5.9 MPa for ZrB_2-18SiC and 5.1 MPa near sample periphery. The macroscopic observation after high-temperature erosion test shows that both ZST0 and ZST10 composites survive such a combination of tensile and erosive loads. This reinforces the thermo-erosive stability of spark plasma sintered ZrB_2-SiC-Ti composites and provides an approximate estimate of stress in high-temperature erosive environment.

14.7 Thermo-Erosive Behavior

The thermo-erosive behavior of ZrB_2-SiC-based composites in the selected high temperature conditions can be discussed. It is reported that intense heat flux and pressure are experienced only in localized regions in supersonic space vehicles, which restricts the temperature and thermally induced stresses [34]. Other important aspect, which governs the stress, is the structural boundary condition.

The ceramic disks of both compositions survived with minimal erosion and cracking, thereby suggesting excellent thermo-erosive stability of the sintered ceramics. The presently investigated composites have inherent residual stress at room temperature due to differential coefficient of thermal expansion (CTE) of various phases. But these stresses tend to relax at higher temperatures [35]. In particular, thermo-structural analysis indicates tensile stress of 39 and 282 MPa near the crater center on the exposed surface of ZST0 and ZST10, respectively. But, the composites withstand such stresses, with maximum material loss of 0.44 wt% for ZrB_2-18 wt% SiC, presumably due to negligible loss in strength during 1200 seconds exposure of high-temperature erosion.

Overall, the understanding obtained from the novel approach of combining experimental results and computational analysis is extremely significant in selecting suitable material for thermal protection systems of supersonic vehicles experiencing erosive wear. The results presented in this chapter demonstrate that a large improvement of 1.7 times in hardness and 2.5 times in fracture toughness obtained in ZST10 results in only 6.6% improvement in erosive wear resistance. Almost one order of magnitude higher stresses is found in the near crater center of ZST10 compared with ZST0. The stresses are lower in the crater region of ZrB_2-18SiC, because of lower thermal gradient. Based on the experimental results obtained in the selected high-temperature erosion conditions and computational analysis, ZrB_2-18SiC-10Ti composite is recommended for thermal protection systems of supersonic aircrafts.

> The reduced effect of mechanical properties on erosive wear resistance can be explained by the large difference in stresses experienced by the composites in selected high-temperature erosion conditions.

14.8 Closure

Summarizing the discussion on a novel approach combining experimental study and computational analysis to understand the thermo-erosive stability of ZrB_2-18SiC-xTi composites ($x = 0$ and 10; composition in wt%), it can be said that the erosive rate of ZrB_2-18SiC-10Ti is 6.6% lower than that of ZrB_2-18SiC composite in the selected high-temperature erosion conditions. One order of magnitude higher stress determined from thermo-structural analysis is attributed to the reduced effect of better mechanical properties, viz. hardness and fracture toughness on erosion loss for ZrB_2-18SiC-10TiC composite. Detailed discussion of the results presented in the current chapter can be found in a published article [1]. Taken together, the present work proposes and demonstrates new guidelines for assessing thermal-erosive-structural stability of materials utilizing coupled experimental and computation analysis.

References

1 Purwar, A., Ariharan, S., Basu, B., and Kumar, B. (2019). Experimental and computational analysis for thermo-erosive stability assessment of ZrB_2-SiC based multiphase composites. *Int. J. Refract. Met. Hard Mater.* 84: 104972.

2 Zimmermann, J.W., Hilmas, G.E., Fahrenholtz, W.G. et al. (2008). Thermophysical properties of ZrB_2 and ZrB_2–SiC ceramics. *J. Am. Ceram. Soc.* 91 (5): 1405–1411.

3 Fahrenholtz, W.G., Hilmas, G.E., Talmy, I.G., and Zaykoski, J.A. (2007). Refractory diborides of zirconium and hafnium. *J. Am. Ceram. Soc.* 90: 1347–1364.

4 Gupta, N., Mukhopadhyay, A., Pavani, K., and Basu, B. (2012). Spark plasma sintering of novel ZrB_2–SiC–$TiSi_2$ composites with better mechanical properties. *Mater. Sci. Eng. A* 534: 111–118.

5 Nisar, A., Ariharan, S., Venkateswaran, T. et al. (2017). Effect of carbon nanotube on processing, microstructural, mechanical and ablation behavior of ZrB_2-20SiC based ultra-high temperature ceramic composites. *Carbon* 111: 269–282.

6 Venkateswaran, T., Basu, B., Raju, G.B., and Kim, D.Y. (2006). Densification and properties of transition metal borides-based cermets via spark plasma sintering. *J. Eur. Ceram. Soc.* 26: 2431–2440.

7 Rezaie, A., Fahrenholtz, W.G., and Hilmas, G.E. (2007). Effect of hot pressing time and temperature on the microstructure and mechanical properties of ZrB_2–SiC. *J. Mater. Sci.* 42: 2735–2744.

8 Zhang, G.J., Deng, Z.Y., Kondo, N. et al. (2000). Reactive hot pressing of ZrB_2–SiC composites. *J. Am. Ceram. Soc.* 83: 2330–2332.

9 Peng, F. and Speyer, R.F. (2008). Oxidation resistance of fully dense ZrB_2 with SiC, TaB_2, and $TaSi_2$ additives. *J. Am. Ceram. Soc.* 91: 1489–1494.

10 Medri, V., Monteverde, F., Balbo, A., and Bellosi, A. (2005). Comparison of ZrB_2-ZrC-SiC composites fabricated by spark plasma sintering and hot-pressing. *Adv. Eng. Mater.* 7: 159–163.

11 Zhang, X., Xu, L., Du, S. et al. (2008). Spark plasma sintering and hot pressing of ZrB_2–SiC_W ultra-high temperature ceramics. *J. Alloys Comp.* 466: 241–245.

12 Purwar, A., Mukherjee, R., Ravikumar, K. et al. (2016). Development of ZrB_2–SiC–Ti by multi stage spark plasma sintering at 1600 °C. *J. Ceram. Soc. Jpn.* 124: 393–402.

13 Mukhopadhyay, A., Basu, B., Bakshi, S.D., and Mishra, S.K. (2007). Pressureless sintering of ZrO_2–ZrB_2 composites: microstructure and properties. *Int. J. Refract. Met. Hard Mater.* 25: 179–188.

14 Sharma, S.K., Kumar, B.V., Lim, K.Y. et al. (2014). Erosion behavior of SiC–WC composites. *Ceram. Int.* 40: 6829–6839.

15 Shipway, P.H. and Hutchings, I.M. (1996). The role of particle properties in the erosion of brittle materials. *Wear* 193: 105–113.

16 Routbort, J.L., Scattergood, R.O., and Turner, A.P. (1980). The erosion of reaction-bonded SiC. *Wear* 59: 363–375.

17 Jianxin, D., Lili, L., and Mingwei, D. (2007). Erosion wear behaviours of SiC/(W, Ti)C laminated ceramic nozzles in dry sand blasting processes. *Mater. Sci. Eng. A* 444: 120–129.

18 Routbort, J.L. and Matzke, H. (1983). On the correlation between solid-particle erosion and fracture parameters in SiC. *J. Mater. Sci.* 18: 1491–1496.

19 Wang, D.F., She, J.H., and Ma, Z.Y. (1995). Effect of microstructure on erosive wear behavior of SiC ceramics. *Wear* 180: 35–41.

20 Wiederhorn, S.M. and Hockey, B.J. (1983). Effect of material parameters on the erosion resistance of brittle materials. *J. Mater. Sci.* 18: 766–780.

21 Kim, J.J. and Park, S.K. (1998). Solid particle erosion of SiC and SiC–TiB_2 composite hot-pressed with Y_2O_3. *Wear* 222: 114–119.

22 Purwar, A., Venkateshwaran, T., and Basu, B. (2017). Experimental and computational analysis of thermo-oxidative-structural stability of ZrB_2-SiC-Ti during arc-jet testing. *J. Am. Ceram. Soc.* 100: 4860–4873.

23 Purwar, A. and Basu, B. (2017). Thermo-structural design of ZrB_2–SiC-based thermal protection system for hypersonic space vehicles. *J. Am. Ceram. Soc.* 100: 1618–1633.

24 Jain, D., Reddy, K.M., Mukhopadhyay, A., and Basu, B. (2010). Achieving uniform microstructure and superior mechanical properties in ultrafine grained TiB_2–$TiSi_2$ composites using innovative multi stage spark plasma sintering. *Mater. Sci. Eng. A* 528: 200–207.

25 Pastewka, L., Moser, S., Gumbsch, P., and Moseler, M. Anisotropic mechanical amorphization drives wear in diamond. *Nat. Mater.* 10: 34–38.

26 Zykova-Timan, T., Ceresoli, D., and Tosatti, E. Peak effect versus skating in high-temperature nanofabrication. *Nature* 6: 230–234.

27 Tegen, I. and Fung, I. (1994). Modeling of mineral dust in the atmosphere: sources, transport, and optical thickness. *J. Geophys. Res. Atmos.* 99 ([D11]): 22897–22914.

28 Manoj Kumar, B.V., Basu, B., Kang, S., and Ramkumar, J. (2006). Erosion wear behavior of TiCN–Ni cermets containing secondary carbides (WC/NbC/TaC). *J. Am. Ceram. Soc.* 89: 3827–3831.

29 Kuryliuk, V., Nadtochiy, A., Korotchenkov, O. et al. (2015). A model for predicting the thermal conductivity of SiO_2–Ge nanoparticle composites. *Phys. Chem. Chem. Phys.* 17: 13429–13441.

30 Mistarihi, Q., Umer, M.A., Kim, J.H. et al. (2015). Fabrication of ZrO_2-based nanocomposites for transuranic element-burning inert matrix fuel. *Nucl. Eng. Tech.* 47: 617–623.

31 Monteverde, F. and Savino, R. (2007). Stability of ultra-high-temperature ZrB_2–SiC ceramics under simulated atmospheric re-entry conditions. *J. Eur. Ceram. Soc.* 27: 4797–4805.

32 Evans, A.G., Gulden, M.E., and Rosenblatt, M. (1978). Impact damage in brittle materials in the elastic-plastic response regime. *Proc. R. Soc. Lond. A* 361: 343–365.

33 Basu, B. and Kalin, M. (2011). *Tribology of Ceramics and Composites*. Hoboken, NJ: Wiley.

34 Purwar, A. and Mahapatra, D.R. (2016). Design of thermal barrier coating system for scramjet using coupled thermo-structural analysis. *Trans. India Ceram. Soc.* 75: 242–249.

35 Watts, J., Hilmas, G., Fahrenholtz, W.G. et al. (2010). Stress measurements in ZrB_2–SiC composites using Raman spectroscopy and neutron diffraction. *J. Eur. Ceram. Soc.* 30: 2165–2171.

15

Erosive Wear of WC-Co Coating

Erosive wear behavior depends on the type and hardness of erodent and on the angle of impingement. Both these aspects are addressed in the present case study on detonation sprayed WC-Co coatings. This chapter summarizes author's published results that the erosive wear rate of coatings increases with hardness of the erodent. More importantly, the dominant erosion mechanism involves plowing and subsurface cracking.

15.1 Introduction

Erosive wear is widely recognized as a major material removal mechanism for thermally sprayed coatings. The influence of type and extent of the binder, porosity, hardness, fracture toughness, and microstructure on the erosive wear behavior of WC-Co coatings has been investigated [1–13]. The use of Co binder, instead of Ni, resulted in better erosive wear performance [2]. For plasma-sprayed WC-Co coatings, Levy and Wang [1] commented on the role of porosity on erosive wear resistance. Wayne and Sampath [8] reported the erosive wear behavior of WC-12Co and WC-17Co coatings obtained by air plasma spray, vacuum plasma spray, and HVOF techniques.

> Co content, porosity, mean free path of the binder, and cuboid size are the important microstructural variables, which are reported to influence erosive wear in WC-Co system.

In an important study, Kim et al. [3] emphasized the role of the inter-splat bond strength, rather than porosity, hardness, surface roughness, extent of carbide retention, on the erosive wear of plasma-sprayed WC-12Co coatings.

Friction and Wear of Ceramics: Principles and Case Studies, First Edition.
Bikramjit Basu, Mitjan Kalin, and B. Venkata Manoj Kumar.

For HVOF-sprayed WC-Co coatings, Kulu et al. [4] commented on the role of hardness on the erosive wear behavior. Wang et al. [5] compared the erosive wear behavior of HVOF-sprayed WC-12Co coatings deposited using three types of feedstock (nanostructured, multimodal, and conventional). The WC-12Co coating with the nanostructure feedstock (50–800 nm) exhibited the lowest erosive wear rate, while the coating with conventional feedstock (2–3 μm) had the highest erosive wear rate. Hawthorne et al. [6] concluded that the coating composition and microstructural integrity are the main factors responsible for erosive wear resistance. Bonache et al. [7] reported higher erosive wear rate of air plasma-sprayed WC-12Co when Al_2O_3 was used as the erodent instead of SiO_2. A number of detonation-sprayed WC-Co coatings evaluated by Sundararajan et al. [13–15] indicate that porosity of less than 1% results in improved erosive wear performance. Some attempts have also been made to study the subsurface cracking behavior (the number, length, and distribution of transverse and parallel subsurface cracks) in HVOF-sprayed WC-Co-Cr coatings subjected to erosion and correlate the same with erosive wear resistance [16].

15.2 Materials and Erosion Experiments

Commercially available sintered and crushed WC-12 wt% Co powder, supplied by Sulzer Metco, USA, was used as the feedstock for the detonation spray coating. The WC-12Co coatings were deposited on grit-blasted mild steel substrate using the detonation spray process and the deposition conditions were so chosen that the coatings were formed under varying ratio of oxygen to acetylene (fuel), i.e. at three OF ratios of 1.16, 1.5, and 2.0, respectively.

The erosive wear behavior of the WC-12Co coatings with three OF ratios, bulk WC-12Co, and mild steel were evaluated using an air jet-type solid particle erosion rig and as described in a previous published report [17]. The erosion test parameters utilized in the present study are provided in Table 15.1.

Table 15.1 Erosion test parameters [17].

Erodents	SiO_2, Al_2O_3, and SiC
Average erodent size (μm)	147 (SiO_2), 206 (Al_2O_3), and 227 (SiC)
Erodent mass feed rate (g/min)	3.1 (SiO_2), 3.6 (Al_2O_3), and 3.8 (SiC)
Nozzle diameter (mm)	4
Stand-off distance (mm)	10
Erodent velocity (m/s)	25 and 45 independent of erodent particle
Impact angles (°)	30 and 90

Source: Reproduced with permission of Elsevier.

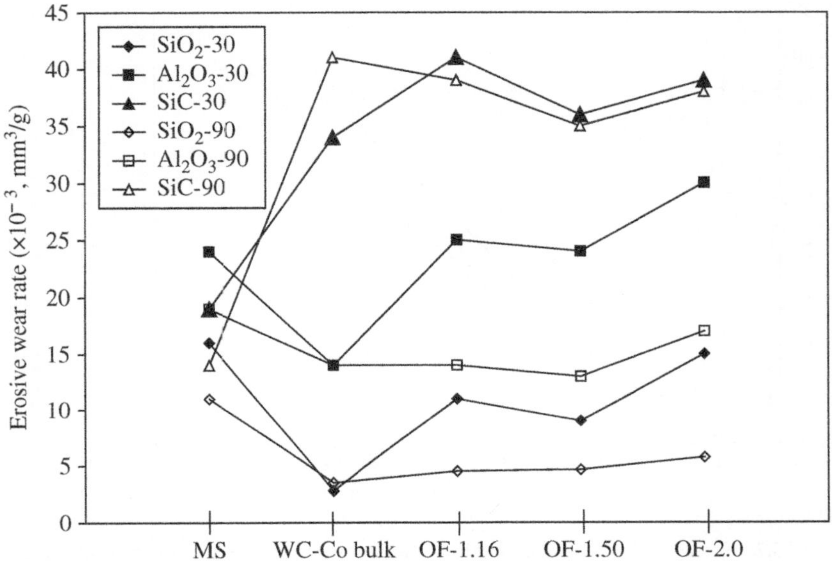

Figure 15.1 Steady-state erosive wear rate of mild steel, bulk WC-Co, and WC-Co coatings for three different types of erodents, which impact at 45 m/s speed at different angles [17]. *Source:* Reproduced with permission of Elsevier.

In particular, three different particles, i.e. SiO_2, Al_2O_3, and SiC, were used as the erodent. Erosion tests, irrespective of the erodents used, were carried out on two impact velocities (25 and 45 m/s) and two impact angles (30° and 90°) for each of the test material. The velocity of the erodent particles prior to their impact on the test sample was measured using a high-speed camera system in conjunction with a laser pulsing system (Spray Watch 2I Scientific, Oseir Ltd., France). It was ensured that the impact velocity was the same (25 and 45 m/s), irrespective of the erodent used by suitably altering the air pressure in the erosion rig.

The erosive wear rate of bulk WC-Co is relatively low compared with WC-Co coatings deposited at three OF ratios (see Figure 15.1). Among all the coatings studied, the coatings deposited at an OF ratio of 1.50 exhibited a relatively low erosive wear rate compared with the coatings deposited at other two OF ratios. The ratio of erosive wear rates at 30° and 90° impact angles (E30/E90), known as erosion ratio, is useful to understand the erosive wear behavior [18, 19]. It needs to be pointed out here that, in general, ductile metals exhibit a maximum erosive wear loss at an impact angle of 30°, while erosive wear rate steadily increases with impact angle to reach a maximum at normal impact (90°) in the case of brittle materials, like ceramics.

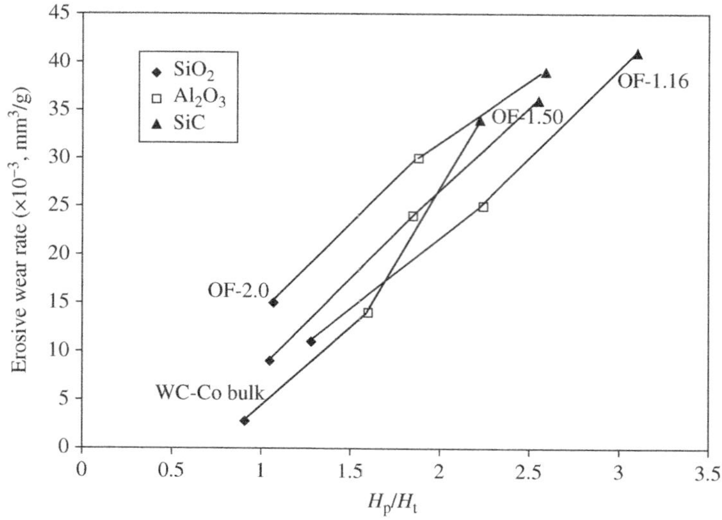

Figure 15.2 Erosive wear rate as a function of erodent-target hardness ratio at 30° impact angle and 45 m/s impact velocity [17]. *Source:* Reproduced with permission of Elsevier.

> The difference in erosive wear loss with impact angle can be attributed to the fundamental difference in the characteristics of erosive wear damage behavior – plastic deformation-dominated damage for metals and brittle fracture-induced damage for ceramics.

Therefore, the value E30/E90 < 1 indicates the brittle nature, while E30/E90 values > 1 indicate the ductile nature and E30/E90 ≅ 1 illustrates the mixed-mode (brittle + ductile) behavior.

The hardness ratio of eroding particle (H_p) and the target (H_t) is also an important parameter that can significantly influence the erosive wear behavior of materials [20, 21]. The erosive wear rate of bulk WC-Co and WC-Co coatings was plotted against H_p/H_t in Figure 15.2. The erosive wear rate of coatings increased with increasing hardness of the erodent. Figure 15.2 clearly illustrates that the H_p/H_t ratio is a critical parameter, capable of explaining the changes in the relative wear rates of different materials. It is clear from Figure 15.2 that an increase in H_p/H_t ratio from 1.0 to 2.5 results in an increase in wear rate by 3.5–4.0 times.

15.3 Erosive Wear Mechanisms (Surface Damage)

In order to understand the erosive wear data presented so far in more detail, SEM analysis of the worn surfaces was carried out. SEM images of eroded surfaces of bulk and coatings using three erodents and two impact angles are shown in Figures 15.3–15.6.

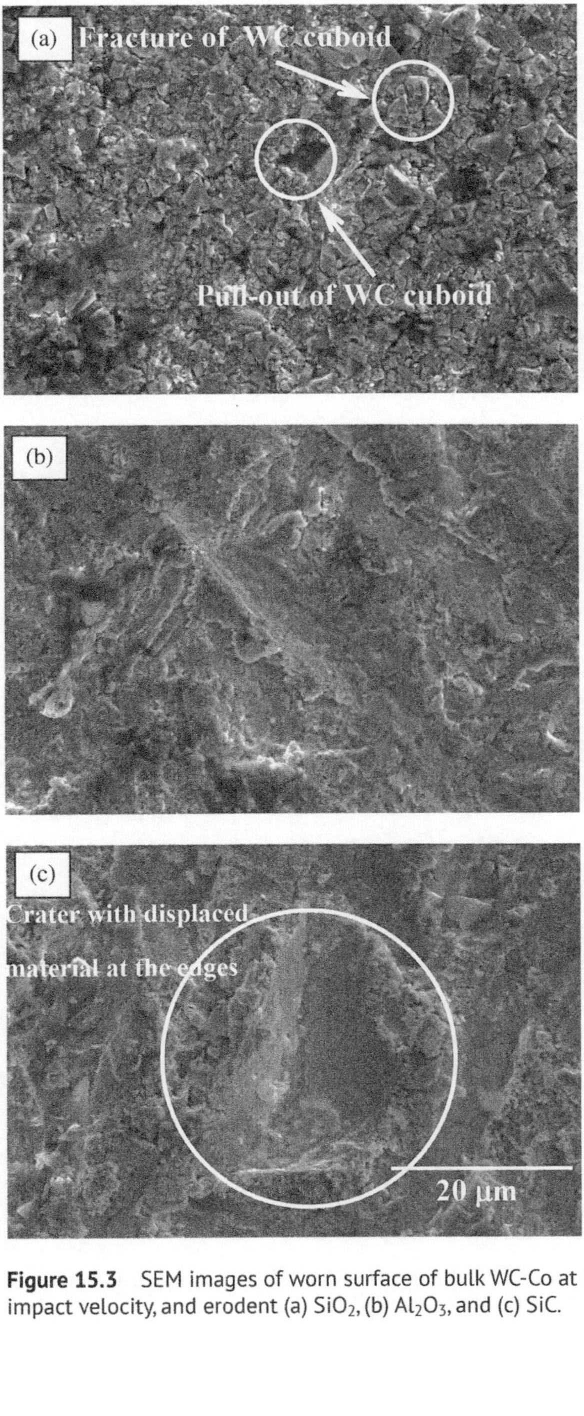

Figure 15.3 SEM images of worn surface of bulk WC-Co at 30° impact angle, 45 m/s impact velocity, and erodent (a) SiO_2, (b) Al_2O_3, and (c) SiC.

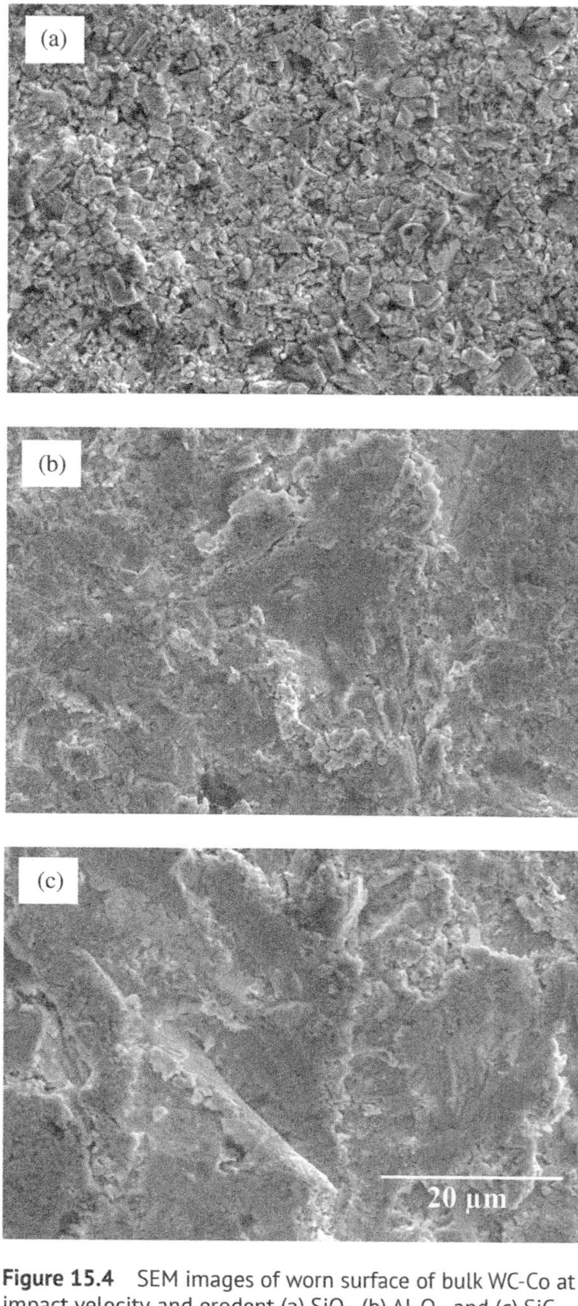

Figure 15.4 SEM images of worn surface of bulk WC-Co at 90° impact angle, 45 m/s impact velocity, and erodent (a) SiO_2, (b) Al_2O_3, and (c) SiC.

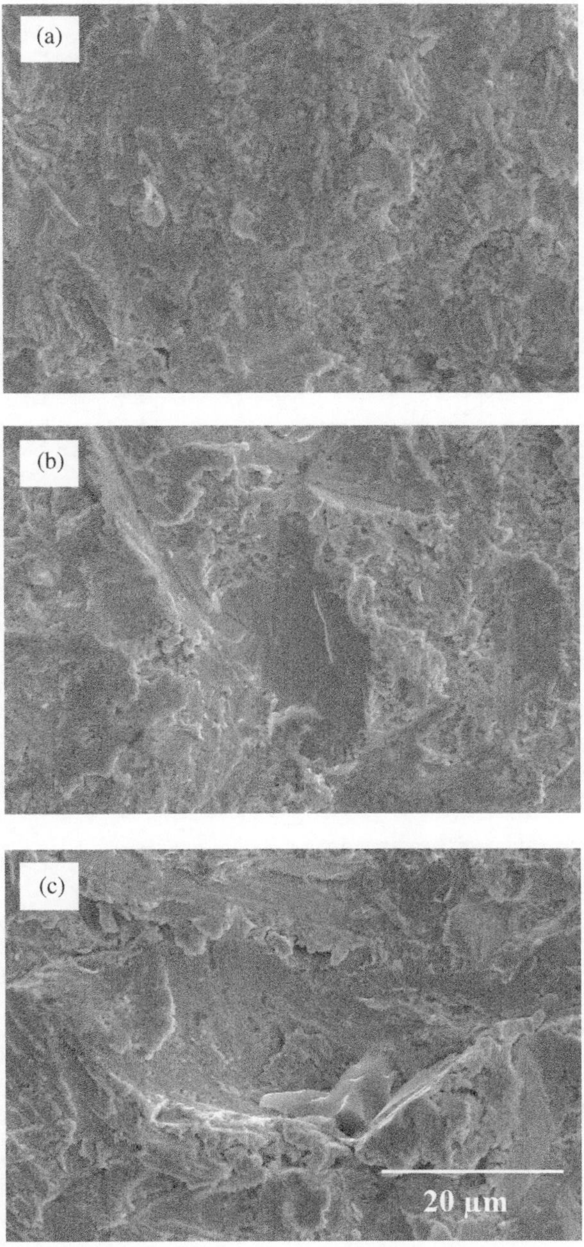

Figure 15.5 SEM images of worn surface of WC-Co coatings deposited at an OF ratio of 1.50 at 30° impact angle, 45 m/s impact velocity, and erodent (a) SiO_2, (b) Al_2O_3, and (c) SiC.

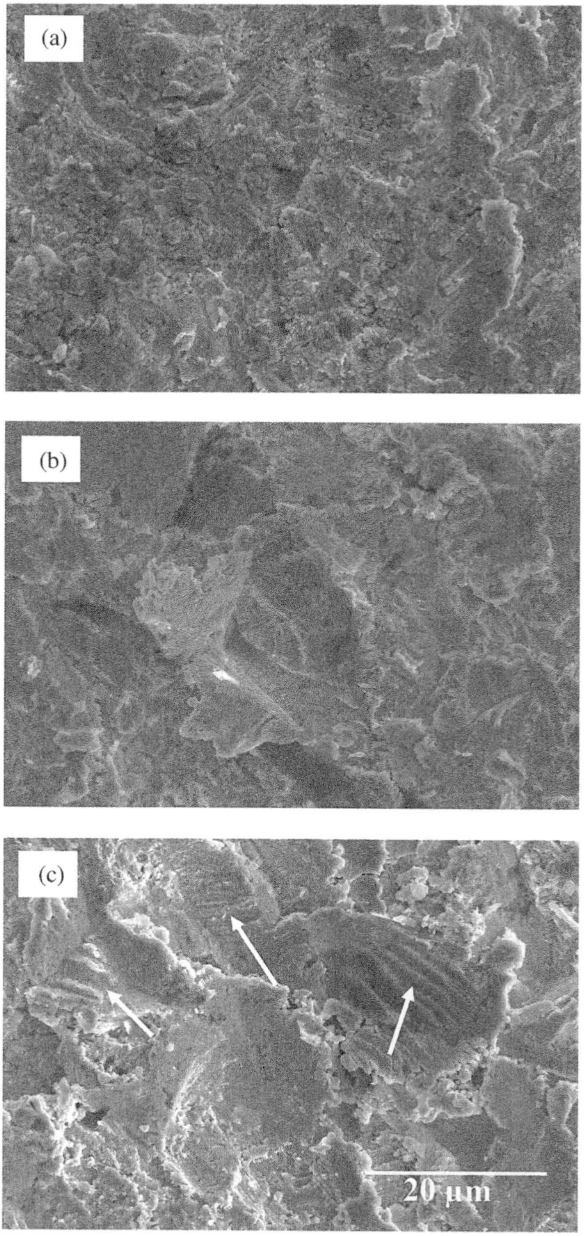

Figure 15.6 SEM images of worn surface of WC-Co coatings deposited at an OF ratio of 1.50 at 90° impact angle, 45 m/s impact velocity, and erodent (a) SiO_2, (b) Al_2O_3, and (c) SiC. The arrows in (c) are evidence of serrations, caused by extensive deformation during erosion impacts.

For bulk WC-Co cermets, eroded with SiO_2 erodent, cracking and pull-out of WC cuboids, leaving behind the eroded surface covered with projected carbides after selective removal of the cobalt matrix, are evident from Figure 15.3. The worn surface of bulk WC-Co cermets after eroded with Al_2O_3 and SiC exhibited the formation of deep craters with displaced material under the impact. The severity of the crater size/displacement increased from Al_2O_3 to SiC erodent due to an increase in erodent hardness, as evident from Figures 15.3 and 15.4. It is also clear from the worn surface images that there is no appreciable difference in the wear mechanism at two widely different impact angles.

Microcutting, plowing, and chipping are the major material removal mechanisms observed in the thermal-sprayed coatings [22]. In the present study, the coatings, deposited at three OF ratios, exhibited chipping, microcutting, plowing, and deformation as the primary material removal mechanisms (Figures 15.5 and 15.6). The material removal was associated with severe deformation resulting in significant spalling, as evident from Figures 15.5 and 15.6. In the case of SiO_2 erodent, the deformation and craters are not visible as the material removal occurs via minor chipping. In addition, the selective removal of cobalt phase and subsequently the fracture and pull-out of WC cuboids were noticed for coatings when eroded with SiO_2. A large-scale deformation on the surface was observed in the coatings, eroded with Al_2O_3 and such behavior is more severe in the case of coatings, eroded with SiC.

> The feature that is distinctly noticeable is that all the coatings without any exception, when eroded with Al_2O_3 and SiC, exhibited a most common behavior of severe plastic deformation resulting in crater/lip formation at both the oblique and normal impact angles.

The relatively higher hardness of Al_2O_3 and SiC erodent over SiO_2 erodent could explain the difference in such behavior. The crater depth is found to increase with increasing particle hardness from Al_2O_3 to SiC.

15.4 Erosive Wear Mechanisms (Subsurface Damage)

SEM analysis of the cross-section of the bulk WC-Co and coatings was also carried out for further understanding about the subsurface material removal mechanisms in the present study and are shown in Figures 15.7–15.9. In case of bulk WC-Co cermets (Figure 15.7), no subsurface crack was noticed, irrespective of erodent type. However, the extent of damaged region was increased in the order $SiO_2 > Al_2O_3 > SiC$. Similarly, the subsurface damage was largely confined to the surface region for the coatings deposited at three OF ratios and eroded with SiO_2 (Figures 15.7–15.9). However, in the case of coatings eroded with Al_2O_3 and SiC, the subsurface damage is quite extensive compared to the subsurface damage

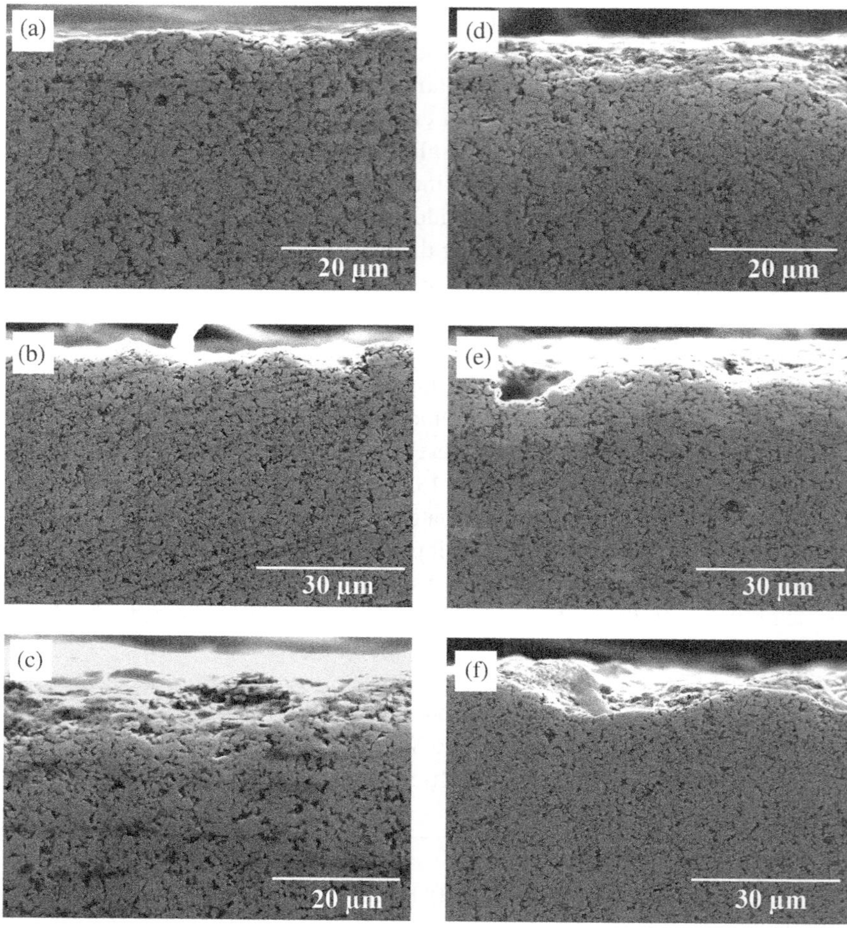

Figure 15.7 Cross-sectional SEM images showing subsurface damage behavior of bulk WC-Co under the erosive wear scars at 30° impact angle (a), (b), (c) and 90° impact angle (d), (e), (f) after tested with SiO_2, Al_2O_3, SiC erodents, respectively. The impact velocity is 45 m/s.

imparted by the SiO_2 erodent (Figures 15.8 and 15.9). In addition to higher subsurface damage, the subsurface cracks were also found in the coatings, deposited at an OF ratio of 2.0 and eroded with Al_2O_3 and SiC (Figure 15.9). This damaged region was extended to around 40–50 μm from the surface, while such subsurface damage region is confined to 5–10 μm in the case of coatings, deposited at an OF ratio of 2.0 and eroded by SiO_2. The crack formation and propagation path were further analyzed by observing the eroded cross-section at higher magnifications under SEM as shown in Figure 15.10. The cracks were found to have propagated

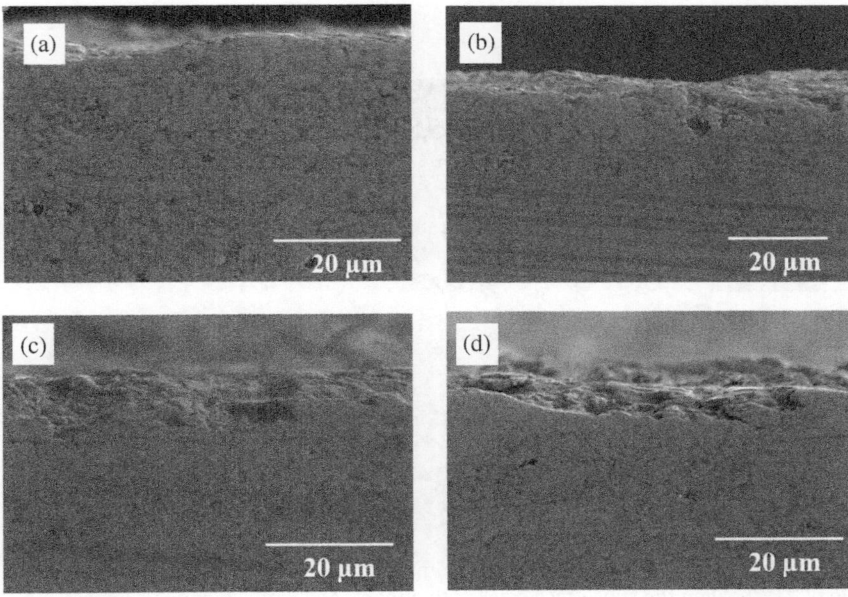

Figure 15.8 Cross-sectional SEM images showing subsurface damage behavior of WC-Co coatings deposited at an OF ratio of 1.50 under the erosive wear scars at 30° impact angle (a), and 90° impact angle (b), (c), (d) after tested with SiO_2, Al_2O_3, SiC erodents, respectively.

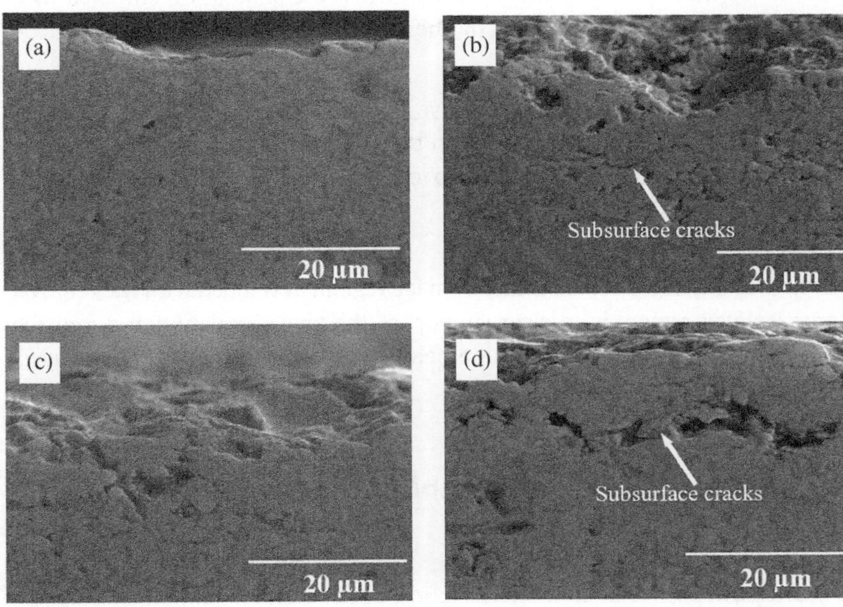

Figure 15.9 Cross-sectional SEM images showing subsurface damage behavior of WC-Co coatings deposited at an OF ratio of 2.0 under the erosive wear scars at 30° impact angle (a), (b), (c) and 90° impact angle (d) after tested with SiO_2, Al_2O_3, SiC erodents, respectively. The impact velocity is 45 m/s.

Figure 15.10 High-magnification BSE SEM image of WC-Co coating deposited at an OF ratio of 2.0 showing the subsurface crack path after eroded with SiC and 90° impact angle, 45 m/s impact velocity [17]. *Source:* Reproduced with permission of Elsevier.

parallel to the coating surface and were predominantly located at the splat boundaries and along the decomposition layer boundaries. In addition to the horizontal cracks, the vertical cracks were also noticed across the band of Co enriched with W and C. This band, being formed by the decomposition of WC cuboids in Co matrix, acts as a brittle region.

> The intersection of vertical cracks with the horizontal cracks creates an isolated region that would easily be removed during subsequent impact during the erosion process.

The higher thickness of subsurface damage region, developed with the removal of large isolated region, is in clear support of the above explanation. From the above observations, the decomposition behavior of WC-Co coatings also plays a vital role in judging the erosive wear rate.

15.5 Correlating Wear Mechanism with Erodent and Coating Properties

The coatings, deposited at all three OF ratios, exhibited microchipping as the primary material removal mechanism after being eroded with SiO_2. Additionally, the displacement of carbide, after each collision and gradual removal of the Co matrix, results in projection of carbides on the entire worn surface. These carbides will fracture and spall from the surface during repeated impacts. However, in the case of coatings deposited at an OF ratio of 2.0, where the volume fraction of the matrix is high due to dissolution of W and C in the Co matrix, plowing in the matrix region was noticed after eroded with SiO_2. The plowing, microcutting, and deformation are the mechanisms involved in the case of coatings eroded with Al_2O_3. The severity of the above material removal mechanisms is high in the case of coatings, eroded with SiC. The highest wear loss in the WC-Co coatings, deposited at an OF ratio of 2.0, is due to the removal of large chunks of material after formation of prolonged subsurface cracks. Highest decomposition as observed in the WC-Co coatings deposited at an OF ratio of 2.0 results in dissolution of considerable amounts of W and C in Co matrix, converting the soft binder phase to be hard and brittle compared with Co binder (without any dissolved W or C). Such brittle phase at weekly bonded inter-splat boundaries leads to a network of subsurface cracks. Therefore, high erosive wear rate can be rationalized for the WC-Co coatings, deposited at an OF ratio of 2.0.

The erosive wear rate increased with increase in the hardness of the erodents, as explained in Figure 15.2. In the case of SiO_2 erodent, the hardness is very close to the hardness of the coatings. The low hardness and low kinetic energy (4.4 μJ), calculated from the test parameters, are sufficient enough to generate the subsurface damage in the case of coatings, deposited at an OF ratio of 2.0. However, the subsurface damage or cracking is confined very close to surface only. In the case of Al_2O_3 erodent, high hardness and high kinetic energy (18 μJ) resulted in substantial penetration of erodent into the coating material, resulting in plowing, microcutting, and removal of the large amount of material. Thereby, substantial subsurface crack growth was extended up to a depth of 45 μm. High hardness and high kinetic energy in the case of Al_2O_3, however, is not sufficient enough to cause extensive subsurface damage in the case of coatings, deposited at OF ratios 1.16 and 1.50. In the case of SiC erodent, though the kinetic energy (15 μJ) is close to that of Al_2O_3, the cracking and delamination were also observed in the coatings deposited at an OF ratio 1.16, because of its high hardness. The difference in hardness of impacting particle along with kinetic energy explains the difference in wear behavior of WC-Co coatings.

Erosion efficiency (η) [23] can be used to characterize the nature and mechanism of erosive wear and is defined as the fraction of material actually removed as the erosion debris out of the total amount of displaced material:

$$\eta = \frac{2EH}{v^2} \tag{15.1}$$

where E is the erosive wear rate at normal impact angle in terms of volume material loss per unit mass of erodent, H is the target hardness, and v is the erodent velocity. The erosion efficiency is equal to zero, if there is no material removal and only displacement (i.e. ideal microplowing). In ductile materials, only a small fraction of displaced material is removed during each impact. Thus, the η value lies <5 in the case of ductile materials.

In the case of brittle materials, erosion occurs via spalling and due to intersection of lateral or radial cracks that result in the material removal in large volume.

Therefore, η is expected to be even greater than 100%. The η value around 500–2000% was noticed in the case of glass [23]. Erosion efficiency of the mild steel, WC-Co bulk, and detonation-sprayed WC-Co coatings was calculated and is listed in Table 15.2. It is clear from the table that η varies from 2.1 to 3.7 in the case of mild steel eroded with three different erosives. The η values of the bulk WC-Co and the three coatings obtained in the present study were in the identical range for a given erodent. The η values with SiO_2 erodent are 4.1–6.2, with Al_2O_3 erodent are 13–18.5, and with SiC erodent are 35–52. This correlates well with the hardness of the erodents, employed in the present study. These η values indicate that the coatings are having a mixed-mode of fracture behavior with larger dominance of "ductile-metal" like behavior. These values also suggest that the material

Table 15.2 Erosion efficiency of investigated materials [17].

Erodent type	Materials	η (%)
SiO_2	Bulk WC-Co	4.4
	MS	2.17
	OF-1.16	4.1
	OF-1.50	5.06
	OF-2.0	6.15
Al_2O_3	Bulk WC-Co	17.76
	MS	3.75
	OF-1.16	12.72
	OF-1.50	14.31
	OF-2.0	18.47
SiC	Bulk WC-Co	52
	MS	2.76
	OF-1.16	35.44
	OF-1.50	38.54
	OF-2.0	41.3

Source: Reproduced with permission of Elsevier.

removed is reasonably larger than the material displaced. However, the above range of η values (compared with η value of 500–2000% for glass) indicates there will be substantial material displacement as well. These observations are in clear support of SEM images, pertaining to worn surfaces, exhibiting lip formation/deformation and crack formation/crack propagation across the inter-splat boundaries. Hence, the η parameter values obtained in the present study commensurate well with the wear rate data and the associated wear rate mechanisms, illustrated through SEM examination of worn surfaces.

15.6 Closure

In summary, the erosive wear of WC-Co coatings is influenced by coatings' microstructure/property, impact angle, and erodent type. Low erosive wear rate is recorded with the coatings decarburized to around 34%.

> The surface and subsurface regions of eroded WC-Co coatings revealed that significant cracking at the splat boundaries resulted in high erosive wear rate for coatings deposited at higher OF ratio.

Also, wear loss is increased with increase in erodent hardness. Detailed understanding on the erosive wear of WC-Co coating is provided in author's published report [17].

References

1 Levy, A.V. and Wang, B. (1988). Erosion of hard material coating systems. *Wear* 121: 325–346.

2 Berger, L.M., Saaro, S., Naumann, T. et al. (2008). Microstructure and properties of HVOF-sprayed chromium alloyed WC-Co and WC-Ni coatings. *Surf. Coat. Technol.* 202: 4417–4421.

3 Kim, H.J., Kweon, Y.G., and Chang, R.W. (1994). Wear and erosion behavior of plasma-sprayed WC-Co coatings. *J. Therm. Spray Technol.* 3: 169–178.

4 Kulu, P., Hussainova, I., and Veinthal, R. (2005). Solid particle erosion of thermal sprayed coatings. *Wear* 258: 488–496.

5 Wang, Q., Chen, Z.H., Ding, Z.X., and Liu, Z.L. (2008). Performance study of abrasive wear and erosive wear of WC-Co coatings sprayed by HVOF. In: *Proceedings: 2nd IEEE International Nanoelectronics Conference (INEC 2008)*, Shanghai, China (24–27 March 2008), 340–344. IEEE.

6 Hawthorne, H.M., Arsenault, B., Immarigeon, J.P. et al. (1999). Comparison of slurry and dry erosion behaviour of some HVOF thermal sprayed coatings. *Wear* 225–229: 825–834.

7 Bonache, V., Salvador, M.D., García, J.C. et al. (2010). Study of erosion behavior of conventional and nanostructured WC-12Co coatings sprayed by atmospheric plasma. *Key Eng. Mater.* 423: 35–40.
8 Wayne, S.F. and Sampath, S. (1992). Structure/property relationships in sintered and thermal sprayed WC-Co. *J. Therm. Spray Technol.* 1: 307–316.
9 De Villiers Lovelock, H.L. and Vanwyk, P. (1998). Effect of powder type and composition on the erosion and abrasion of HP/HVOF deposited WC-Co coatings. In: *Proceedings: 15th International Thermal Spray Conference*, Nice, France (25–29 May 1998), 153–158. ASM International.
10 Wood, R.J.K., Mellor, B.G., and Binfield, M.L. (1997). Sand erosion performance of detonation sprayed tungsten carbide/cobalt-chromium coatings. *Wear* 211: 70–83.
11 Wood, R.J.K. (2010). Tribology of thermal sprayed WC-Co coatings. *Int. J. Refract. Met. Hard Mater.* 28: 82–94.
12 Suresh Babu, P., Basu, B., and Sundararajan, G. (2008). Processing–structure–property correlation and decarburization phenomenon in detonation sprayed WC–12Co coatings. *Acta Mater.* 56: 5012–5026.
13 Babu, P.S., Rao, D.S., Rao, G.V.N., and Sundararajan, G. (2007). Effect of feedstock size and its distribution on the properties of detonation sprayed coatings. *J. Therm. Spray Technol.* 16: 281–290.
14 Sundararajan, G., Sen, D., and Sivakumar, G. (2005). The tribological behaviour of detonation sprayed coatings: the importance of coating process parameters. *Wear* 258: 377–391.
15 Sundararajan, G. and Suresh Babu, P. (2009). Detonation sprayed WC-Co coatings: unique aspects of their structure and mechanical behavior. *Trans. Indian Inst. Met.* 62: 95–103.
16 Barber, J., Mellor, B.G., and Wood, R.J.K. (2005). The development of sub-surface damage during high energy solid particle erosion of a thermally sprayed WC-Co-Cr coatings. *Wear* 259: 125–134.
17 Babu, P.S., Basu, B., and Sundararajan, G. (2011). The influence of erodent hardness on the erosion behavior of detonation sprayed WC-12Co coatings. *Wear* 270: 903–913.
18 Bhushan, B. (1999). *Principles and Applications of Tribology*. New York: Wiley.
19 Hutchings, I.M. (1992). *Tribology: Friction and Wear of Engineering Materials*. Burlington, MA: Butterworth, Heinemann Publication.
20 Feng, Z. and Ball, A. (1999). The erosion of four materials and seven erodents – towards an understanding. *Wear* 233–235: 674–684.
21 Hussainova, I., Kubarsepp, J., and Pirso, J. (2001). Mechanical properties and features of erosion of cermets. *Wear* 250: 818–825.
22 Kingswell, R., Rickerby, D.S., Scott, K.J., and Bull, S.J. (1991). Comparison of erosion wear behavior of vacuum plasma sprayed and bulk alumina. In: *Thermal Spray Research and Applications* (ed. T.T. Berneda), 179–185. ASM International.
23 Sundararajan, G., Roy, M., and Venkataraman, B. (1990). Erosion efficiency – a new parameter to characterize the dominant erosion micromechanism. *Wear* 140: 369–338.

Section V

Case Study: Machining-Induced Wear of Cermets

16

Crater Wear of TiCN Cermets in Conventional Machining

In a number of chapters, the published results correlating the wear resistance with ceramic microstructure are presented on the basis of tribological testing on lab-scale coupons. Considering the significant application potential in machining, this chapter presents the summary of the published study from author's group to highlight the machining-induced damage of a series of TiCN-Ni cermets with various combinations of secondary carbides. A major emphasis has been placed on discussing the competition between abrasion/adhesion and chemical wear. Importantly, a comparison will be made to assess the performance among conventional and spark plasma sintered cermets with the commercial cermet.

16.1 Introduction

The tool inserts with superior performance are of high demand in manufacturing of an engineering component. Online monitoring of cutting forces is recommended for continuous evaluation of tool performance and for avoiding product rejection. Owing to the unique combination of high hardness, superior wear resistance, and good surface finish, TiCN-based cermets are considered as new-generation cutting tool inserts. In particular, TiCN-based cermets are reported to be promising for high-speed machining and finishing operations on difficult-to-machine materials [1–3]. In general, powder mixtures of TiCN and low melting point binder materials (like Co, Ni, Mo, etc.) are sintered in a vacuum or inert atmosphere by liquid-phase-assisted sintering. The dissolution and reprecipitation process in liquid-phase sintering results in a typical core-rim-binder structure [4, 5]. The sintered cermets of hard TiCN grains and soft binder phases provide combination of good wear resistance and toughness.

Friction and Wear of Ceramics: Principles and Case Studies, First Edition.
Bikramjit Basu, Mitjan Kalin, and B. Venkata Manoj Kumar.

Furthermore, transition metal carbides are generally added to improve mechanical and thermal properties of TiCN cermets [6–11]. In particular, the addition of NbC or TaC is reported to enhance hot hardness and thermal shock resistance and thereby improving the performance in interrupted cutting operations. The addition of VC is reported to enhance fatigue strength of TiCN cermets [8–11]. On the other hand, the addition of ZrC or HfC results in reduction of strain misfit between different phases [11]. Also, WC addition results in improved densification in sintering and fracture toughness of sintered TiCN cermets [12–15]. It is to note that sintering and microstructural studies of TiCN-based cermet materials have been reported in a considerable extent, but reports on understanding the wear behavior in extreme tribological conditions are limited. In this regard, performance of TiCN-Ni-based cermets in various tribological conditions [15–18] has been studied in our research group.

Regarding performance assessment of TiCN-based cermet tool inserts, limited research studies are available on material removal mechanisms in machining [13, 19]. Such investigations on the machining performance assessment and its relation with wear mechanisms are of primary concern. In this context, TiCN-based cermets were studied for their machining performance and related with their microstructure and wear mechanisms in the present chapter. The first part of the present chapter reports the effect of WC addition on tool performance of TiCN-20 wt%Ni-based cermets while machining boiler-grade steel in dry orthogonal cutting conditions [20]. In the second part of the chapter, the machining performance of Ti(CN)-5 wt%WC-20 wt%Ni and Ti(CN)-5 wt%WC-10 wt%Ni-10 wt%Co-5 wt%TaC cermets is compared with that of the commercially available carbide tool [21]. The major material removal mechanisms of rake face of the tools are elucidated.

16.2 TiCN Cermets and Machining Conditions

Two sets of cermets were used as cutting tool inserts. The first set, i.e. TiCN-20 wt.% Ni-(0–15 wt%)WC cermets, was sintered at 1510 °C for one hour in a vacuum. The cermets were used to machine a boiler-grade A192 steel pipe in orthogonal cutting test conditions. A piezoelectric dynamometer was used to monitor the cutting forces during machining.

The second set of cermets, Ti(CN)-5 wt%WC-20 wt%Ni and Ti(CN)-5 wt%WC-10 wt%Ni-10 wt%Co-5 wt%TaC, was conventionally sintered at 1550 °C in flowing argon for two hours, and spark plasma sintered in argon at 70 MPa and 1400 °C for three minutes. The second set of cermets was subjected to dry turning at 133 rpm for three minutes or 435 rpm for one minute against a stainless steel workpiece. Turning operations were performed three to five times for 0.9 mm depth of cut at 0.5 mm/rev feed rate. Cutting force was studied as a function of cermet composition, cermet processing type, and cutting speed. The important mechanisms of material removal from rake face of the TiCN-based tools and commercial tool are elucidated using SEM-EDS analysis.

16.3 Wear Mechanisms of TiCN-WC-Ni Cermets

In order to assess the tool life, continuous monitoring of cutting force was done and results in terms of cutting force versus machining time for the investigated cermets are presented in Figure 16.1. Overall, a three-stage behavior is observed for the given TiCN-Ni-based cermets. The sharp cutting edge of the tool is rapidly broken down and a finite wear land is established in the first stage. Also, almost negligible contact area in the initial stage causes high normal pressures and leads to subsurface plastic flow. The flank surface temperature becomes sufficiently high when the contact area reaches a critical value and that results in the catastrophic failure of the tool. The subsurface plastic flow is again possible in these conditions due to sufficient normal pressures. In the later stage, the linear curve

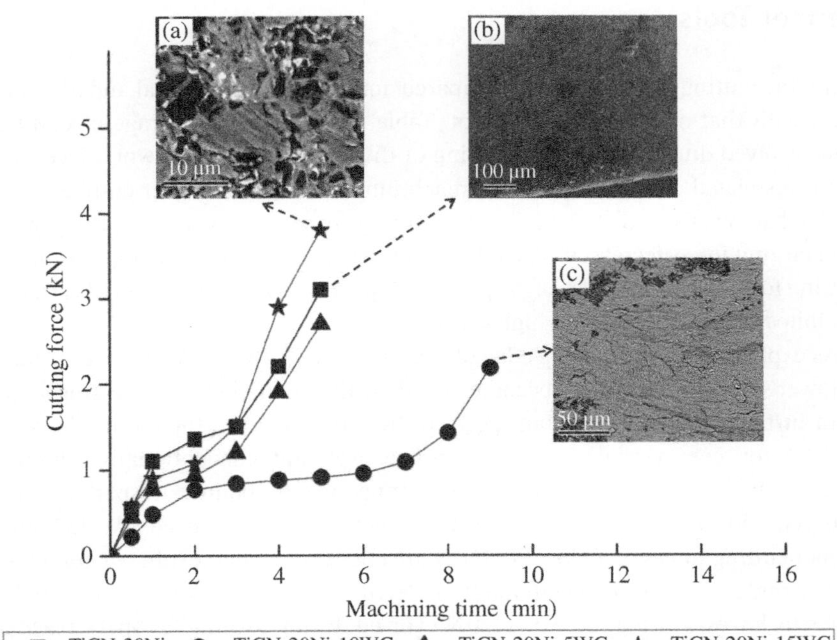

Figure 16.1 Cutting force versus machining time for the investigated cermets. Insets (a–c) represent typical SEM images of worn surfaces of respective cermets [20]. *Source:* Reproduced with permission of Elsevier.

with uniformly increased cutting force is observed. This particular stage of form-stability determines the tool life [22]. Among the investigated cermets, TiCN-20 wt%Ni-10 wt%WC cermet yields maximum tool life and minimum cutting force.

Typical SEM images of the investigated cermets post machining are presented as insets in Figure 16.1. The baseline TiCN-20 wt%Ni cermet tool exhibits severe damage and distinctive debris near cutting edge of rake face (see inset a in Figure 16.1). With the addition of WC, the tools of TiCN-Ni cermets show certain differences on worn surfaces. Overall, the adhesion of a layer is observed in all cermets. With the addition of 10 wt%WC in TiCN-20 wt%Ni cermet, the worn surface shows noticeably contrast tribolayer (see inset b in Figure 16.1). This layer reduces the tool wear and cutting forces and results in improved performance. As per the EDS analysis, tribolayer is found to be rich with iron and oxygen (not shown in the figure). Further, the severe pull-outs are found on the worn surface of TiCN-20 wt%Ni-15 wt%WC cermet (see inset c in Figure 16.1). The extensive intergranular fracture results in severe wear, which agrees with higher cutting force.

16.4 Machining with TiCN-WC-TaC-Ni-Co Cermet Tools

Here, the cutting performance is compared among the conventional and SPS cermets with that of commercial cermet. Table 16.1 shows the average of cutting force evolved during orthogonal cutting of the 304 stainless steel workpieces for the investigated cermets in selected machining operations. Higher cutting force evolved at lower speed (133 rpm) and as cutting speed increases (435 rpm), generated cutting force decreased. The effect of cermet processing and composition on cutting force is significant at high speed (435 rpm). Spark plasma sintered cermets exhibited less cutting force at high speed.

As explained earlier, cutting edge of the tool plows onto the workpiece surface at lower speed, while cutting becomes steady with a reduction in the cutting force with further increase in cutting speed. In the present study, the cutting force is highest for commercial cemented carbide tool and conventionally sintered Ti(CN)-5WC-20Ni cermet, while lowest cutting force is obtained in spark plasma sintered Ti(CN)-5WC-10Ni-10Co-5TaC cermet at a given speed. At 133 rpm, higher cutting forces of 1270 and 1250 N are obtained during turning of commercial carbide tool and conventionally sintered Ti(CN)-5WC-20Ni cermet tools, whereas lowest cutting force of 1210 N evolved during turning for spark plasma sintered TiCN-10Ni-10Co-5WC-5TaC cermet (Table 16.1). Compared with the cutting forces at 135 rpm, less cutting forces are evolved at 435 rpm. At 435 rpm, higher cutting forces of 1065 and 1060 N evolved during turning for commercial carbide tool and conventionally sintered Ti(CN)-5WC-20Ni cermet, whereas

Table 16.1 Machining (turning) parameters and generation of cutting force [21].

		Stainless steel rod (304 grade) (25 mm diameter) at 0.9 mm depth of cut, 0.5 mm/rev feed rate					
Cermet	Processing technique	d_i (mm)	d_f (mm)	N rpm, for 180 s	F_c, N at 133 rpm	N rpm, for 60 s	F_c, N at 435 rpm
TiCN-20Ni-5WC	Conventional	24.9	24	133	1250 ± 15	435	1060 ± 10
TiCN-10Ni-10Co-5WC-5TaC		24.9	24	133	1220 ± 10	435	1020 ± 5
TiCN-20Ni-5WC	SPS	24.9	24	133	1245 ± 15	435	1040 ± 10
TiCN-10Ni-10Co-5WC-5TaC		24.9	24	133	1210 ± 10	435	990 ± 5
Commercial carbide tool	Not known	24.9	24	133	1270 ± 15	435	1065 ± 10

Source: Reproduced with permission of FLOGEN STARTS OUTREACH.

lowest cutting force of 990 N is evolved during turning for spark plasma sintered TiCN-10Ni-10Co-5WC-5TaC cermet (Table 16.1). At higher speed, spark plasma sintered TaC added TiCN-WC-Ni/Co cermet tool is capable of reducing the cutting force significantly. Improved mechanical properties are attributed for spark plasma sintered TiCN-10Ni-10Co-5WC-5TaC cermet tool. Refined carbide size is attributed to improved mechanical properties with the addition of TaC in TiCN-WC-Ni/Co cermet. Further, it is reported that TaC addition improves thermal shock resistance and hot hardness, and thereby enhances the performance of cermets in interrupted cutting operations [9, 10]. Among various transition metal carbides, TaC exhibits consistent mechanical behavior and better crack propagation resistance [10, 23, 24]. Thus, the addition of TaC improves the cutting performance of TiCN-WC-Ni/Co cermet.

After turning at 133 and 435 rpm, worn tool edges were examined using scanning electron microscope and typical images of worn rake surfaces are shown in Figure 16.2a–h. The worn tool surface of conventionally sintered Ti(CN)-5WC-20Ni cermet shows cracks, grain pull-out, and fracture (Figure 16.2a) at 133 rpm and an increase in the intensity of grain pull-out, cracking, and fracture at 435 rpm (Figure 16.2b).

The hard contact asperities in machining generate cracks and subsequent grain pull-outs and fracture in cermet tools. Friction-induced heat at the nose is also high at high speed. Thermal cracking is therefore highly possible at high speeds. Worn tool surface of spark plasma sintered Ti(CN)-5WC-20Ni cermet shows abrasive tracks over most of the contact length and grain pull-out with the presence of shallow grooves at lower speed of 133 rpm (Figure 16.2c) and grain pull-out in clusters and abrasion at higher speed of 435 rpm (Figure 16.2d). The fractured material is prone to oxidation at contact temperatures. Thus, oxides of constituents of cermet or steel are likely to evolve at the contact surface. Harder tungsten oxide formed during machining plow onto the workpiece. Abrasive ridges or grooves formed in Figure 16.2d are attributed to high stresses at the chip–tool interface in machining. Grooves were also reported on the surface of Al_2O_3–TiCN tool during machining hardened steel [25]. The conventionally sintered Ti(CN)-5WC-10Ni-10Co-5TaC cermet tool worn surface shows shallow grooves and grain pull-out at 133 rpm (not shown as figure) while increased grain pull-out regions are observed at 435 rpm (Figure 16.2e). The SPSed Ti(CN)-5WC-10Ni-10Co-5TaC cermet tool worn surface revealed grain pull-out on the tool face and discontinuous adhesion beneath the tool face at 133 rpm (not shown as figure), while grain pull-out and continuous adhesion beneath the tool face are observed at 435 rpm (Figure 16.2f). The adhesion is believed to occur primarily due to the transfer of material through the diffusion process during machining. The presence of adhered layer underneath the tool face is said to protect the tool from further damage by continuous rubbing with workpiece and chips. The TaC added TiCN-WC-Ni/Co

Figure 16.2 SEM images of worn tool surfaces of (a, b) Ti(CN)-5WC-20Ni cermet processed via conventional sintering at 133 and 435 rpm, (c, d) Ti(CN)-5WC-20Ni cermet processed via SPS at 133 and 435 rpm, (e, f) Ti(CN)-5WC-10Ni-10Co-5TaC cermet processed via conventional sintering and SPS at 435 rpm, and (g, h) cemented carbide at 133 and 435 rpm. ["Use of figure 4 from the *Sustainable Industrial Processing Summit, Volume 8 Surfaces & Interfaces, Composite, Ceramic and Nanomaterials*. Edited by F. Kongoli, I. Braems, V. Demange, J.M. Dubois, M. Pech-Canul, C.L. Patino, O. Fumio, FLOGEN, 2017. pp. 210–222," under the permission of FLOGEN STARTS OUTREACH (www.flogen.org)].

cermet tool resulted in resistance of crack or fracture during machining. The characteristic wear mechanisms observed during turning of Ti(CN)-5WC-10Ni-10Co-5TaC cermet are grain pull-out whereas those observed in TiCN-20Ni-5WC cermet are grain pull-out, cracking, and fracture. Figure 16.2g and h are images of worn tool surfaces of commercial cemented carbide showing severe abrasion with grain pull-outs and cracking at 133 rpm, and grain pull-out and severe abrasion ridges at 435 rpm. The tendency for plowing into the workpiece results in severe abrasion during machining. Cutting forces generated at the nose interface are also high for commercial cemented carbide whereas less forces observed for the investigated cermets. Among the TiCN-based cermets, spark plasma sintered Ti(CN)-5WC-10Ni-10Co-5TaC is promising by having decreased severity in wear.

The presence of minor peaks of Ti, Wi, and Ta and dominant peaks of Fe and O in the EDS analysis of cermet tool surfaces after machining (not shown in the figure) suggest the material transfer from workpiece and its oxidation. Typical images of chips (not shown as figure) collected after turning operation shows that at lower cutting speed of 133 rpm, the width of chip is higher than that obtained at 435 rpm. It is possible that the tool plows onto the workpiece to a large extent at low cutting speed resulting in increased width.

The present study essentially indicates superior performance of the TiCN-based cermet tools than commercial cemented carbide tools. Based on the results obtained in machining test, it can be safely said that spark plasma sintered Ti(CN)-5WC-10Ni-10Co-5TaC cermet is preferred for cutting tool material for machining hard steel.

16.5 Correlating Cermet Composition, Microstructure, and Wear During Machining

It is well documented that machining is a complicated material removal process with simultaneous occurrence of variety of mechanisms. Nevertheless, it can be understood from the present study that material loss in tool occurs by the dominant process of adhesion on both the clearance and rake faces, and by chipping the cutting edge. As discussed previously, the motion of the chip across the rake face causes friction and results in material loss, named as crater wear. Based on the results, it can be said that rake surfaces of TiCN-Ni-WC cermet tools are largely influenced by the abrasion and adhesion in machining of boiler-grade steel pipe. In other words, the material removal mechanisms for the investigated cermet tools can be categorized as mechanical wear and chemical wear. However, the extent of contribution from each mechanism toward overall material loss is determined by the WC amount in cermet composition.

Furthermore, the abrasion is reported to be the mechanically activated form of wear and depends largely on material properties. For example, the resistance against abrasion is increased with increase in hardness. This is also in line with the cutting force data (see Figure 16.1). While machining, hard wear particles or asperities develop variety of cracks that on further propagation and intersection results in pull-out of cermet grains. The large stress concentrations developed at the interface of tool and chip are responsible for the formation of grooves or abrasive ridges. The formation of abrasive ridges is also found on the surface of alumina-TiCN tool while machining of steel [25].

On the other hand, the adhesive wear process can be elucidated in two possible ways: (i) loose debris generation through stick-slip phenomenon and (ii) material transfer through diffusion. At the microscopic level, asperity junctions are pulled out of the contact surface and form a large amount of debris particles. In case of metals, adhered asperity junctions are likely to stick in initial stages, and subsequently crack along each surface and slips/rolls over either surface [26]. In the case of heterogeneous materials, poor bonding between different phases leads to severe adhesion and large amount of loose debris particles. However, only few appearances of debris particles on the tool surfaces suggest insignificant contribution from stick-slip phenomenon in the present study. Therefore, the adhesion observed in the investigated cermets is possible by material transfer through the diffusion process.

> The diffusion can be referred to as atomic transfer at asperity contacts of chip and tool interface promoted by high temperature and pressure during machining.

Accordingly, the diffusion-mediated tribochemical reactions are likely to take place at elevated temperatures. Trigger and Chao [27] also observed diffusion in adhesive wear on crater surfaces of cemented carbide tools. Also, severe oxidation of transferred material resulted in the formation of a tribolayer rich in iron and oxygen in the present study. It can be understood that chemical affinity between the cermet tool and the steel workpiece accelerates temperature at the chip–tool interface and results in the material transfer by diffusion and subsequent formation of tribolayer.

However, the nature or stability of the tribolayer is influenced by WC content in the TiCN-20Ni cermet. The material loss is less due to the less severity of oxidation compared with abrasion. On the other hand, the continuous rubbing results in the oxide layer removal and exposes fresh subsurface to wear. Thus, the layer formation rate has to be dominant than the layer removal rate for the cermet to achieve lower wear. At a given instant, the oxide layer must be intact at the surface to overcome the loss due to continuous rubbing. The worn surfaces of cermet tools revealed increased formation of tribolayer with an increase in WC content up to 10 wt% in the baseline TiCN-20Ni cermet. Accordingly, TiCN-20Ni-10WC cermet is dominantly protected by adhered thick tribolayer. Also, larger WC content implies larger fraction of interface boundaries and thus increased stress concentration sites. Thus, pull-outs or pits observed on the worn surfaces of cermet containing large WC content (beyond 10 wt%) are responsible for higher wear.

16.6 Closure

In summary, the material loss in TiCN-Ni-WC cermet tools during machining is greatly influenced by two challenging mechanisms: mechanical wear and chemical wear. When WC content is limited to 10 wt%, the results indicate dominant material transfer through high-temperature diffusion. On the other hand, the material removal is dominated by grain pull-outs and grooves in cermets containing WC content beyond 10 wt% during cutting against steel. Detailed discussion on machining characteristics of TiCN-Ni-WC cermet is available in author's publication [20].

The discussion on machining performance of a new set of TiCN-WC-TaC-Ni-Co cermet against steel workpiece implies that at lower speed (133 rpm), the tool cutting edge is likely to plow onto the surface of workpiece. However, steady cutting is possible with a reduction in cutting force as the cutting speed increases to 435 rpm. While lower cutting force is found in spark plasma sintered Ti(CN)-5WC-10Ni-10Co-5TaC cermet, higher cutting force is observed in commercial cemented carbide tip tool and Ti(CN)-5WC-20Ni cermet tool. The conventionally sintered Ti(CN)-5WC-20Ni cermet tool worn surface reveals severe grain pull-out, cracking, and fracture at higher cutting speed. While plowing of tungsten oxide onto the workpiece leads to deeper abrasion in case of commercial carbide tool, wear debris particles generate cracks and grain pull-outs on the surface of TiCN-based cermet tool. The presence of adhered layer beneath the tool face of spark plasma sintered Ti(CN)-5WC-10Ni-10Co-5TaC cermet provided them the best damage resistance during machining. Detailed analysis of results presented on machining performance of TiCN-WC-TaC-Ni-Co cermets is presenred in one of the author's published research report [21].

References

1 Watanabe, T., Dostu, T., and Nakanshi, T. (1996). Sintering properties and cutting-tool performance of Ti(C,N)-based ceramics. *Key Eng. Mater.* 114: 189.

2 Pastor, H. (1988). Titanium carbonitride-based hard alloy for cutting tool. *Mater. Sci. Eng. A* 105–106: 401–409.

3 Zhang, S. (1998). Material development of titanium carbonitride-based cermets for machining applications. *Key Eng. Mater.* 138–140: 521–543.

4 Ahn, S. and Kang, S. (2001). Effect of various carbides on the dissolution behavior of $Ti(C_{0.7}N_{0.3})$ in a $Ti(C_{0.7}N_{0.3})$-30Ni system. *Int. J. Refract. Met. Hard Mater.* 19: 539–545.

5 Ahn, S. and Kang, S. (2000). Formation of core/rim structures in Ti(C,N)-WC-Ni cermets via a dissolution and precipitation process. *J. Am. Ceram. Soc.* 83 (6): 1489–1494.

6 Ettmayer, P., Kolaska, H., Lengauer, W., and Dreyer, K. (1995). Ti(C,N) cermets-metallurgy and properteis. *Int. J. Refract. Met. Hard Mater.* 13: 343–351.

7 Zhang, S. (1993). Titanium carbonitride-based cermet: processes and properties. *Mater. Sci. Eng. A* 163: 141–148.

8 Qi, F. and Kang, S. (1998). A study on microstructural changes in Ti(CN)–NbC–Ni cermets. *Mater. Sci. Eng. A* 251: 276–285.

9 Rolander, U., Winel, G., and Zwinkls, M. (2001). Effect of Ta on structure and mechanical properties of (Ti, Ta, W)-Co cermets. *Int. J. Refract. Met. Hard Mater.* 19: 325–328.

10 Kang, S. (1996). Some issues in Ti(CN)–WC–TaC cermets. *Mater. Sci. Eng. A* 209: 306–312.

11 Mun, S. and Kang, S. (1999). Effect of HfC addition of microstructure of Ti(CN)-Ni system. *Powder Metall.* 42 (3): 251–256.

12 Ahn, S.Y. and Kang, S. (2000). Formation of core/rim structure in Ti(C,N)–WC–Ni cermets via a dissolution and precipitation process. *J. Am. Ceram. Soc.* 83 (6): 1489–1494.

13 Kwon, W.T., Park, J.S., Kim, S.W., and Kang, S. (2004). Effect of WC and group IV carbides on the cutting performance of Ti(C,N) cermet tools. *Int. J. Mach. Tools Manuf.* 44: 341–346.

14 Ahn, S.-Y., Kim, S.-W., and Kang, S. (2001). Microstructure of Ti(C,N)–WC–NbC–Ni cermets. *J. Am. Ceram. Soc.* 84 (4): 843–849.

15 Sarkar, D., Manoj Kumar, B.V., Ahn, S. et al. (2004). Fretting wear behavior of Ti(CN)-based advanced cermets. *Key Eng. Mater.* 264–268: 1115–1118.

16 Manoj Kumar, B.V., Basu, B., Kang, S., and Ramkumar, J. (2006). Erosion wear behavior of TiCN-Ni cermets containing secondary carbides (WC/NbC/TaC). *J. Am. Ceram. Soc.* 89: 3827–3831.

17 Manoj Kumar, B.V., Balasubramaniam, R., and Basu, B. (2007). Electrochemical behavior of TiCN-Ni based cermets. *J. Am. Ceram. Soc.* 90: 205–210.

18 Manoj Kumar, B.V., Ramkumar, J., Basu, B., and Kang, S. (2007). Electro-discharge machining performance of TiCN-based cermets. *Int. J. Refract. Met. Hard Mater.* 25: 293–299.

19 Bellosi, A., Calzavarini, R., Faga, M.G. et al. (2003). Characterization and application of titanium carbonitride-based cutting tools. *J. Mater. Proc. Technol.* 143-144: 527–532.

20 Manoj Kumar, B.V., Ram Kumar, J., and Basu, B. (2007). Crater wear mechanisms of TiCN-Ni-WC cermets during dry machining. *Int. J. Refract. Met. Hard Mater.* 25: 392–399.

21 Verma, V. and Manoj Kumar, B.V. (2017). TaC-containing Ti(CN)-WC-Ni/Co cermets for the improved machining performance. In: *Sustainable Industrial Processing Summit, Volume 8 Surfaces & Interfaces, Composite, Ceramic and Nanomaterials* (eds. F. Kongoli, I. Braems, V. Demange, J. M. Dubois, M. Pech-Canul, C. L. Patino, O. Fumio), 210–222. FLOGEN.

22 Lal, G.K. (1994). *Introduction to Machining Science*, 1e. New Delhi: New Age International Ltd.

23 Rowcliffe, D.J. and Warren, W.J. (1970). Structure and properties of tantalum carbide crystals. *J. Mater. Sci.* 5: 345–350.

24 Rowcliffe, D.J. and Hollox, G.E. (1971). Hardness anisotropy, deformation mechanisms and brittle-to-ductile transition in carbide. *J. Mater. Sci.* 6: 1270–1276.

25 Senthil Kumar, A., Raja Durai, A., and Sorankumar, T. (2003). Machinability of hardened steel using alumina based ceramic cutting tools. *Int. J. Refract. Met. Hard Mater.* 21: 109–116.

26 Greenwood, J.A. and Tabor, D. (1957). The properties of model friction junctions. *Inst. Mech. Eng. Proc. Conf. Lubr. Wear* 92: 314–318.

27 Trigger, K.J. and Chao, B.T. (1956). The mechanism of crater wear of cemented carbide tools. *Trans. Am. Soc. Mech. Eng.* 78: 1119.

17

Wear of TiCN-Based Cermets in Electrodischarge Machining

In a preceding chapter, the fundamental principles of one of the advanced manufacturing techniques, i.e. die sinking electro-discharge machining (EDM), is discussed. This chapter describes the wear damage of TiCN-20 wt% Ni cermets containing 10 wt% of different secondary carbides: WC, NbC, and TaC, when they were used as EDM tool in making a cavity of 1 mm depth in a commercial high-speed steel (HSS). The published results from the author's prior work, discussed in this chapter, would allow one to realize that it is possible to obtain smooth surface and least tool wear ratio for TiCN-20Ni-TaC cermet, while maximum material removal rate for baseline TiCN-20Ni cermet. Concerning the material removal mechanisms, surface melting and binder evaporation as well as pull-out and spalling of carbonitride/carbide grains were emphasized as the major wear mechanisms. The mechanistic analysis does reveal the formation of recast layer and/or material transfer from workpiece as the physical processes involved in cermet tool wear.

17.1 Introduction

As discussed in an earlier chapter of this book, a subset of ceramics and cermets with electrical conductivity larger than the percolation threshold can be manufactured using advanced machining, like electro-discharge machining (EDM). In general, metallic material systems like varieties of steels, tungsten and its alloys, copper and its alloys, graphite, etc., are used as EDM electrodes [1–5]. In order to minimize the electrode wear, materials with high melting points and good electrical and thermal conductivities like SiC, TiCN, Si_3N_4-TiN, Al_2O_3-SiC_W-TiC, ZrO_2-TiN, etc., have been recommended [6–13]. As described in the previous chapter of this book, TiCN-Ni-based cermets have been developed as alternative materials to WC-Co hard materials for cutting tool insert applications [14–22].

Friction and Wear of Ceramics: Principles and Case Studies, First Edition.
Bikramjit Basu, Mitjan Kalin, and B. Venkata Manoj Kumar.

In the above backdrop, this chapter describes some of the key results from the author's published work to illustrate the EDM performance of a series of TiCN-20Ni-XC (where X:W/Nb/Ta) cermets in identical EDM conditions to make cavity in a high-speed tool steel (HSS) as workpiece [23]. In addition to assessing the qualifying parameters of EMD like tool wear ratio (TWR), material removal rate (MRR), and surface roughness (R_a and R_z), an attempt is made to illustrate the dominant material removal mechanisms for the investigated cermets in the selected EDM conditions.

17.2 Materials and EDM Tests

Details of sintering, microstructural characteristics, and mechanical properties of the TiCN-20Ni-XC (where X:W/Nb/Ta) cermets are already described in a previous chapter of this book. The standard four-probe test at room temperature reveals minimum electrical resistivity of $43\times10^{-5}\,\Omega\cdot$m for baseline TiCN-20Ni cermet, and higher electrical resistivity of $158\times10^{-5}\,\Omega\cdot$m and $163\times10^{-5}\,\Omega\cdot$m for TiCN-20Ni-10WC and TiCN-20Ni-NbC, respectively. TiCN-20Ni-TaC cermet exhibits electrical resistivity of $74\times10^{-5}\,\Omega\cdot$m. Polished specimen from a commercially available (AISI M2 grade) HSS steel bar was used as workpiece (as cathode) and respective cermet tool (as anode) was used in die sinking machining tests on an EDM machine (Electra R-35, Electronic, India) with kerosene oil as dielectric medium. All EDM experiments were done to make a depth of cut of 1.0 mm in the HSS workpiece in normal polarity conditions. The weight loss in the cermet tool electrode after each test was noted and the plot is shown in Figure 17.1.

17.3 Wear of TiCN-Cermets During EDM

The performance of the investigated cermets is assessed in terms of three attributes: MRR, TWR, and roughness (R_a and R_z) of EDM surface as shown in Figure 17.1. The diameter of the eroded hole and depth of cut were used to determine the volume of the material removed from HSS workpiece. The volumetric loss of material per unit time is represented as MRR, while the ratio of percentage of volumetric loss of material removed from tool to the workpiece is represented as TWR. As the surface quality of the tool after EDM is also an important parameter to qualify the tool materials, the surface roughness in terms of arithmetic roughness average (R_a) and roughness depth (R_z) is measured.

Among the investigated cermets, the baseline TiCN-20Ni cermet electrode exhibits the highest MRR of 7.65 mm^3/min due to the lowest electoral resistivity. However, a small difference in resistivity resulted in a considerable difference in

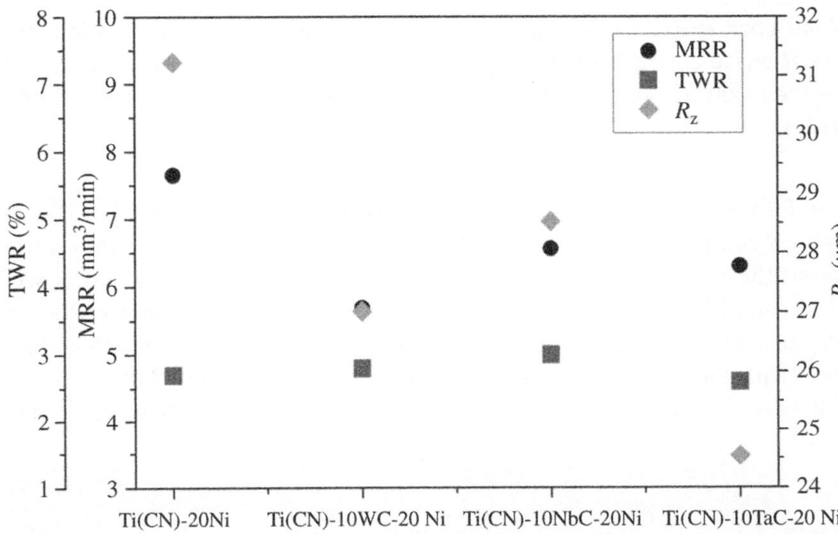

Figure 17.1 EDM performance test results for the investigated TiCN-Ni-based cermet tools [23]. *Source:* Reproduced with permission of Elsevier.

the MRR of TiCN-20Ni-10WC (5.69 mm^3/min) and TiCN-20Ni-10NbC (6.56 mm^3/min) cermets. Compared with TiCN-20Ni-10WC and TiCN-20Ni-10NbC cermets, TiCN-20Ni-10TaC cermet with lower resistivity exhibits higher MRR of 6.30 mm^3/min. With regard to tool wear results, TiCN-20Ni-10TaC cermet exhibits the lowest %TWR of 2.6, whereas TiCN-20Ni-10NbC cermet exhibited highest %TWR of 3.0. Further, the baseline TiCN-20Ni cermet and TiCN20Ni-10WC cermets show higher %TWR of ~2.7.

Post EDM surface analysis indicates a smooth surface with R_a: 4.3 μm and R_z: 24.5 μm for TiCN-20Ni-10TaC cermet, whereas the most rough surface with R_a: 5.2 μm and R_z: 31.2 μm is obtained for baseline TiCN-20Ni cermet. On the other hand, the HSS workpiece surface became extremely rough with R_a: 12–17 μm. Based on the three performance attributes, it can be confirmed that the TiCN-20Ni-10TaC cermet is a preferred composition with lower %TWR, lower surface roughness, and moderate MRR.

17.4 Mechanisms of Material Removal During EDM

Typical representative SEM images of the post EDM surfaces of investigated TiCN-20Ni-based cermets are presented in Figure 17.2. The EDM surface of baseline TiCN-Ni cermet shows channel-type longer cracks, grain draining, and recast smooth surface.

While channel-type cracking is indicative of melting and evaporation of binder phase during the initial part of the EDM process, recast smooth layers are the result of molten ceramic phase at high temperatures during the EDM process.

The formation of recast layers was also reported in die sinking-type EDM of Si_3N_4-TiN composites [10]. The presence of flaky debris observed in Figure 17.2a suggests the non-protective nature of the recast layer. With the addition of secondary carbides in the baseline cermet composition, roughness characteristics of EDM surfaces are changed. TiCN-20Ni-10WC cermet exhibits less rough surface in line with the roughness values as shown in Figure 17.1. In addition, severe cracking of the surface indicates extensive damage of TiCN-20Ni-10WC cermet

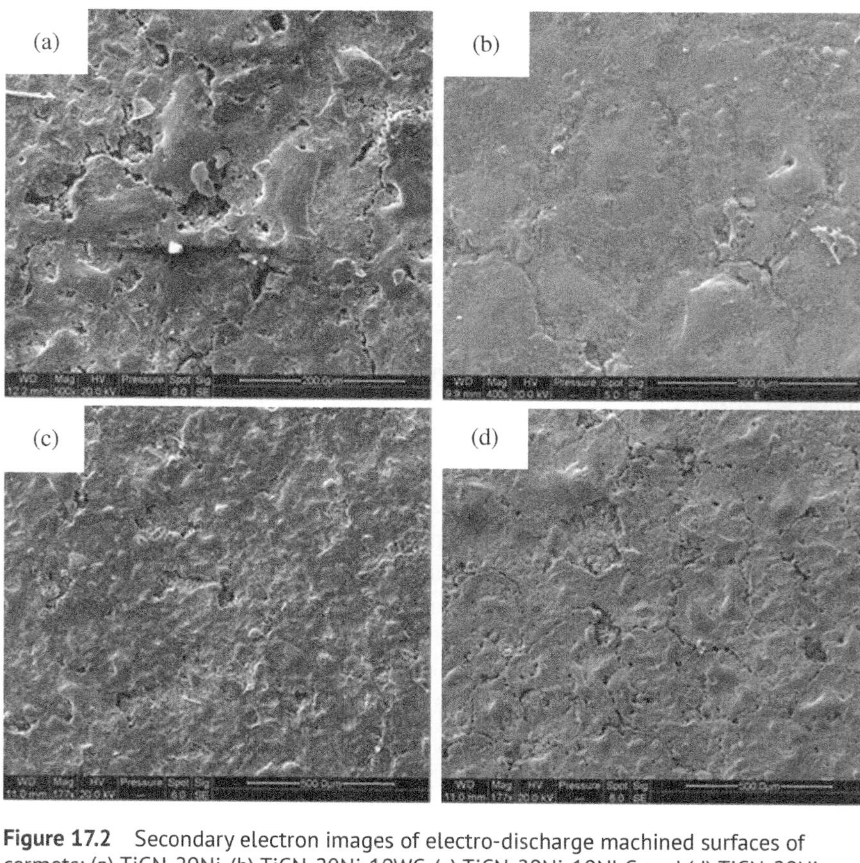

Figure 17.2 Secondary electron images of electro-discharge machined surfaces of cermets: (a) TiCN-20Ni, (b) TiCN-20Ni-10WC, (c) TiCN-20Ni-10NbC, and (d) TiCN-20Ni-10TaC [23]. *Source:* Reproduced with permission of Elsevier.

(see Figure 17.2b). This is also in accordance with higher %TWR as shown in Figure 17.1. The overall increase in roughness can be observed in the case of the EDM surface of TiCN-20Ni-10NbC cermet (see Figure 17.2c) compared with that of TiCN-20Ni-10WC cermet. In the case of EDM surfaces of TiCN-20Ni-10TaC cermet, the surface is predominantly covered with a smooth recast layer (see Figure 17.2d). The roughness data in Figure 17.1 also show lower R_Z: 24.5 μm for TiCN-20Ni0-10TaC cermet.

The overall micromechanisms of material removal in EDM of TiCN-based cermets can be realized. During initial stages, the heat produced by sparks between tool and die preferentially melts the low melting point binder-phase materials. The failure of the binder phase results in the formation of a non-protective skeleton against material removal. This further leads to subsequent decohesion/debonding of TiCN-rich core and (Ti, W/Nb/Ta)CN-rich rim grains. On the other hand, the concurrent formation of transfer or recast layer retards material loss and leads to smoother surface with less roughness value. The spalling is noticeably less in the case of TaC containing cermets with high toughness.

17.5 Closure

The published [23] EDM performance test results, discussed in this chapter necessarily indicate the beneficial effect of secondary carbide addition in baseline TiCN-20Ni cermet composition. While the TiCN-20Ni-10TaC possesses the lowest TWR, good surface finish, and moderate MRR, baseline TiCN-20Ni cermet exhibits the highest MRR. The material removal occurs with the following sequence of micromechanisms: melting and evaporation of low melting point binder phase, decohesion/debonding of core and rim phases. Interestingly, the material removal process at an interim stage during EDM is lowered with the formation of recast smooth layer.

References

1 Petrofes, N.F. and Gadalla, A.M. (1988). Processing aspects of shaping advanced materials by electrical discharge machining. *Adv. Mater. Manuf. Process.* 3: 127–153.

2 Arthur, A., Dickens, P.M., and Cobb, R.C. (1996). Using rapid prototyping to produce electrical discharge machining electrodes. *Rapid Prototyp. J.* 2 (1): 4–12.

3 Wang, W.F. (1997). Effect of tungsten particle size and copper content on working behaviour of W–Cu alloy electrodes during electrodischarge machining. *Powder Metall.* 40 (4): 295–300.

4 Johnson, J.L. and German, R.M. (1994). Chemically activated liquid phase sintering of tungsten-copper. *Int. J. Powder Metall.* 30 (1): 91–102.

5 Tsai, H.C., Yan, B.H., and Huang, F.Y. (2003). EDM performance of Cr/Cu-based composite electrodes. *Int. J. Mach. Tools Manuf.* 43: 245–252.

6 Li, L., Wong, Y.S., Fuh, J.Y.H., and Lu, L. (2001). EDM performance of TiC/copper-based sintered electrodes. *Mater. Des.* 22: 669–678.

7 Luis, C.J., Puertas, I., and Villa, G. (2005). Material removal rate and electrode wear study on the EDM of silicon carbide. *J. Mater. Process. Technol.* 164–165: 889–896.

8 Bellosi, A., Guicciardi, S., and Tampierie, A. (1992). Development and characterization of electroconductive Si_3N_4-TiN composites. *J. Eur. Ceram. Soc.* 9: 83–93.

9 Pitman, A. and Huddleston, J. (2000). Electrical discharge machining ZrO_2/TiN particulate composite. *Br. Ceram. Trans.* 99: 77–84.

10 Lauwers, B., Kruth, J.P., Liu, W. et al. (2004). Investigation of material removal mechanisms in EDM of ceramic composite materials. *J. Mater. Process. Technol.* 149: 347–352.

11 Jiang, D., Vleugels, J., Van der Biest, O. et al. (2005). Electrically conductive and wear resistant Si_3N_4-based composites with $TiC_{0.5}N_{0.5}$ particles for electrical discharge machining. *Mater. Sci. Forum* 492–493: 27–32.

12 Sanchez, J.A., Cabanes, I., Lopez de Lacalle, L.N., and Lamikiz, A. (2001). Development of optimum electro discharge machining technology for advanced ceramics. *Int. J. Adv. Manuf. Technol.* 17 (12): 897–905.

13 Matsuo, T. and Oshiwa, E. (1992). Investigation on the optimum carbide content and machining conditions for wire EDM of zirconia ceramics. *Ann. CIRP* 41 (1): 231–234.

14 Mahdavinejad, R.A. and Mahdavinejad, A. (2005). ED machining of WC–Co. *J. Mater. Process. Technol.* 162–163: 637–643.

15 Zhang, S. (1993). Titanium carbonitride-based cermets: process and properties. *Mater. Sci. Eng. A* 163A: 141–148.

16 D'Errico, G.E., Buglisosi, E., and Guglielmi, E. (1998). Tool-life reliability of cermet inserts in milling tests. *J. Mater. Process. Technol.* 77: 337–343.

17 Kwon, W.T., Park, J.S., Kim, S.-W., and Kang, S. (2004). Effect of WC and group IV carbides on the cutting performance of Ti(CN) cermet tools. *Int. J. Mach. Tools Manuf.* 44: 341–346.

18 Ettmayer, P., Kolaska, H., Lengauer, W., and Dreyer, K. (1995). Ti (CN) cermets: metallurgy and properties. *Int. J. Refract. Met. Hard Mater.* 13: 343–351.

19 Ahn, S. and Kang, S. (2001). Effect of various carbides on the dissolution behavior of $Ti(C_{0.7}N_{0.3})$ in a $Ti(C_{0.7}N_{0.3})$-30Ni system. *Int. J. Refract. Met. Hard Mater.* 19: 539–545.

20 Qi, F. and Kang, S. (1998). A study on microstructural changes in Ti(CN)–NbC–Ni cermets. *Mater. Sci. Eng. A* 251: 276–285.

21 Ahn, S. and Kang, S. (2000). Formation of core/rim structures in Ti(C,N)-WC-Ni cermets via a dissolution and precipitation process. *J. Am. Ceram. Soc.* 83 (6): 1489–1494.

22 Anstis, G.R., Chantikul, P., Lawn, B.R., and Marshall, D.B. (1981). A critical evaluation of indentation techniques for measuring fracture toughness: I. Direct crack measurements. *J. Am. Ceram. Soc.* 64: 533.

23 Kumar, B.V.M., Ramkumar, J., Basu, B., and Kang S. (2007). Electro-discharge machining performance of TiCN-based cermets; *Int. J. Refract. Met. Hard Mater.* 25: 293–299.

Section VI

Future Scope

18

Perspective

In this concluding chapter, the accelerated development for next generation wear-resistant materials is mentioned. The need for design and manufacturing of tribo-testers with high-temperature testing capability has been emphasized. The requirement to characterize the surface and subsurface damage using sophisticated *in situ* diagnostic tools is also highlighted in order to improve understanding on micromechanisms of material removal. Further, the need for modeling coupled with experiments to develop predictability in wear of ceramics is highlighted. At the end, it will be emphasized that the "tribomaterialomics" approach will significantly aid in development of new generation wear-resistant materials, whose predictive performance would be closely tracked by "digital twins."

18.1 Innovation Cycle for Wear-Resistant Materials

In the present book, considerable efforts have been endowed to elucidate the wear micromechanisms of ceramics and ceramic composites. Many of the ceramic materials have low density and possess the unique ability to retain strength and hardness at high temperatures. Based on these characteristics, ceramics are preferred in several tribological applications.

In the past, ceramic materials had found few specialized applications, where they were almost irreplaceable. Owing to low density and noncatastrophic failure mode, silicon nitride ceramics have been widely used for high-speed precision ball bearings and spindle hybrid bearings. On the other hand, being a wear resistant and corrosion resistant, silicon carbide has been successfully used over a wide range of sliding applications such as mechanical axial face seals. High hardness

Friction and Wear of Ceramics: Principles and Case Studies, First Edition.
Bikramjit Basu, Mitjan Kalin, and B. Venkata Manoj Kumar.

and superior wear resistance make alumina ceramics candidate materials for use in dry and water lubricated sliding conditions. Alumina has also been used for joint prosthesis due to its additional qualification of biocompatibility.

> Poor thermal shock resistance and poor fracture toughness are major drawbacks for ceramics, which restrict wider usage in engineering applications.

The application range of ceramics can be widened if the limiting properties are enhanced by a suitable processing and microstructural engineering approach. On the other hand, ceramics are engineers' best choice for components subjected to extremely severe structural and tribological conditions.

> Since tribology is a system-dependent property, any claim on industrial applicability of a new material requires testing under closely simulated conditions with different counterbody materials over a range of operating variables.

In addressing unmet technological needs, there has been a continuous surge toward development of new wear-resistant materials, mostly using an empirical approach. A typical innovation cycle is shown in Figure 18.1. This approach involves lab-scale processing using conventional and/or advanced manufacturing techniques, wherein the process parameters are intuitively selected and tailored to meet the requirement of the spectrum of physical and mechanical properties. Subsequently, researchers use tribological test protocols to simulate application-specific conditions to assess wear resistance. This phase of research is scientifically the most exciting phase and researchers use several characterization tools to determine the changes in friction and wear properties. Although not reported widely, it is important to simulate flash temperature or contact stress distribution at the tribological interface, using computational analysis. The last and most important step in the innovation cycle is the prototype manufacturing and testing using performance-qualifying test protocols. In the entire innovation cycle (see Figure 18.1), the quantum of data progressively decreases as one goes from lab-scale development to product prototype manufacturing, while the application significance of the data increases in the reverse manner. Therefore, it can be perceived that the effort required for innovation of wear-resistant materials not only involves the integration of ideas and concepts from multiple disciplines but also involves significant time and cost. Importantly, the above-described innovation cycle also generates significant data, as shown in Figure 18.1.

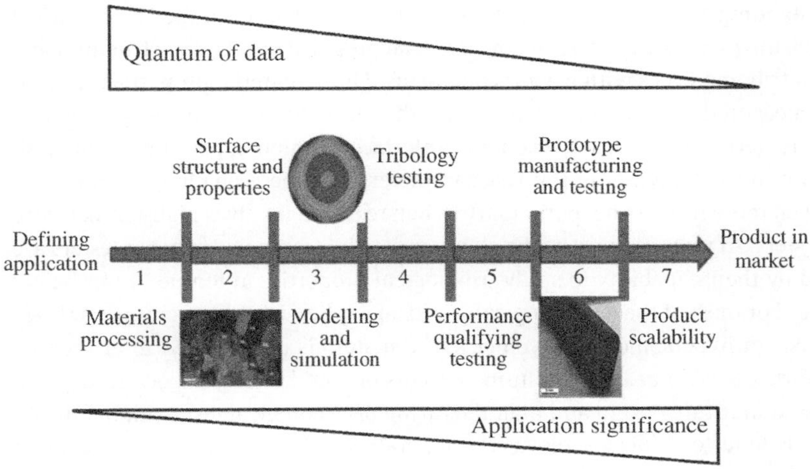

Figure 18.1 Innovation cycle for wear-resistant materials.

18.2 *In Situ* Diagnosis of Tribological Interaction

The creation and maintenance of a large research infrastructure, spanning from materials processing to tribology facilities, including surface characterization tools are important challenges. Considering the requirement of distinctly different skill sets/expertise, the maintenance of such facilities has often been a major issue. The in situ diagnostics of the tribomechanical wear mechanisms and tribochemical wear phenomenon can provide significant insights into the evolution of damage mechanisms or the onset of chemical reaction at tribological surfaces. The present practice is the microscopic observation of worn surfaces at the end of long test duration or the spectroscopic evaluation of the worn regions to assess the formation of different tribochemical reaction products. Alternatively, the researchers have been using interrupted test designs, i.e. to stop the tests after predetermined number of test cycles and then to use similar diagnostic tools, to understand "what's going at the worn surfaces?" If the field of tribology has to grow in future, the tribology equipment manufacturer should introduce a new test setup with appropriate in situ diagnostic tools. This will help the researchers to investigate temporal evolution of wear mechanisms at contacting interfaces during testing.

18.3 High-Temperature Wear Testing

The field of tribology has largely been based on experimental data generation of friction and wear rate. This is corroborated with the microscopic observations, often with spectroscopic evaluation of the worn surfaces, which often aid in understanding the underlying wear mechanisms. Also, a number of empirical relationships are used to correlate the material properties and microstructural features of tribomaterials with wear mechanisms. Those material properties are generally measured at room temperature and therefore do not necessarily represent "true" material properties at frictional interfaces, which experience high flash temperatures depending on the test parameters. All these challenges make this field even more interesting, particularly when research involves high-temperature tribotesters. At present, most of the commercial high-temperature wear testers are limited by their capability to study tribological properties at temperatures below 1000 °C. For most of the technologically critical applications, like scramjet or leading edges in hypersonic space vehicles, the materials are exposed to erosion or other forms of wear at a temperature in excess of 1000 °C. This aspect needs to be addressed in future design and manufacturing of tribological test equipments. As materials tend to oxidize at elevated temperatures, the situation at tribocontact becomes more complex with oxide-scale properties governing the performance.

> The published literature reveals limited studies on high-temperature tribology of materials. This is mainly because of the unavailability of testing facilities that can be used at temperatures above 1000 °C.

It is to note that the tribology data obtained in room temperature conditions are not applicable for high-temperature applications. This is because of the fact that surface reactions with the surrounding environment at high temperature can cause transition in dominant mechanisms of material removal and often increase the wear rate.

> A comprehensive understanding of the surface/subsurface damage and tribochemical layer or debris particles using advanced characterization techniques like Fourier transform infrared spectroscopy (FT-IR), X-ray photo electron spectroscopy (XPS), Raman spectroscopy, etc., is essential in order to improve our knowledge on the wear mechanisms and to assess performance of new material systems under extreme tribological conditions. Focused ion beam (FIB) machining coupled with SEM is useful to study the fine-scale deformation or fracture of materials at different spatial regions across the worn surface.

18.4 Modeling and Simulation in Tribology

Regarding modeling, two types of models, tribochemical and tribomechanical wear models, are generally reported in the literature [1]. However, the validation of such wear models needs one to conduct a set of control experiments with materials having properties determined over a broader range of operating parameters, like sliding velocity, load counterbody, etc. Furthermore, experiments have to be conducted on different materials with identical set of operating parameters in order to verify the functional dependence of the wear volume on material parameters. For tribochemical wear model, lack of experimental information of contact surfaces makes the absolute comparison of experimentally determined wear volume values with the model-predicted values almost impossible. Also, no existing model is able to distinguish the transition in wear regime from mild wear to tribochemical wear. More rigorous models with properly defined contact conditions are required to discern such transition. The other concerning issue in modeling and simulation is the contact temperature estimation.

In view of the absence of any largely acceptable model, researchers use a variety of models to estimate contact temperature. Often, estimated contact temperatures differ over a significant range for a given tribomaterial pair.

The need for strong computational analysis can be realized in performance assessment of components in extreme conditions of tribology, such as elevated temperatures, high-velocity impingement, etc. As high-temperature wear is driven not only by temperature but also by material yielding, stress evolution in the contact region becomes important to assess the material loss. While determining stress levels in high-temperature worn regions is not possible by experimental techniques, computational analysis using finite element (FE) simulation can provide necessary insights in estimating the complex nature of stress and temperature distribution in the worn regions. In fact, a novel approach of a combination of experimental erosion testing and FE analysis is demonstrated in one of the chapters of the book to assess the true potential of ZrB_2-SiC-based multiphase ceramic composites for critical components in thermal protection systems for supersonic space vehicles. Such an approach should be followed in similar studies to evaluate wear performance of new tribomaterials.

18.5 Tribomaterialomics – A New Concept

Recent years have also witnessed the emergence of data science as the fourth major paradigm in driving scientific discovery after experiments, theory, and computations [2]. Data science has already had a major impact on many aspects of the human endeavor, from social networking and search engines to e-commerce and self-driving cars [3]. In essence, data science deals with automated aggregation of data in accessible repositories and its systematic analyses using a variety of statistical learning tools.

Building on early successes in social networking and e-commerce, data science is now making major inroads into the traditional scientific approaches and workflows employed in hard sciences (i.e. engineering and science disciplines). Already, data science has had a significant impact in fields such as genetics [4], astronomy [5], and weather predictions [6], among many others. The recent announcement of the Materials Genome Initiative (MGI) [7] by the US federal government has brought attention to the potential benefits of data science approaches to accelerated materials design, development, and deployment in emerging advanced technologies with positive impact on economy, sustainability, national security, and the quality of human life.

In the field of Materials Science, the process-structure-property (PSP) linkages constitute the essential *materials knowledge* needed to drive efforts of materials innovation. Despite its tremendous importance, a rigorous mathematical framework for the systematic capture of this knowledge base is not available. The central impediment comes from the fact that the material structure exhibits rich heterogeneity at a hierarchy of length scales spanning from the atomistic to the macroscale. Capturing all details of the complex internal structure of most materials requires an extremely large number of structure parameters. However, in practice, one hopes to identify a small set of salient measures of the material internal structure that could be tracked through the processing history imposed on the material and at the same time capture the most dominant influences on the overall properties of interest (dictated by the selected application). These correlations are generically referred to as PSP linkages, and provide the essential scientific and engineering pathways for the optimized design of the material for any selected application.

Early efforts at establishing PSP linkages employed grossly simplified measures of the material structure. For example, the yield strength of a polycrystalline metal is inversely proportional to the square root of the average grain size [8, 9]. However, the adoption of such simplistic measures often limit the applicability of the extracted PSP linkages to small windows in materials and process spaces (i.e. the model forms and parameter values change substantially with changes in chemical composition and process history). In line with such understanding, one representative example of PSP linkage for tribological materials is shown in Figure 18.2.

Despite several decades of research, we still do not have proven capability to design tribological materials with predictable wear resistance properties. In line with the existing concepts of "Materials Genome," we define here a new concept "Tribomaterialomics".

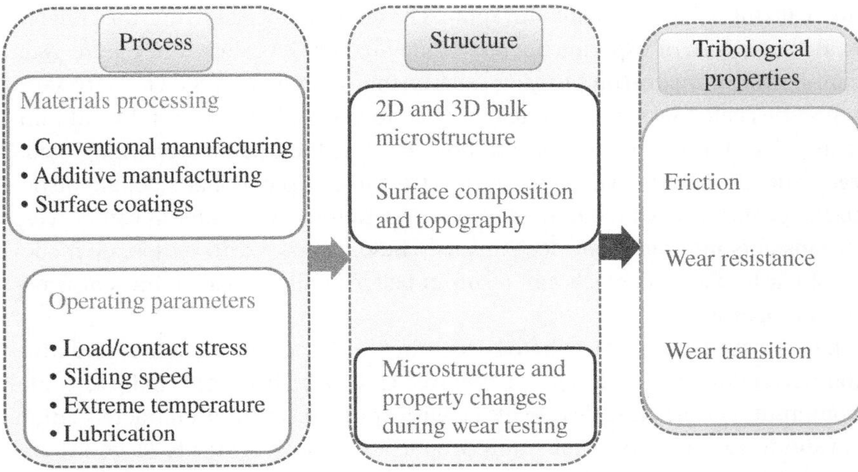

Figure 18.2 Correlation among tribological properties, process, and structure of materials.

> Tribomaterialomics is defined as "Integration of high-throughput tribology test results, computational tools, and large databases to explore multiscale tribological properties in correlation with basic material elements and combine them to discover and design new materials for tribological applications."

The new concept of "tribomaterialomics" is based on the application of the theory and technology of materials genomics to tribological materials (see Figures 18.2 and 18.3).

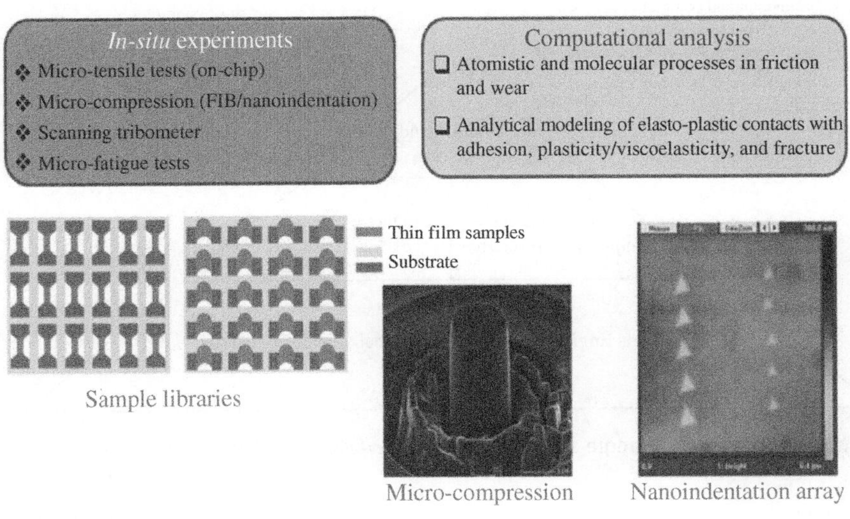

Figure 18.3 High-throughput characterization of materials properties. *Source:* Adapted from Zhang and Xiang [10].

In realizing tribomaterialomics approach, one has to adopt the nonconventional approaches of experiments and modeling need. As shown in Figure 18.3, micro-tensile or micro-compression experiments and micro-fatigue tests for sample libraries generate significant data related to materials properties. In addition, the use of scanning tribometers with predetermined test parameters can provide large data sets of tribological properties. On the computational side, atomistic- and molecular-level simulations, involving friction of new material can provide more insights into the wear mechanisms. These together with mesoscale modeling of elasto-plastic contacts can result in better predictive capabilities of tribological properties.

The concept triangle of tribology together with various elements to define tribomaterialomics is shown in Figure 18.4. As discussed before, high-throughput sample libraries together with multi-resolution characterization to evaluate surface topography and properties would constitute large experimental database.

In a number of studies, attempts are also made to establish quantitative relationships between wear resistance and system parameters using empirical or semi-empirical equations. Multiscale models like FE analysis, MD simulation, and discrete Fourier transform (DFT) analysis pursued to reveal atomistic-/molecular-level interactions at tribological interface. It is expected that data science approaches involving machine learning and data analytics tools will be used in future to develop better predictive capability of the wear resistance for the next-generation tribomaterials.

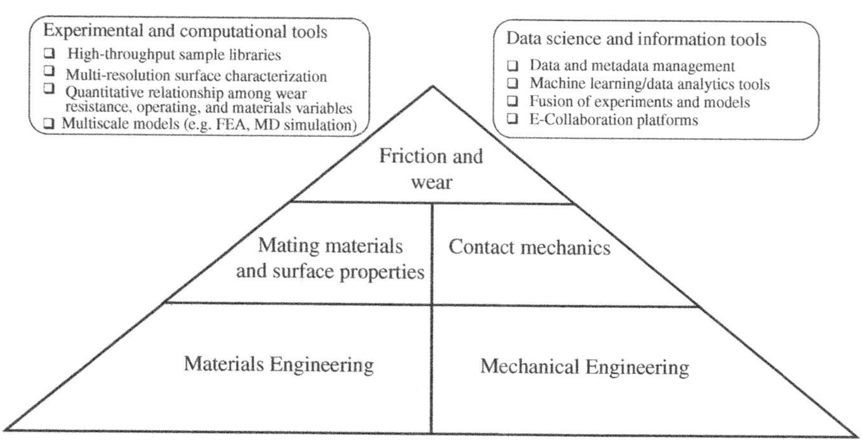

Figure 18.4 Concept triangle and elements of a new concept, "Tribomaterialomics."

18.6 Education and Mentoring of Next-Generation Researchers

The interdisciplinary nature of the field of tribology, in general, clearly emerges from several chapters of this book. This field integrates the fundamental concepts of Mechanical Engineering, Materials Science, Chemistry, and Physics. Over the last few decades, this subject has been largely taught in the domain of mechanical engineering with a few examples in Materials Science. Therefore, one of the major challenges is the lack of widespread academic programs in this subject and effective training of young researchers in this interdisciplinary field. As a result, many young researchers do not have much idea about the basic concepts of tribology and its significant impact on engineering applications.

It is important to mention that like Mechanical Engineering concepts, the fundamentals of Materials Science do play an important role in developing new wear-resistant materials and evolving new insights into friction and wear mechanisms. In the last few decades, it has been widely observed that more researchers with undergraduate or core education from Materials Science background significantly contributed to the field of tribology. As the field is extended from Mechanical Engineering more toward Materials Science, this subject should be treated with equal emphasis on Materials fundamentals as well. Currently, only very few Materials Science departments in the world offer tribology courses at the post-graduate level. Also, this subject is not taught as a compulsory course in engineering science curriculum in most universities around the world, despite its significance in technological applications.

In the backdrop of the above discussion, it is also recommended to include "Tribology" as a core/compulsory course for freshmen/junior undergraduate students across all the disciplines in engineering. Further, active researchers in tribology are recommended to make efforts through outreach activities like public lectures in engineering colleges/high schools, online Massive Open Online Courses (MOOC), etc. towards popularizing this important and socially relevant research area.

An extensive network of universities/national labs/companies needs to evolve for training and knowledge dissemination in the field of tribology. The cumulative outcome can lead to accelerated development of wear-resistant materials and deeper insights into friction and wear mechanisms.

References

1 Tiwari, A., Basu, B., and Bordia, R.K. (2009). Model for fretting wear of brittle ceramics. *Acta Materialia* 57: 2080–2087.

2 Hey, T., Tansley, S., and Tolle, K. (2009). *The Fourth Paradigm. Data-Intensive Scientific Discovery*. Microsoft Research.

3 https://www.wired.com/story/zoox-self-driving-car-video-san-francisco (30 September 2019).

4 Excoffier, L., Laval, G., and Schneider, S. (2005). Arlequin (version 3.0): an integrated software package for population genetics data analysis. *Evolutionary Bioinformatics* 1: 47–50.

5 Bell, G., Hey, T., and Szalay, A. (2009). Beyond the data deluge. *Science* 323 (5919): 1297–1298.

6 Boehm, J., Niell, A., Tregoning, P., and Schuh, H. (2006). Global Mapping Function (GMF): a new empirical mapping function based on numerical weather model data. *Geophysical Research Letters* 33 (7) https://doi.org/10.1029/2005GL025546.

7 National Science and Technology Council (NSTC) (2011). *Materials Genome Initiative for Global Competitiveness*. Washington, DC: NSTC.

8 Petch, N.J. (1953). Cleavage strength of polycrystals. *Iron and Steel Institute – Journal* 174 (Part 1): 25–28.

9 Hall, E.O. (1951). The deformation and ageing of mild steel III. Discussion of results. *Proceedings of the Physical Society. Section B* 64: 747–753.

10 Zhang, X. and Xiang, Y. (2017). *Journal of Materiomics* 3: 209–220.

Appendix: Appraisal

A.I Multiple Choice Questions

Select the most appropriate answer from the given choices:

1 Ceramics – a class of materials:
A With low thermal conductivity (<30 W/m·K)
B With high hardness (>15 GPa)
C With low CTE ($<4\times10^{-6}\,K^{-1}$)
D With inherent brittleness

2 Typically, with today's advanced materials, very high cutting/machining speed is achievable, which is as high as
A 1000 m/min
B 1000 m/s
C 10 m/min
D 100 m/s

3 For a glass material, XRD spectrum will show:
A Diffuse peak
B Strong peak
C Only background intensity
D None of the choices

4 By definition, high temperature materials are those that are used at temperatures above
A 100 °C
B 300 °C
C 500 °C
D None of the above

Friction and Wear of Ceramics: Principles and Case Studies, First Edition.
Bikramjit Basu, Mitjan Kalin, and B. Venkata Manoj Kumar.

5 Traditional ceramics, compared with engineering ceramics
- **A** have lower flexural strength
- **B** have high E-modulus
- **C** are much costlier
- **D** None of the choices

6 The yield strength in compression for ceramics is
- **A** Equal to the same in tension
- **B** Half than that in tension
- **C** Approximately eight times that in tension
- **D** Just double than the same under tension

7 The processing of ceramic materials involves
- **A** Melting and casting
- **B** Sintering
- **C** Hot rolling
- **D** Hot extrusion

8 During grain growth in ceramics, those grains will shrink for which the number of grains edges is
- **A** Less than 6
- **B** More than 6
- **C** Equal to 6
- **D** Less than 10

9 As Weibull modulus increases within finite values, the strength distribution becomes:
- **A** Wider
- **B** Narrower
- **C** A straight line
- **D** None of the choices

10 In order to eliminate total porosity in a solid-state sintered ceramic without losing mechanical properties,
- **A** Pores should move along with grain boundaries
- **B** Pores should be located at the central region of grains
- **C** Sintering should be done for longer time at the final stage
- **D** Pores should be coarser/larger sizes

11 Advanced sintering processes include:
- **A** Hot isostatic pressing
- **B** Hot pressing

C Spark plasma sintering
D All of the above

12 The grain size increases, when (in reference to sintering conditions to reach fully dense ceramics)
A Sintering time is longer
B Sintering time is shorter
C Sintering temperature is lower
D None of the choices

13 During sintering of ceramics,
A Only pore size changes
B Only pore shape changes
C Both pore size and pore shape changes
D Maximum neck growth occurs by vapor-phase transport

14 Spark discharges during sintering is produced during
A Spark plasma sintering
B Hot pressing
C Hot isostatic pressing
D Normal

15 In case of mode II loading, the crack faces will largely experience
A Tensile force
B Compressive force
C Shear force
D Twisting force

16 In a brittle ceramic, stress concentration at crack tip will be higher for
A Longer blunt crack
B Longer sharp crack
C Shorter but blunt crack
D Shorter but sharp crack

17 If a spherical WC ball indents a polished Al_2O_3 sample, the following crack morphology will develop:
A Median cracks
B Radial cracks
C Cone cracks
D Lateral cracks

18 Toughening by crack bridging is most effective in:
A SiAlON
B ZTA
C WC-Co
D MgO-doped Al_2O_3

19 Ceramics are characterized by Weibull modulus in the range of:
A 5–15
B 30–50
C More than 50
D More than 100

20 Crack in a crystalline material represents discontinuity in
A Interatomic bonds
B Function
C Shape
D None of the above

21 Hardness is a measure of resistance against
A Fracture
B Crack growth
C Permanent deformation
D Pore growth

22 If the crack growth resistance increases, then
A fracture toughness will decrease
B fracture toughness will increase
C hardness will increase
D hardness will decrease

23 Strength of a material describes
A load bearing capability
B resistance against permanent deformation
C scratch resistance
D corrosion resistance

24 Stress concentration at the crack tip
A will have linear dependency on crack length
B will have inverse square root dependency on crack length
C is independent of crack tip radius of curvature
D will be proportional to square root of ratio of crack length and crack root radius of curvature

25 Friction is conceptually a:
A Gravitational force
B Pressure
C Resistive force
D Movement

26 Asperities in the context of tribology represent
A Lubricants in contacts
B Surface irregularities
C Air absorbed by surfaces in relative motion
D None of the above

27 Local welding at sliding contacts is also known as:
A Compression
B Asperition
C Friction
D Adhesion

28 Plowing contribution would be expected to be more
A During the initial contact of a spherical wear particle
B During the increased penetration of the spherical asperity
C All the time during the deformation by spherical asperity
D For all the above cases

29 Adhesional contribution to friction will be
A larger for self-mated couple
B larger for dissimilar material couple
C smaller for self-mated couple
D independent of material combination

30 For higher work of adhesion, the contact temperature will be
A Higher
B Lower
C Same as for lower work of adhesion
D Unpredictable

31 The coefficient of friction of material combinations, during running-in-period, typically:
A increases
B decreases
C remains constant
D none of the above

32 The coefficient of friction of ceramics is typically less than that of metals, because of

A Superior thermal properties
B Better mechanical properties
C Better physical properties
D None of the above

33 Frictional energy dissipation is

A Directly proportional to sliding velocity
B Inversely proportional to sliding velocity
C Independent of sliding velocity
D Increases and decreases with sliding velocity

34 The coefficient of friction of known engineering materials:

A cannot be >1.0
B generally 0.3
C can be zero, in most cases
D can be as low as 0.05

35 For brittle materials like ceramics,

A the coefficient of friction should be dependent on toughness
B the coefficient of friction should be independent on toughness
C the contact will be essentially plastic
D cracks expected at center of Hertzian contact

36 For metallic materials, the coefficient of friction will always

A increase with load
B decrease with load
C be constant
D none of the above

37 The coefficient of friction of HDPE composite/alumina couple in SBF typically lies around

A 0.1
B 1.0
C 0.001
D 0.0001

38 Friction of graphite in vacuum is:

A Higher than that in ambient air
B Lower than that in ambient air

C The same as in ambient air
D None of the above

39 Flash temperatures for different engineering material combinations in tribocontact:
A are always higher than the bulk temperature
B are always lower than the bulk temperature
C are the same as bulk temperature
D can be same as bulk temperature for a particular class of material

40 Which of the following tribosystems is expected to have larger contact temperature:
A Self-mated steel
B Self-mated hardmetal (WC-Co)
C Self-mated zirconia
D Self-mated alumina

41 Which of the following parameters cannot play any significant role in friction and wear of engineering materials:
A Electrical resistivity
B Hardness
C Elastic modulus
D Thermal conductivity

42 The tip radius of diamond stylus on a profilometer
A ~2 nm
B ~2 mm
C ~2 μm
D ~2 pm

43 "Pop-in" event occurs in nanoindentation during
A Loading only
B Unloading only
C Both loading and unloading
D None of the above

44 At loaded contacts under identical conditions, which is/are expected to experience elastic behavior:
A TiN-coated steel against alumina ball
B Al-alloys against steel
C Self-mated alumina
D None of the above

45 Under identical contact conditions, which one (self-mated) has lowest Hertzian contact dimension:

A Alumina
B Zirconia
C Titanium boride
D PTFE

46 Which of the following does not always decrease up to a certain depth below the tribosurfaces:

A Contact temperature
B Hardness
C Principal shear stress
D Normal stress

47 The material removal of soft surface due to grooving of sliding surface is known as:

A Tribochemical wear
B Pitting wear
C Abrasive wear
D Adhesive wear

48 Cold local welding is referred to as:

A Adhesive wear
B Tribochemical wear
C Abrasive wear
D Erosive wear

49 The material removal as a result of stresses below endurance limit is:

A Erosive wear
B Tribochemical wear
C Fatigue wear
D Pitting wear

50 The material removal as a result of electrochemical reactions is:

A Erosive wear
B Corrosive wear
C Adhesive wear
D Fatigue wear

51 In an articulating joint with hard and soft materials experiencing relative motion

A Hard material wears soft material
B Soft material wears hard material
C Strength of the interface increases
D None of the above

52 Based on the theory of tribology, which of the following (self-mated) will have lowest friction:

A HDPE
B Ti6Al4V
C SS 316L
D Zirconia

53 Wear debris particles can cause

A Inflammatory reaction
B Toxicity
C Systemic response
D All of the above

54 For the wear of brittle materials, one of the following crack patterns plays the most significant role:

A Radial-median cracks
B Lateral cracks
C Cone cracks
D None of the above

55 Wear involves

A Linear relative motion
B Motion due to impact
C Motion due to application of torque to the loading system
D All of the above

56 For abrasive wear, the wear rate

A will be typically higher for two-body than three-body
B two-body case is always equal to that of three-body case
C is inversely proportional to load
D does not depend on the abrasive shape/size

57 Fretting wear
- **A** Increases with displacement amplitude
- **B** Involves linear relative displacement of small amplitude
- **C** Does not depend on normal load
- **D** Does not involve abrasive/adhesive wear mechanism

58 For materials with higher wear rate,
- **A** Higher will be the coefficient of friction irrespective of material combinations
- **B** Lower will be the coefficient of friction irrespective of material combinations
- **C** The coefficient of friction can be lower for specific material class
- **D** None of the above

59 The contact temperature is expected to be most related to
- **A** thermal conductivity
- **B** hardness
- **C** elastic modulus
- **D** electrical resistivity

60 During tractive rolling, the system has
- **A** Nonzero slip velocity
- **B** Zero slip velocity
- **C** Zero rotational speed
- **D** None of the above

61 The wear due to solid particle impingement is:
- **A** Pitting wear
- **B** Fretting wear
- **C** Erosive wear
- **D** Tribochemical wear

62 For erosive wear, the wear rate
- **A** will always increase with impact angle
- **B** will always decrease with impact angle
- **C** is independent of impact angle
- **D** will be higher for materials with low hardness

63 With increasing abrasive particle size, the abrasive wear rate will always
- **A** Increase to a certain level
- **B** Increase and then decrease
- **C** Remains steady state
- **D** None of the given choices

64 In some cases, unlubricated wear may not necessarily involve
A Relative motion
B Measurable material loss
C Relative linear tangential displacement
D Asperity contact

65 The major aim of lubrication is
A to eliminate friction and wear
B to accumulate heat at the contact
C to maintain high friction and wear
D none of the given choices

66 The lubricating characteristics of a lubricant include:
A Viscosity
B Surface tension
C Both a and b
D None of the above

67 In lubricated medium, the coefficient of friction is
A independent of surface roughness
B dependent on sliding velocity
C constant irrespective of sliding velocity
D always higher for smoother surfaces

68 The coefficient of friction under oil lubrication is typically low, because of
A Tribodegradation of oil
B Pronounced abrasion
C Less plastic deformation
D None of the above

69 The coefficient of friction of a given tribocouple is expected to be lowest for
A Rolling friction
B Boundary lubrication
C Water lubrication
D Mixed lubrication

70 In which of the following cases, asperity contact does not occur:
A Hydrostatic lubrication
B Elastohydrodynamic lubrication (EHD)
C Mixed lubrication
D Boundary lubrication

71 External source-driven fluid pressure is associated with:
A hydrostatic lubrication
B solid–liquid lubrication
C solid lubrication
D liquid lubrication

72 The major characteristics of lubricating oil shall include:
A cloud point
B pour point
C viscosity index
D All of the above

73 Apart from lubrication, material loss can be minimized by
A Suitable design of machine components
B Near net machining
C Appropriate selection of materials
D All of the above

74 In case of aerodynamic bearings, lubricant is:
A Paraffin oil
B Water
C Synthetic oil
D Pressurized air

75 The friction coefficient is expressed as
A Ratio of tangential frictional force to normal force
B Ratio of normal force to tangential frictional force
C Ratio of normal force to normal frictional force
D None of the above

76 The roughness parameters of a surface can be obtained using
A Surface profilometer
B Atomic force microscope
C Transmission electron microscope
D a and b

77 Choose proper example from the following where wear is not desirable:
A Writing with pencil
B Machining

C Shaving
D Bearings

78 Tribological behavior of a material can be studied by
A Tensile testing
B Compressive testing
C Friction and wear testing
D Creep testing

79 Choose correct order of materials for the respective property
A Elastic modulus: ceramics > polymers > metals
B Hardness: ceramics > metals > polymers
C Compressive strength: metals > polymers > ceramics
D Fracture toughness: ceramics > polymers > metals

80 The microstructural characterization technique SEM stands for
A Scanning energy microscopy
B Scanning electron microscopy
C Secondary energy microscopy
D Standard electron microscopy

81 The analytical technique XRD used for phase analysis stands for
A X-ray diffraction
B X-ray diffusion
C X-ray deformation
D X-ray deflection

82 Which surface roughness parameter will be preferred for recording finer variations in surface asperities?
A R_a
B R_q
C R_p
D None of the above

83 In Hertzian contact, choose subsurface level at which shear stress is maximum
A 0.48 times contact radius
B 0.38 times contact radius
C 0.78 times contact radius
D equal to contact radius

84 From the following, select dominant mechanism of material removal in an abrasive wear study of detonation sprayed WC-Co coating:

A Pull-out of WC cuboids in coatings
B Delamination as a result of intra-splat cracking in coatings
C WC cracking in coatings
D Delamination as a result of inter-splat cracking in coatings

85 In Stribeck curve, the coefficient of friction is plotted against

A (Viscosity × Pressure/Velocity)
B (Velocity × Pressure/Viscosity)
C (Viscosity/Pressure × Velocity)
D (Viscosity × Velocity/Pressure)

86 Select proper statement for abrasive wear rate of detonation sprayed WC-12Co coating

A Abrasive wear rate increases with an decrease in abrasive hardness
B Abrasive wear rate increases with an increase in abrasive hardness
C Abrasive wear rate has no relation with abrasive hardness
D None of the above

87 Choose conformal contact from the following:

A Pin-on-flat
B Flat bottom block-on-ring
C Ball-on-disc
D Round bottom pin-on-flat

88 Choose appropriate shape of debris resulting from adhesion wear,

A Wedge-shaped fragments
B Equiaxed-shaped particle
C Needle-shaped particle
D All of the above

89 Choose appropriate shape of debris resulting in abrasive wear,

A Chips shaped
B Wedge shaped
C Equiaxed-shaped particle
D All of the above

90 Abrasive wear rate of metals depends strongly on

A Only hardness
B Only toughness

C Both hardness and toughness
D None of the above

91 In metallic systems, rapid oxidation in sliding wear occurs in
A High temperature and high speed
B Low temperature and low speed
C Low temperature and high speed
D High temperature and low speed

92 Higher wear rate can be observed in erosion of
A Ceramics in normal incidence
B Ceramics in shallow angles
C Metals in normal incidence
D Rubber in normal incidence

93 For zirconia ceramic, choose major toughening mechanism
A Phase transformation
B Crack deflection
C Ductile metal bridging
D Fiber reinforcement

94 For WC-Co cermets, choose the major toughening mechanism from the following:
A Phase transformation
B Crack deflection
C Ductile metal bridging
D Fiber reinforcement

95 Choose dominant cracking type from the following relevant to wear of ceramic materials:
A Cone cracks
B Radial median
C Lateral crack
D All of the above

96 From the following, select specific characters of ceramics that make conventional fabrication technique not suitable for ceramic processing:
A High melting point
B Low temperature stability
C High self-diffusion coefficient
D None of the above

97 Select suitable processing technique for ceramics
A Casting
B Forging
C Powder metallurgical processing
D Rolling

98 Sintering refers to the consolidation process of powder compacts by diffusional mass transport possible at a temperature (in absolute scale)
A >0.5 melting temperature
B <0.5 melting temperature
C =0.5 melting temperature
D > melting temperature

99 Select the most suitable processing technique for processing TiCN-Ni cermets
A Liquid-phase sintering
B Solid-state sintering
C Casting
D All of the above

100 Select correct arrangement of sintering techniques in the increasing order of heating rates (SPS: spark plasma sintering; HP: hot pressing)
A SPS, HP, pressureless sintering
B HP, pressureless sintering, SPS
C SPS, pressureless sintering, HP
D Pressureless sintering, SPS, HP

101 Wear of SiC ceramics in dry unlubricated sliding conditions increased at high loads due to
A Increased fracture
B Increased deformation
C Increased deformation and fracture
D None of the above

102 The grain pull-out of ceramic grains during wear necessarily indicates
A Dominant intergranular fracture
B Dominant transgranular fracture
C Mixed mode of inter and transgranular fracture
D None of the above

103 The significant presence of cylindrical debris on the surface of SiC worn in highly humid sliding conditions:

A Tribochemical wear
B Abrasive wear
C Adhesive wear
D Fatigue wear

104 Compared with the size of wear debris obtained from dry unidirectional sliding of SiC ceramics, size of wear debris obtained from dry reciprocating sliding

A is reduced
B is increased
C remains same
D has no relation

105 In a dry unlubricated sliding study of SiC-WC composites against SiC counterbody, select the most possible result from the following:

A The wear damage in reciprocating sliding conditions is similar in extent to the wear damage in unidirectional sliding conditions.
B The wear damage in reciprocating sliding conditions is more than the wear damage in unidirectional sliding conditions.
C The wear damage in reciprocating sliding conditions is less than the wear damage in unidirectional sliding conditions.
D None of the above

106 Advanced sintering processes, typically

A Lowers the overall densification time
B Involves faster heating rate than conventional sintering
C Involves slower heating rate
D Both a and b

107 From the following, choose WRONG statement related to major issues in fabrication of ceramic nanocomposites:

A Grain growth inhibition while achieving high density
B Nanosized grains retention in sintered microstructure
C Green compaction of nanoparticles without density gradient
D Unlimited grain growth while achieving full density

108 Choose possible ceramic nanocomposites that show possible improvement in machining

A Nano BN dispersed Si_3N_4 composite
B SiC-WC composites
C B_4C-SiC composites
D Al_2O_3-SiC composites

109 From the study of fretting wear behavior of TiCN-Ni cermets against steel ball in dry ambient conditions, choose the most possible observation on dominant wear mechanism

A Abrasion at high load
B Tribochemical wear at high load
C Abrasion and tribochemical wear at high load
D Tribochemical wear at low load

110 For a study of zirconia fretted agilest steel in a pure tribochemical wear condition, select the most possible observation from the following:

A Intergranular fracture and delamination wear dissipates large amount of energy that results in high friction
B Intergranular fracture and delamination wear dissipates large amount of energy that results in low friction
C Intergranular fracture and delamination wear dissipates small amount of energy that results in high friction
D Intergranular fracture and delamination wear dissipates small amount of energy that results in low friction

111 Replacing Co with ZrO_2 ceramic in WC-Co provides

A Resistance to degradation in high-temperature conditions
B Resistance to failure under sudden change in loading
C Improved chemical stability at high temperature
D Superior wear resistance and oxidation resistance
E All of the above

112 For ceramic composites, select the correct statement from the following:

A High hardness results in less erosion wear
B High elastic modulus results in high erosion wear
C High fracture toughness results in less erosion wear
D None of the above

113 With an increase in erodent hardness, the extent of erosive wear of SiC-WC composites

A Decreases
B Increases
C Does not change

114 As the hardness ratio of coating to erodent approaches to one,
A Brittle to ductile transition observed
B Ductile to brittle transition observed
C No change from brittle behavior
D No change from ductile behavior

115 Choose major influencing factors for detonation spray coating microstructure
A Particle temperature
B Particle velocity
C Particle and substrate material properties
D All of the above

116 Choose major bonding mechanisms of thermal spray coatings
A Mechanical bonding
B Metal to metal bonding
C Chemical bonding
D All of the above

117 The responsible mechanism(s) for the improved fracture toughness in ZrB_2-SiC composites with large content of SiC is/are
(i) Crack bridging
(ii) Ductile to metal bridging
(iii) Transformation toughening
(iv) Crack deflection
A i and iv
B ii and iv
C ii and iii
D i and iii

118 From the following, select the possible result in high-temperature (800 °C) erosion wear of ZrB_2-SiC composites with large SiC content:
A Large extent of borosilicate phase reduces material loss
B Large extent of borosilicate phase increases material loss
C Small extent of borosilicate phase reduces material loss
D Small extent of borosilicate phase increases material loss

119 The major tribochemical compound responsible for reducing friction and wear of Mg-SiC composites fretted against steel ball in ambient conditions

A Magnesium oxide
B Silicon oxide
C Dense hydrous magnesium silicate
D None of the above

120 From the following, choose the most preferable test configuration to study the friction and wear of materials for articulating joint applications:

A Sliding wear
B Fretting
C Erosion
D All of the above

121 From the following, select the most possible result from a dry unlubricated fretting wear study of SiAlON ceramics against steel, SiAlON, and Al_2O_3:

A Wear rate is the lowest against steel ball
B Wear rate is the lowest against Al_2O_3
C Wear rate is the lowest against SiAlON
D Wear rate is the highest against steel

122 In a fretting wear test against WC-Co counterbody, YTZP ceramics with high fracture toughness in ambient humid or dry condition exhibited lower wear resistance due to

A Microcracking as a consequence of tetragonal to monoclinic ZrO_2 phase transformation
B Microcracking as a consequence of monoclinic to tetragonal ZrO_2 phase transformation
C Monoclinic to tetragonal ZrO_2 phase transformation due to microcracking

123 Wear volume of ceramics in abrasion conditions is

A Increased with increase of dissipation energy
B Increased with increase of abrasion parameter
C No relation with dissipation energy
D Both a and b

124 With increase in erosion test temperature from 25 to 800 °C, the erosion wear depth for NbB_2-ZrO_2 composites

A Increases
B Decreases
C Remains constant
D None of the choices

125 Two steel balls of radii 10×10^{-3} m and 15×10^{-3} m are in dry unlubricated sliding conditions at a normal force of 5 N; If Poisson's ratio and elastic modulus for steel balls are 0.3 and 2.1×10^{11} Pa, respectively, the contact radius (in micron meters) is:

A 5.8
B 58
C 11.6
D 116

126 After sliding of brass pin-on-steel flat for 30 km at 10 N normal load, the mass losses of the steel and brass are found to be 2 and 34 mg, respectively. The hardness of the steel and brass are 2.5 and 0.8 GPa, respectively, and the density of steel and brass are 8 and 8.5 g/cm^3, respectively. The respective wear coefficients for the brass pin and the steel flat are

A 1.07×10^{-2} and 2.08×10^{-3}
B 1.07×10^{-5} and 2.08×10^{-6}
C 1.07×10^{7} and 2.08×10^{8}
D 1.07×10^{10} and 2.08×10^{11}

127 The following is true for fatigue wear:

A Surface cracks lead to irregular-shaped wear debris
B Subsurface cracks lead to delaminated sheet wear debris
C Both a and b
D Surface cracks lead to delaminated sheet wear debris

128 Select proper arrangement of materials in increasing order of friction-induced temperature in self-mated dry unlubricated sliding conditions,

A Metals, ceramics, polymers
B Metals, polymers, ceramics
C Ceamics, metals, polymers
D Ceramics, polymers, metals

129 Select the polymer processing technique from the following:

A Compression molding
B Powder injection molding
C Extrusion
D All of the above

130 In erosive wear study of WC-Co coating, as the hardness ratio of coating to erodent approaches to one,

A Transition from ductile to brittle behavior is expected
B Transition from brittle to ductile behavior is expected
C No change from brittle behavior
D No change from ductile behavior

A.II Select the Appropriate Combination

1 Select the appropriate combination of material pair and flash temperature rise in dry unlubricated sliding conditions

Material pair	Flash temperature rise (in K)
(1) Steel vs. copper in ambient	(i) 473
(2) Steel vs. steel in vacuum	(ii) 1273
(3) Steel vs. steel in ambient	(iii) 773

A 1-i
B 1-ii
C 2-iii
D 3-ii

2 Choose correct combination for an abrasive wear condition

(1) Ductile material	(i) Chipped out wear particle
(2) Brittle materials	(ii) Ribbon-like wear particle

A 1-i, 2-ii
B 1-ii, 2-i
C 1-i, 2-ii
D 1-ii, 2-i

3 Chose correct combination of erosive wear condition with major worn surface observations

(1) Low-velocity liquid erosive wear	(i) Lips and crater
(2) High-velocity liquid erosive wear	(ii) Holes or pits

A 1-i, 2-ii
B 1-ii, 2-i
C 1-i, 2-i
D 1-ii, 2-ii

4 Match the correct mechanism for solid particle erosion

(1) Low angle of impingement	(i) Plastic deformation/fracture
(2) High angle and low speed	(ii) Melting
(3) High angles and medium speed	(iii) Abrasion
(4) High angles and high speed	(iv) Fatigue

A 1-i, 2-ii, 3-iii, 4-iv
B 1-iii, 2-iv, 3-i, 4-ii

C 1-ii, 2-iii, 3-i, 4-iv
D 1-iv, 2-iii, 3-i, 4-ii

5 Choose the proper reason from the following for brittleness in ceramics:
(i) Ionic bonding
(ii) Covalent bonding
(iii) Narrow dislocation core width
(iv) More than five independent and active slip systems
A i
B ii
C iii
D i and iii
E i, ii, and iii
F i, ii, iii, and iv

6 Select correct statement for ceramic materials
(i) Compressive stress conditions delay fracture
(ii) Tensile stress conditions delay fracture
(iii) Tensile stress condition accelerate fracture
(iv) Compressive stress conditions accelerate fracture
A i and ii
B ii and iii
C i and iii
D i and iv

7 Match the following

(1) Hardness	(i) $MPa \cdot m^{(0.5)}$
(2) Fracture toughness	(ii) N/m^2
(3) Strength	(iii) kgf/mm^2

A 1-i, 2-ii, 3-iii
B 1-iii, 2-i, 3-ii
C 1-ii, 2-iii, 3-i
D 1-i, 2-iii, 3-ii

8 Match the following combination of temperature and particle velocity with the coating type

(1) APS	(i) Low velocity and low temperature
(2) ARC	(ii) Medium velocity and high temperature
(3) DSC	(iii) High velocity and medium temperature
(4) Cold spray	(iv) High velocity and medium temperature

A 1-ii, 2-i, 3-iii, 4-iv
B 1-i, 2-ii, 3-iii, 4-iv
C 1-ii, 2-i, 3-iv, 4-iii
D 1-iv, 2-iii, 3-i, 4-ii

9 Match the effect of materials addition with the most possible improvement of properties in TiCN-based cermets

(1) MoS_2, WC	(i) Hardness
(2) (Nb, Ta)C	(ii) Densification
(3) Ni/Co	(iii) Hot hardness (thermal shock resistance)
(4) TiC	(iv) Toughness

A 1-i, 2-ii, 3-iii, 4-iv
B 1-ii, 2-iii, 3-iv, 4-i
C 1-iv, 2-iii, 3-ii, 4-i
D 1-ii, 2-iii, 3-iv, 4-i

10 Select appropriate combination of COF value and material pair in the dry unlubricated sliding conditions

(1) 0.5	(i) Steel vs. steel in dry unlubricated sliding conditions
(2) 0.2	(ii) Steel vs. graphite in dry unlubricated sliding conditions
(3) 0.1	(iii) Steel vs. steel in the presence of a thick lubricated film
(4) 0.02	(iv) Steel vs. zirconia ceramics in dry unlubricated sliding conditions

A 1-iii
B 1-i
C 3-ii
D 4-iv

A.III Fill in the Blanks with the Most Appropriate Response

1) The type of force that decreases the length and increases the diameter of the material is called _______.
2) The type of force that increases the length and decreases the diameter of the material is called _______.
3) The ceramics are best characterized by the combination of _______ bonding.
4) Instrumented indentation can be used to measure both _______ and _______.
5) The Hooke's law is valid upto the point of ______ limit.

6) If the fracture surfaces of a specimen under tension are flat, it can be concluded that it has undergone ______.
7) In the elastic region, shear stress and shear strain are related as ______.
8) The mode II loading involves the application of ______ force at the crack faces.
9) Welding occurring at sliding contacts is also known as ______.
10) In Stribeck curve, the coefficient of friction is plotted against ______.
11) The coefficient of friction of ceramics under oil lubrication is typically low, because of ______.
12) A little increase in the coefficient of friction during hydrodynamic lubrication stage is due to ______.
13) In mixed lubrication, the coefficient of friction ______ with an increase in speed.
14) From the given set of coefficient of friction (COF) values 0.02, 0.1, and 0.5, write suitable values for the mating materials in a sliding contact.

Mating materials	COF
Steel vs. steel	______
Steel vs. steel in the presence of a thick lubricated film	______
Steel vs. graphite	______

15) The wear involving relative impact motion is known as ______.
16) Pin-on-disc tester is used to evaluate ______ of materials.
17) In fretting wear, the sliding speed is typically much ______ than that in case of pin-on-disk wear tests.
18) The dependence of radial crack length on load follows the relationship: ______.
19) The erosive wear rate for brittle materials has the strongest dependence on ______.
20) Tribology is a ______-dependent property.

A.IV Mention the Appropriate Material/Equipment in the Blank

1) Bearing component to be used in high-temperature applications also requiring high strength retention at elevated temperature: ______.
2) A ceramic material experiencing high wear rate in tribocontact with SiAlON: ______.
3) To improve the lubrication efficiency under load bearing conditions: ______.
4) To improve the wear performance of existing bearing components in environmentally green technology: ______.

5) For phase identification in a tribochemical layer, in the case when XRD is not available: _______.
6) To obtain detailed nanoscale surface topography measurement: _______.
7) To measure mechanical properties (E, H) of an ultrathin solid lubricated layer, deposited on steel substrate: _______.
8) To determine the abrasive wear rate: _______.
9) To obtain bearing area curve (nondestructive route), when asperity heights in microns range: _______.
10) A contact temperature measuring probe, physically not attached to contacting body: _______.

A.V Identify Whether the Following Statements are True/False

1) The most reliable method of evaluating fracture toughness of ceramics is through single edge V-notch bend test.
2) Resistance to crack growth propagation in a material is called fracture toughness.
3) ZTA ceramics are preferred for biomedical applications mainly due to a better combination of biocompatibility and tribological properties.
4) Wear debris at tribocontact causes three-body wear scenario.
5) Diamond-like carbon (DLC) coating on steel reduces EHL friction in lubricating oil contacts.

A.VI Short Review Questions and Descriptive Questions

1 Suggest the best usage of specific brittle materials for the following technological applications:
- **A** Load bearing hip joints
- **B** Cutting tool inserts
- **C** Extrusion dies
- **D** Coatings on Ni-based super alloys
- **E** Mechanical seals

2 Support or contradict the following statements with proper reasoning:
- **A** Ceramics typically have much higher elastic modulus than metals.
- **B** Ceramics are inherently brittle, i.e. dislocation movement is impossible at room temperature.

3 Schematically illustrate the following:

A Two different methods of toughening mechanisms (one each from process zone and bridging zone)

B Fracture toughness measurement of ceramic implants using indentation cracking and SEVNB technique

C Difference between Hardness and Strength (definition and measurement technique) of ceramics

D The differences between polymer, ceramic, and ductile metals w.r.t. tensile stress–strain response.

E Different modes of loading and displacements at crack faces in ceramics (mode I, II, and III).

F Evolution of cone cracks in alumina implant under indentation loading.

G Stress–strain response and the crack propagation in ceramics under uniaxial compression.

4 Illustrate the following with appropriate sketches/graphs:

A Loading configuration for three-point flexural vs. four-point flexural test of ceramics

B Compression failure of ceramics

C Indentation cracking-based toughness measurement

5 Explain briefly the guidelines to be followed for the following property measurements in case of tribological materials:

A Hardness

B Fracture toughness

C Strength

6 Fracture is an important performance-limiting property of ceramics. Explain Griffith's model of fracture to define the critical crack size. In this context, define stress intensity factor (mode I). How can one define fracture toughness?

7 On a single stress–strain plot, distinguish the difference in deformation behavior of a brittle ceramic, ductile metal, and thermoplastic polymer, when all experience independently uniaxial tension.

8 For desired application, you need to fabricate the sintered monolithic materials with uniform density (98% or more) on industrial scale production. The application requires minimum property combination, as follows: Hardness – 11 GPa, toughness – 9 MPam$^{1/2}$; E – 200 GPa, and application temperature – up to 500 °C (dry air atmosphere). The properties should be isotropic on a macroscopic scale. Given the above information,

A What would be your material (composition) of choice?

B What are the stages in production, i.e. processing steps to ensure uniform densification?

C What are the experimental techniques that you adopt to characterize the microstructure (grain size) and properties (H_v, K_{Ic}, E, 6_f) of the sintered materials?

D Do you really require a precise/close microstructural control in terms of grain size or material composition to achieve desired toughness? If so, explain the reasons from the fundamental thermodynamic point of view.

E Now, if you want to increase the hardness, what would be your cost-effective strategy?

F What are the factors and mechanisms that would contribute to toughness, while improving hardness?

9 Out of the surface engineering techniques, which is the technique that takes place at lowest temperature than other techniques and with the lowest depth of hardened layer? What is the mechanism and variables in that technique?

10 List down at least four different surface modifications typically used to improve the wear performance of steel. Write down the steps that you need to take in order to analyze tribological problems and to come up with a feasible technological solution.

11 Differentiate between Flame Spraying vs. Plasma Spraying (use sketches wherever required).

12 Explain chemical and metallurgical processes involved in thermal spraying of WC-Co powders on a metallic substrate.

13 What would be the different surface layers on a vertical c/s microstructure of a Ni-based superalloy used as a gas turbine component? Explain the selection of materials at different layers? Which process is usually adopted to deposit the topmost layer and would that layer be desired to be porous or dense and why?

14 Define Tribology.

15 Mention four examples of situations where (i) low friction is desirable and (ii) friction should be high. Justify in case example why high or low friction is important.

16 Support or contradict the following with proper reasoning:

A Friction and wear is not a material property, but a system property!
B Tribology is considered as an important research area as far as nation's economy is concerned!

17 Support or contradict the following with proper reasoning:
A Surface profilometry is an essential tool to characterize surface profiles, when other characterization tools are available.
B R_q is a more significant parameter than R_a.

18 What is the information one needs to know in order to understand the behavior of rough surfaces?

19 Differentiate (with sketches and mathematical expressions, wherever applicable) for the following:
A Kurtosis and skewness
B Autocorrelation function and cumulative height distribution function

20 If R_q and R_a values are insufficient, then what parameters/functions one needs to know in order to distinguish two tribosurfaces?

21 Draw schematically the stress distribution both at the surface and subsurface region for the circular Hertzian contact. From the stress distribution, can you comment on the yielding of the contact? If material A has a higher hardness than material B, which one will yield first at the contact of A & B and why?

22 Graphically show how the plowing contribution to friction increases with the groove width for spherical hard asperity. How will be the relationship for conical asperities? Why plowing contribution is inadequate to explain experimentally obtained friction values?

23 Distinguish between the following, use sketch wherever necessary:
A Roughness and waviness
B Height distribution curve and bearing area curve, with special reference to the importance to the construction of bearing area curve

24 Graphically show the load-dependent frictional behavior for the three major classes of engineering materials. Just mention the major mechanism on each graph.

25 Given the properties of metals, ceramics, and polymers, what would be the fundamental mechanistic differences in the dissipation of frictional energy for the self-mated metals, ceramics, and polymers?

26 Draw a bearing area curve and explain how it is constructed from a given surface profile. Indicate the important bearing curve parameters and mention the physical significance of each parameter.

27 What are the basic assumptions made in the Hertzian contact analysis?

28 What are the two basic material parameters that need to be optimized (and how?) in order to prevent yielding at the Hertzian contact?

29 Distinguish between contained and uncontained plasticity. How the contact behavior depends on the material parameters: show graphically.

30 Write basic working principles of the following: (a) AFM and (b) Stylus profilometer.

31 List the different techniques for measuring the contact temperature. What is the fundamental equation that needs to be solved to compute the contact temperature? Why measurement of the flash temperature with any of the available technique is difficult?

32 Briefly specify the reasons. Why solid lubricated materials and polymers show low COF compared with other engineering materials?

33 What would be the preventive measures that you will consider to reduce the unlubricated adhesive wear of materials, when the contact is predominantly either (a) plastic or (b) elastic?

34 Briefly explain the term "junction growth" and comment on its importance in the sliding friction of metals.

35 Draw useful plots to show how wear rate varies with

A displacement amplitude in case of fretting wear

B mechanical properties (H_V and K_{IC} combination) in case of abrasive wear

C abrasive particle size in case of metals (for both two-body as well as three-body)

36 Justify the following:

A Cracks are commonly observed at the edges of Hertzian contacts of brittle materials.

B For self-mated metals, plastic contact is more common, while elastic contacts for self-mated ceramics.

37 Graphically show how does the elastic–plastic transition depend on (a) sharpness of conical indenter and (b) size of spherical indenter?

38 Justify the following:

A Typically wear debris particles are coarser for metallic materials, while finer for ceramics.

B Rolling friction is always less than the kinetic friction in sliding.

39 Graphically show how the erosive rate of brittle materials is dependent on the angle of impingement and mechanical properties. How does the erosive wear mechanisms change with the angle of impingement?

40 Alumina is tested in a high-temperature wear apparatus ($T \sim 1200\,°C$). Would you expect any difference in the wear behavior under similar sliding conditions compared with that under room temperature?

41 With schematic sketches, show how the material fragment gets detached from the worn surface due to plastic shear crack growth?

42 Differentiate in terms of the process/mechanisms (use sketches wherever required):

A Crossed cylinder contact vs. four ball tribocontact

B Erosive wear by laminar flow and turbulent flow

C Oxidative wear vs. tribochemical layer

D tribochemical and lubricated wear

43 Why oxidation is enhanced during sliding?

44 Justify the following observations:

A In a cylindrical steel pin-on-steel disc sliding wear test in air, the coefficient of friction (COF) reached peak value of 0.7 in initial stages that reduced to 0.5 during further course of sliding. High level of oxygen found on the wear track that is surrounded by black debris.

B High COF and wear volume are found when the identical test to that mentioned in (a) is conducted in a vacuum. The worn surface of the disc is silvery and rough.

C Worn surface is found to be black, shiny, and smooth when the identical test to that mentioned in (a) is conducted in air at 300 °C. High level of oxygen is found on the wear track while no wear debris particles are found outside the wear track.

D With an increase in load, the wear rate increases by two orders of magnitude and wear track becomes rough in a sliding test of alumina ball against an alumina disc in air.

E In general condition of laboratory, low COF and wear rates are found in a sliding test of silicon nitride disc against a silicon nitride ball, whereas rapid failure occurred when the sliding test is done in an actuator on a satellite in space.

F The wear rates are found to be low when a polymer disc slides against a steel ball. With an increase in sliding velocity or normal load, very different coefficient of friction and one order of magnitude higher specific wear rates are found.

45 A ductile material A is to be worn against steel on a pin-on-disk machine to evaluate its potential as a tribomaterial. In order to establish wear mechanisms and wear transition,

A What would be the parameters to characterize the unworn surface of "A" and by what technique?

B How will you design experiments, i.e. what would be the variables in your wear tests?

46 What are the different characterization tools you will use to measure wear rate and wear mechanism including the analysis of wear debris particles and subsurface wear damage? For each tool, mention the basic mechanism (one liner) and characteristic output/information that you expect to obtain. What characterization tool is normally used to confirm the formation of hydroxide tribolayer?

47 Draw useful plots/sketches to illustrate the following:

A Abrasive particle size in case of metals (for both two-body as well as three-body)

B The abrasive wear mechanism of ceramics by fracture

48 With schematics, illustrate the stages and mechanisms involved in the wear of ZrO_2(Y-TZP)-20 vol% SiC_p composites against steel for the sliding conditions when tribomechanical and tribochemical both significantly contribute to wear process.

49 Comment on the possible wear mechanisms

A SiAlON ceramics against alumina
B TiB_2 materials against steel under unlubricated conditions
C Self-mated zirconia in high humidity
D SiAlON-TiB_2 materials against steel under unlubricated conditions

50 Write a brief account on the how the toughness can influence the wear behavior of transformation-toughened zirconia ceramics.

51 Summarize all the factors influencing the general wear behavior of metals and ceramics with a special reference to the latter. Write a one line brief statement about their influence against each factor.

52 A vertical cross-section of the worn surface for the following materials is investigated under TEM: (a) pure Al and (b) Al_2O_3. What characteristic microstructural differences would you expect when all are worn under severe wear conditions under ambient conditions? How would be the sub-surface hardness change for (a)?

53 If a steel material is tested at −260 °C in a cryogenic liquid, what change would you expect in the COF and wear mechanism when all are worn under similar severe sliding conditions?

54 How one can obtain E-modulus and toughness of a thin (~0.2 μm) ceramic coating? Plot and compare the response of hard and soft coatings against a nanoindenter.

55 Derive Archard's equation for volume of the material removed in adhesive wear contacts.

56 What is abrasive wear? Explain two-body abrasive wear and three-body abrasive wear, with neat sketches.

57 Using Griffith theory, derive relation for fracture stress for a ceramic plate with central through-thickness crack.

58 (a) Differentiate between material removal in unidirectional sliding and reciprocatory sliding contacts.
(b) Identify the respective wear test configurations for the following schematic representations:

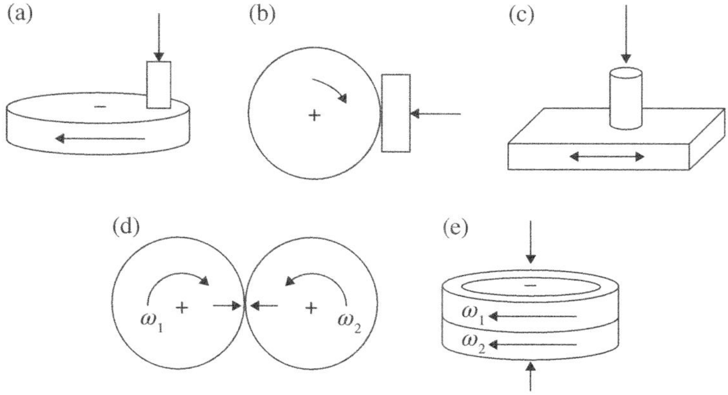

59 In separate erosive wear studies of SiC ceramics by SiC erodent and low carbon steel by SiC erodent, maximum wear obtained at higher angle of incidence of erodent. Support or contradict the finding with proper reasoning.

60 Identify major wear mechanism(s) from each worn surface image.

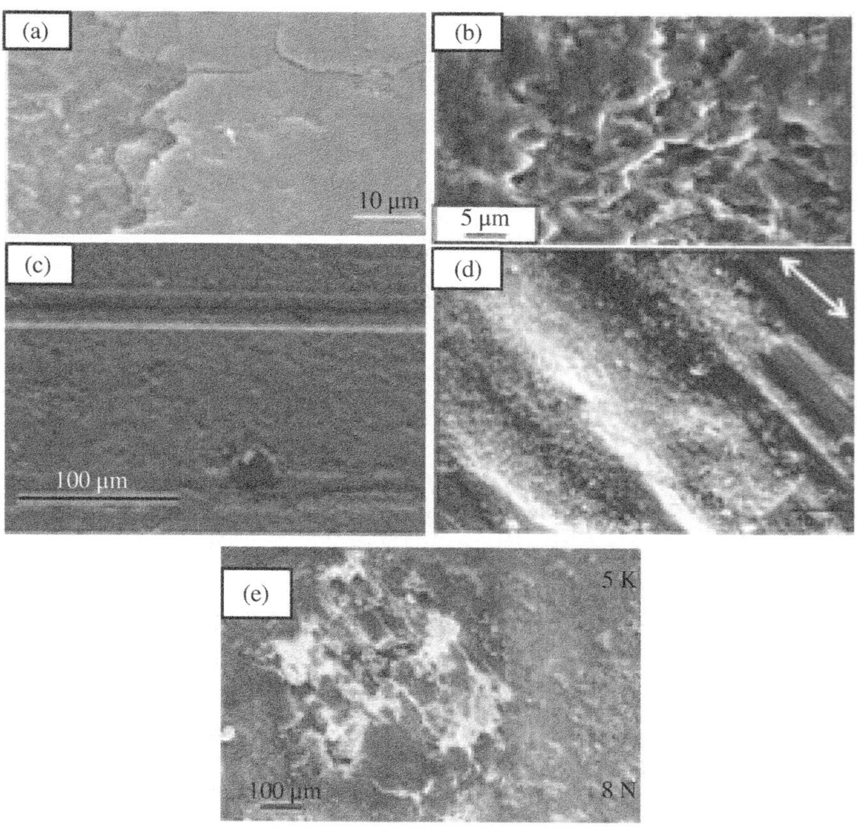

61 Derive a suitable relation for volume of the material removed in abrasive wear by plastic deformation.

62 With neat sketches, explain radial-median and lateral cracking patterns that are relevant for removal of brittle solid material in abrasive wear conditions.

63 What is fretting wear? Explain stick, partial slip, and gross slip regimes in elastic–plastic contact situations of fretting.

64 Draw typical erosion wear vs. impingement angle plots for ductile and brittle materials, and explain reasons for the difference in erosive behavior.

A In separate studies, SiC-WC composites were eroded by stream of silicon carbide (SiC) particles and stream of alumina (Al_2O_3) particles. With proper reasoning, identify the case that resulted in higher wear of composites.

65 In a research study, WC-Co coatings on steel substrate were subjected to abrasion by SiO_2, Al_2O_3, and SiC particles. Write possible common mechanisms of material removal. Also, comment on the abrasive wear rate with respect to given abrasive particles.

A.VII Analytical Questions

1) A thin wire of a shape memory alloy-based cardiovascular stent with a diameter of 0.75 mm experiences a tensile load of 250 N and at this load, it is found that the 100 m wire elongates by 5 cm. Calculate the engineering stress and engineering strain at this point. After the load is removed, the length of the wire is found to be 100.03 m. Calculate the yield strength, if the elastic modulus is given to be 200 GPa.
2) A plate made of an unknown oxide bioceramic has a fracture toughness of $2\,MPa.m^{1/2}$ and a plate of alumina with a fracture toughness of $5\,MPa.m^{1/2}$. Due to difference in manufacturing techniques used to produce these two materials, the longest surface crack is found to be 3 μm in the unknown bioceramic, whereas it is 7 μm in alumina. Determine which plate is to be used, if the maximum allowable stress is 250 MPa in the direction perpendicular to the crack.
3) The plain strain fracture toughness of a hypothetical biomedical grade Ti-alloy is $K_{IC} = 24\,MPam{\cdot}m^{1/2}$. Assume Y and tensile yield strength are 1.5 and 500 MPa, respectively. Determine the size of the smallest crack that can lead to brittle. If the elastic modulus is 71 GPa, what is the surface energy? (Hint: Use the crack length obtained in the previous section and apply the Griffith's theory for brittle fracture).

4) What would be the allowance crack size for Al_2O_3 ($K_{IC} = 3\,MPa{\cdot}m^{1/2}$) and how it will be different for ZrO_2 ($K_{IC} = 10\,MPa{\cdot}m^{1/2}$), when both experience 100 MPa tensile stress? Explain each term in the equation that you use. Assume, $Y = 1$.
5) Determine the load at which zirconia ceramic implant with a critical crack size of 1 μm ($K_{IC} = 10\,MPa{\cdot}m^{1/2}$, $E = 200\,GPa$) will fracture under four-point loading. What will be the radius of curvature at fracture for the same implant sample (square cross-section with thickness 5 mm and span length 40 mm).
6) Determine the fracture strength of an alumina ceramic with a critical surface crack size of 1 μm ($K_{IC} = 4\,MPa{\cdot}m^{1/2}$, $E = 400\,GPa$). What will be the radius of curvature at fracture for the same alumina bar (rectangular cross-section with thickness 5 mm) under flexural loading?
7) An alumina ball of 10 mm diameter is pressed against a nominally steel plate. (a) What normal load needs to be applied in order to initiate yield in the plate; (b) what would be the contact radius at the point of yield; and (c) at what depth the yield will first occur? Assume Hertzian contact behavior and hardness of steel is 7 GPa.
8) For the given mean peak radius of 10 μm and standard deviation of peak heights of 0.4 μm for lapped steel plate surfaces, determine the expected asperity deformation (elastic or plastic)? Will the result depend on the normal load, if yes, then how?
9) Determine parameters of contact for two steel balls: The radii of the balls are 10×10^{-3} m and 15×10^{-3} m; the normal force is 5 N; the Young's modulus for both balls is 2.1×10^{11} Pa; Poisson's ratio of steel is 0.3.
10) Determine parameters of contact for a steel ball on a flat steel plate. The radius of the ball is 10×10^{-3} m; the normal force is 5 N; the Young's modulus for ball and plate is 2.1×1011 Pa; Poisson's ratio of steel is 0.3.
11) For the given mean peak radius of 10 μm and standard deviation of peak heights of 0.2 μm for lapped alumina plate surfaces, determine the expected asperity deformation (elastic or plastic)? Will the result depend on the normal load, if yes, then how?
12) In sliding of a soft and flat surface against hard ball at two different loads, the frictional coefficient is 0.25 and 0.35 and the groove widths are 1.0 and 1.5 mm. Calculate the adhesive component of friction, provided two major sources of friction are additive in nature.
13) Two alumina plates (identical composition) each with standard deviation of peak heights of 0.4 μm and mean peak radius of 5 μm are in tribocontact under a load of 8 N. What would be the predominant wear mode and compute the volume worn away after sliding takes place for an hour, given the sliding velocity of 50 mm/s? Assume a typical value of wear coefficient of 10^{-8}. Also, write the major assumption that you considered in calculation as far as the contact behavior is concerned.

14) Calculate the expected surface crack size for a dental crown material if the fracture toughness, K_{IC}, is 0.7 MPam$^{1/2}$ and the applied stress determined from FEM analysis is 21 MPa.
15) Two zirconia plates (identical composition) each with standard deviation of peak heights of 0.4 μm and mean peak radius of 5 μm are in tribocontact under a load of 8 N. Compute the volume worn away after sliding takes place for an hour, given the sliding velocity of 50 mm/s? Assume a typical value of wear coefficient of 10^{-6}.
16) Under 10 N normal load, a cylindrical bronze pin of radius 1 mm rests on a steel disk at a mean radius of 25 mm. If the disk rotates at 300 rpm for 10 hours, mass losses of the disk and pin are 3 and 50 mg, respectively. Calculate the wear coefficient of steel disk.
17) The material removal rate of SiC in grinding is 2×10^{-6} μm^3/h. Determine the material removal rate if SiC with 1.5 times fracture toughness and same hardness and same elastic modulus is used in grinding.
18) Coefficients of friction of 0.4 and 0.2, respectively, are obtained when a conical indenter and a flat body of same material are independently slid against a soft surface. Determine the coefficient of friction for a conical indenter of same material with a half of the semi-angle. The cross-sectional area of flat body is large enough for negligible plowing contributions.
19) A ceramic component has internal cracks of 0.001 cm in length. The plane strain fracture toughness of the component is 4.5 MPam$^{0.5}$ and the tensile strength is 660 MPa. Comment whether the cracks cause the component to fail before the tensile strength is reached.
20) In a dry unlubricated sliding test, a SiC ball of radius 10 mm rests under 10 N normal load on a SiC disk at a mean radius of 1.5 mm. The disk rotates at 500 rpm for 30 minutes. The average frictional force in steady state was found to be 3 N. The average width and average depth of wear track were 100 and 10 μm, respectively. Determine the wear volume of the disk and average steady-state coefficient of friction.
21) Determine mean Hertzian stress in contact of steel ball with a flat steel plate. The radius of the ball is 5×10^{-3} m. The normal force is 10 N. The Poisson's ratio and Young's modulus for the steel are 0.3 and 2.1×10^{11} Pa, respectively.
22) The strength and fracture toughness of a silicon nitride component are 300 MPa and 3.6 MPa.m$^{0.5}$, respectively. Determine the maximum size of internal crack that causes no fracture in the component. Assume $Y = 1$.
23) In a dry unlubricated sliding test, a steel ball of radius 10 mm rests under 5 N normal load on an aluminum alloy disk at a mean radius of 3 mm. The disk rotates at 250 rpm for 60 minutes. The average frictional force in steady state was found to be 3 N. The average width and average depth of wear track were 200 and 20 μm, respectively. Determine the wear volume of the disk and average steady-state coefficient of friction.

24) Determine maximum Hertzian (initial) stress in contact of steel ball with a flat steel plate. The radius of the ball is 10×10^{-3} m. The normal force is 5 N. The Poisson's ratio and Young's modulus for the steel is 0.3 and 2.1×10^{11} Pa, respectively.
25) The alumina pin mass loss after sliding against silicon carbide disc rotated at 60 rpm in a spiral track for one minute is 0.2 g. Estimate the mass loss for a soda-lime glass pin if abraded in above-mentioned test conditions. Density of alumina: 3900 kg/m^3; density of soda-lime glass: 2500 kg/m^3; hardness of alumina: 1600 HV; hardness of soda-lime glass: 540 HV; fracture toughness of alumina: 4.0 MPa·m$^{1/2}$; fracture toughness of soda-lime glass: 0.75 MPa·m$^{1/2}$; Young's modulus of alumina: 400 GPa; Young's modulus of soda-lime glass: 75 GPa.

A.VIII Model Answers

Multiple Choice Questions

1) D
2) A
3) A
4) C
5) A
6) C
7) B
8) A
9) B
10) A
11) D
12) A
13) C
14) A
15) C
16) B
17) C
18) A
19) A
20) A
21) C
22) B
23) A
24) B
25) C
26) B
27) D
28) B
29) A
30) A
31) A
32) A
33) A
34) D
35) D
36) A
37) A
38) A
39) A
40) C
41) A
42) C
43) C
44) C
45) C
46) B
47) C
48) A
49) C
50) B
51) A
52) A
53) D
54) B
55) D
56) A
57) B
58) C
59) A
60) A
61) C
62) D
63) A
64) B
65) A
66) C
67) B
68) A
69) A
70) B
71) A
72) D
73) C
74) D
75) A
76) D
77) D
78) C
79) B
80) B
81) A
82) B
83) A
84) D
85) D
86) B
87) A
88) A

89) A
90) A
91) A
92) A
93) A
94) C
95) C
96) A
97) C
98) A
99) A
100) A
101) A
102) A
103) A
104) A
105) C
106) D
107) D
108) A
109) B
110) A
111) E
112) C
113) B
114) B
115) D
116) D
117) A
118) A
119) C
120) B
121) D
122) A
123) D
124) B
125) B
126) B
127) C
128) B
129) D
130) B

Select the Appropriate Combination

1) A
2) B
3) B
4) B
5) D
6) C
7) B
8) A
9) B
10) B

Fill in the Blanks with the Most Appropriate Response

1) compression
2) tension
3) ionic and covalent
4) hardness, elastic modulus
5) elastic
6) brittle fracture
7) $\tau = G \cdot \gamma$
8) shear
9) adhesion
10) $(\eta v/F)$
11) tribochemical reaction
12) internal viscous friction
13) decreases
14) 0.5, 0.02, and 0.1
15) erosion
16) tribological properties
17) less
18) $P \propto C_r^{3/2}$
19) angle of impingement
20) system

Mention the Appropriate Material/Equipment in the Blank

1) Si_3N_4
2) Si_3N_4
3) Additives (EP)
4) Solid lubricants (DLC/graphite/MoS_2)
5) XPS or FT-IR
6) AFM
7) Nanoindentation
8) Pin-on-disk or Ball crater tester
9) Stylus profilometer
10) Thermal Camera (Pyrometer)

Identify Whether the Following Statements are True/False

1) True
2) True
3) True
4) True
5) True

Index

a

abrasion 13, 14, 17, 85, 156, 159, 161–163, 165, 167, 168, 170, 171, 181, 190, 198, 199, 203, 225, 227–232, 297, 302–306
- two-body 162
- three-body 162

abrasive 9, 13, 32, 158, 159, 161, 162, 164, 167–170, 173, 196, 220, 224, 227–234, 302, 305

acetabular socket 216 *see also* implant(s)

additive 19, 63, 64, 134, 175, 178–181, 187–192, 216,240, 325

adhesion 14, 39–41, 121–123, 128, 130, 133, 139, 156, 159, 160, 165, 177, 181, 220, 237, 238, 241, 249, 297, 300, 302–305, 325
- friction 140 *see also* friction
- interfacial 160

agglomerates 206, 208, 261, 262

agglomeration 25

Al_2O_3 7, 8, 10, 13, 24, 29, 64, 82, 129, 169,174, 215–218, 220–224, 227, 228, 230–233, 239, 242, 256–258, 266, 280, 281–292, 302, 305, 320

alloy 11, 42, 53, 57, 65, 100, 128, 130, 152, 131, 309
- metallic 7
- intermetallic 128
- superalloy 9

alloying 61, 64

alumina *see* Al_2O_3
- zirconia-toughened alumina (ZTA) 215–224

Amonton's laws 120, 121

amorphous phase 45, 46, 49
- anisotropy 7, 136, 137, 223

applications
- high temperature 265, 322
- tribological 3, 6–11, 14, 30, 57, 71, 79, 84, 88, 99, 187, 195, 255, 319, 325

Archard
- equation 162
- model 147, 149
- wear theory 160

aspect ratio 89

asperity 105, 113, 114, 119, 121–123, 125, 130, 134, 140–142, 144–146, 149, 150, 159, 160, 162, 172, 180, 181, 184, 305

atomic force microscopy (AFM) 108, 109, 134, 150, 181

Friction and Wear of Ceramics: Principles and Case Studies, First Edition.
Bikramjit Basu, Mitjan Kalin, and B. Venkata Manoj Kumar.

b

B_4C 8, 64
basal planes 132
bearing area 111–113
  curve 111,113
bending 109
  four-point 89, 90, 96
  three-point 89, 90, 95
binder 24, 43, 45, 46, 227, 231, 232, 279, 291, 297, 309, 312, 313
bioceramics 215
biomaterial(s) 84, 131
bond 23, 32, 37, 39–41, 80–82, 123, 128, 132, 137, 159, 160, 171, 181, 228–231, 242, 279, 291
  cohesive 23
  covalent 8, 78, 130, 181
  ionic 78, 130
  metallic 130
  van der Waals 78, 130 *see also* force(s)
boundary interface 238 *see also* layer
brittle materials 84, 95, 96, 114, 161–164, 168, 170, 203, 269, 281, 292
brittleness 22, 80, 87

c

carbide(s)
  cemented 42, 206, 300, 303–306
  secondary 297, 309, 312, 313
case study
  machining-induced wear of cermets 295
  erosive wear of SiC ceramics 253
  sliding wear of SiC ceramics 185
  WC-Co cermets 42
ceramics
engineering 8, 64, 215
  monolithic 11, 86
  non oxide 8, 11, 173, 174, 192, 193, 239, 242, 266
  oxide 8, 11, 215, 239, 242
  structural 3, 8, 9, 11, 17, 130, 239
  traditional 8
cermet(s) 13,16, 43, 296, 203, 287, 297–306, 309–313
TiCN 13, 297–306, 309–313
WC-Co 43–45, 47, 48, 287
chemical 5, 6, 12, 39, 47, 121, 141, 155, 177, 265, 304–306, 324
  reaction 171, 321
  stability 127
chemistry 5, 6, 171, 173, 179, 181, 327
chipping 170, 198, 256, 257, 258, 287, 290, 304
coating(s) 11, 14–17, 21, 22, 30, 31–37, 39, 40–49, 53, 56, 111, 134, 160, 178, 227–234, 240, 279–282, 285–293, 325
  diamond-like carbon (DLC) 11, 237, 239, 242–249
coefficient of friction 121, 123, 123, 127–132, 134, 135, 140, 142, 146, 147, 148, 174–178, 188–190, 197, 202, 205, 208, 220, 223, 237, 246–248 *see also* friction
coefficient of thermal expansion 223, 274
composite(s) 3, 5, 8, 10, 12, 13, 17–18, 30, 43, 58, 64–65, 67, 69–70, 86, 96, 118, 132, 142, 146, 153, 162
  ceramic 5, 8, 11, 12, 14, 16, 17, 30, 64, 86, 162, 198, 223, 265–270, 319, 323
  metal matrix 7
  particulate 7, 8, 12, 43
  SiC-WC 16, 195–210, 255–262. 266
  ZrB_2-SiC 16, 265–275, 323
composition 24, 38, 42, 178, 198, 203, 215–217, 227, 230, 267, 269–271, 274, 280, 299, 300, 304, 311–313, 324, 325
conduction 114, 146, 270

contact
  elastic 115, 166
  plastic 105, 125, 166, 325, 326
  single-asperity 134, 150
contaminant 123, 160
convection 114, 146
Coulomb's law 121, 131
crack(s)
  cone 87, 163
  deflection 219
  lateral 85, 161–163, 170, 171, 187, 191, 258, 261
  length 82, 85, 87, 96, 98
  median 98, 161, 170, 171
  primary 81, 164
  propagation 80–82, 88, 161, 162, 195, 203, 218, 223, 293, 302
  short, method 96
crack tip 80, 82–84
crater wear 297, 304
crown 13
cryogenic environment 146
cutting tools 11, 167, 187, 255

d

data science 16, 323, 324, 326
  debonding 315
deformation 22, 26, 42, 77–80, 88, 95, 99, 105, 112–114, 116, 122–123, 126, 131, 140, 162, 164, 169, 170, 191, 195, 202, 237, 256, 260, 272, 282, 286, 287, 291, 293, 322
  elastic 123, 125, 126, 126, 166, 177, 261
  plastic 77, 87, 125, 141, 147, 160–162, 175, 256, 261, 282, 287
  viscoelastic 130
delamination 116, 130, 165, 202, 224, 233, 291 *see also* wear
densification 23, 24, 27–29, 49, 64, 298
density 7, 9–12, 22–24, 26–28, 30, 43, 95, 129, 130, 148, 157, 168, 174, 216, 228, 229, 255, 270, 319
dental 12, 172
deposition 21, 30, 32–37, 49, 53, 62, 227, 228, 265, 267, 268, 270, 276, 280
  detonation spray 32, 33, 36–39, 41–45, 227, 228, 233, 279, 280, 292
  flame spray 33–35, 39
  high-velocity oxy-fuel (HVOF) 33, 36, 37, 39, 43, 44, 47, 49, 279, 280
  plasma spray 33, 34, 43, 44, 49, 64, 228, 279, 280
  wire arc spray 35, 36
dislocation 164
dispersion 249
dissociation 38, 137
dissolution 46–48, 65, 227, 291, 297
drilling 3, 54, 167
ductile metals 168, 281
ductility 42, 130
durability 6

e

elastic modulus 7, 9, 10, 15, 43–45, 79, 80, 91–95, 98, 99, 113, 115, 143, 158, 164, 167, 181, 191, 203, 224, 270
  see also mechanical properties
elastomers 6, 7, 129, 131
elastohydrodynamic lubrication 176, 177, 237, 242
electrical conductivity 58, 309
electrochemical 63
electro-discharge machining (EDM) 53, 57–66, 309–313
  die sinking 57, 58, 65, 309, 310, 312
  wire 30, 34–36, 57–59, 134
enamel 172
energy
  dissipation 146
  kinetic 42, 167, 170, 267–268, 291

environment 5, 6, 14, 15, 17, 27, 122, 134, 146, 149, 156, 171, 175, 206, 265, 266, 270, 273, 322
erosion 5, 14, 15, 61, 62, 156, 159, 167–171, 174, 175, 181, 255–262, 265–275, 279–281, 290, 292, 322, 323 *see also* wear
  brittle materials 84, 95, 96, 114, 161–164, 168, 170
  ductile materials 116, 169, 292
erosive
  thermo-erosive 265, 266, 268–270, 272–274
exfoliation 174

f

fatigue 63, 123, 159, 162, 164, 165, 181, 205, 298, 325, 326 *see also* wear
Fe_2O_3 174, 201
FEM 40
femoral ball 13 *see also* implant(s)
fiber 7, 12, 29, 85, 86, 256
  reinforcement 86
film(s) 5, 11, 16, 56, 105, 122, 123, 127, 134, 156, 175, 177, 179, 180, 237 *see also* layer
fluid flow 167
force(s)
  adhesive 130
  attractive 249
  friction 116, 119–123, 138, 139, 180, 237
  interatomic 130
  repulsive 109
  van der Waals 132
fracture
  brittle 80, 82, 161, 162, 174, 175, 228, 256, 282
  intergranular 204, 300
  mechanics theory 84–85 *see also* theory
  mixed-mode 282, 292
  resistance 85, 215
  transgranular 188, 204
fracture strength 83, 89 *see also* strength
fracture toughness 9, 10, 15, 43–44, 71, 84, 85, 95–97, 99, 162, 164, 167, 174, 188, 191, 196, 198, 202, 203, 206, 210, 215, 217–219, 224, 230, 255, 259, 270, 274, 275, 279, 298, 320 *see also* mechanical properties; toughness
fretting 5, 111, 112, 140, 145, 146, 159, 163, 165, 166, 181 *see also* wear
  corrosion 166
  fatigue 63, 123, 159, 162, 164, 165, 181, 205, 298, 325, 326
  wear 158, 159, 163, 165, 166
friction, *see also* coefficient of friction
  force 120–122, 138, 188 *see also* force(s)
  heating 139, 140
  laws 119, 120
  loss 114, 127
  mechanisms 119, 237
  static 121–123, 125, 126
  steady-state 123, 124, 127, 130, 143, 146, 150, 189, 197, 205, 219, 230, 244, 256, 257, 262, 267, 281
fullerenes 129, 134

g

glass 6, 7, 82, 92, 131, 256, 292, 293
  soda-lime 85, 163
  borosilicate 266
grain 11, 22, 23, 130
  boundary 23, 63, 223
  elongated 7, 11, 12, 40, 188
  equiaxed 11, 12, 23, 188, 196, 216
  growth 24, 27, 29, 30, 43
  pull-out 86, 287, 300, 304, 306–308
  size 14, 196, 198, 216, 218, 223, 224, 259, 324

graphite 28, 31, 57, 129, 132–134, 309
Griffith's theory 83
grinding 3, 5, 24, 54, 59, 130, 156

h
hardening 6, 105, 160
hardness 7–11, 13, 15, 42–45, 56–59, 87, 88, 95, 99, 113, 149, 158, 160–165, 167, 168, 170, 181, 187, 191, 195, 196, 198, 202, 203, 217, 218, 224, 227–233, 259, 262, 265, 270, 274, 279, 280, 282, 287, 297, 298, 302, 305, 319 *see also* mechanical properties
heat 15, 23, 27, 38, 42, 63, 114, 119, 121, 140–149, 155, 178, 270 *see also* friction, heating
  dissipation 146
  generation 119, 146, 155
  steady-state flow 143
Hertzian theory 115, 118
HfC 298
high temperature wear 321–323
Hooke's law 109
Humidity 15, 130, 134, 174, 181
hydroxyapatite (HAp or HA) 13

i
impact angle 268, 280–291, 293
implant(s) 12, 131
indentation
  cracking 85, 97, 217
  fracture theory 170
  toughness 45, 96, 174,
  Vickers hardness 11, 43, 44, 88, 198, 217
indenters 84, 85
inertness 15
  chemical 265
interaction(s) 4–6, 105, 123, 125, 126, 133, 139, 155, 161, 237–239, 248, 249, 326
  multi-asperity 141
  physicochemical 180
  solid-liquid 181, 237–239, 241, 242, 248, 249
interatomic
  distance 83
  force 130 *see also* force(s)
interfacial properties 17
interlocking 11, 39, 49, 121, 123, 130
iron oxide 200

j
junction growth 128 *see also* adhesion; friction, mechanisms

l
layer 39, 41, 42, 62–64, 121, 125, 129, 130, 132, 134, 137, 140, 156, 159, 172–175, 180, 181, 198, 199, 202, 222, 238, 266, 290, 300, 302, 305, 306, 309, 312, 313, 322 *see also* tribochemical layer
  hydrated 175, 190, 206, 208
  hydroxide 156
  low-shear 134
  protective 175, 202, 204, 266
  sacrificial 181
  wear debris 5, 131, 160, 165, 172, 178, 201, 206, 208, 209, 306
lubricant 5, 15, 16, 56, 101, 122, 127, 129, 131, 132, 134, 150, 156, 160, 175–178, 180, 181, 237, 238, 240, 242, 244, 246, 248, 250
  solid 129, 131, 151
lubrication 3–6, 14–17, 105, 134–136, 155, 156, 158, 160, 162, 164, 166, 168, 170, 172, 174–182, 237, 242, 325
  boundary lubrication (BL) 135, 177
  elastohydrodynamic (EHD) 176, 177, 237, 242
  mixed lubrication (ML) 135, 177, 237
  regimes 6, 15, 155, 175, 176, 178, 181

m

machine learning 326
machining 5, 11, 13–16, 53, 54, 56–66, 297–300, 302, 304–306, 309–310, 312, 322
machining-induced damage 297
maintenance 156, 321
material(s)
  metallic 7, 8, 91, 127, 130, 142, 309
  properties 6, 15, 42, 121, 145, 156, 181, 215, 270, 304, 322
  transfer 304–306, 309
materials engineering 5, 6, 326
materials science 22, 324, 327
matrix 7, 12, 24, 46, 86, 131, 164, 206, 208, 224, 255, 256, 262, 287, 290, 291
mechanical engineering 5, 6, 21, 326, 327
mechanical properties 8, 9, 15–17, 45, 71, 72, 74, 76, 78, 80, 82, 84, 86–88, 90, 92, 94, 96, 98–100, 113, 131, 134, 141, 150, 157, 158, 191, 196, 203, 216, 217, 259, 260, 265, 266, 274, 302, 309, 320
melting point 7, 9, 22, 42, 62, 64, 131, 137, 139, 265, 297, 309, 313
metallurgy 42
MgO 216, 217
microcracking 86, 191
microcracks 39, 64
microfracture 97, 192, 261
microstructure 6, 7, 11, 14, 15, 17, 21, 27, 39, 40–42, 45, 46, 48, 49, 64, 80, 81, 95, 97, 141, 156, 167, 171, 188, 196, 209, 215, 216, 218, 222, 228, 266, 279, 293, 297, 298, 304, 325
  cubic 67, 220, 221
  tetragonal 215, 216, 216, 224
milling 24–26, 54, 56, 59, 156
model
  Archard's 149 *see also* contact temperature
  Kong-Ashby 148, 149
MoS_2 129, 132, 134–136

n

nanocomposite(s) 178
alumina 13, 29, 64, 174, 215–218, 220–224, 239, 257, 266, 305, 320
nanocrystalline 45, 49
  ceramics processing 14
  reinforcement 7, 64, 86, 173, 206, 255, 259, 266
  nanotubes 266
NbC 64, 298, 309–313
nitrogen (dry) 24, 38, 239
nitrogen atmosphere 188
noise 119, 126, 179

o

oil 27, 65, 134, 177–180, 237, 241, 242, 247
  lubrication 105, 153, 175, 238, 239, 244–246, 248, 249
  viscosity 175, 177, 178, 240, 242, 248
optical interferometry 108, 110
oxidation 9, 12, 46, 49, 65, 66, 121, 141, 148, 174, 175, 256, 266, 302, 304, 305
  high-temperature 12
  layer formation 198, 199, 305
  reaction 46
  resistance 266
  wear 159, 174, 181 *see also* wear
oxide 8, 11, 39, 40, 160, 164, 169, 174, 192, 198, 199, 204, 215, 239, 242, 302, 305, 322

p

particle(s) 22–24 28, 30–32, 34, 35, 37, 39, 42, 43, 46, 49, 61, 133, 157, 162, 165, 169–172, 173, 178, 192, 196, 198, 199, 202, 204–210, 227, 228, 232, 255–259, 261, 262, 265, 267, 268, 281, 322
  hardness 231, 289
  nano 135, 175
  size 167, 168

strength 94
wear 123, 125, 131, 132, 161, 305, 306 *see also* wear debris
particulate(s) 7, 8, 12, 43, 266, 267
Peclet number 147
physical properties (data, values) *see* mechanical properties
physics 4–6, 27, 49, 53, 54, 57, 66, 145, 173, 327
plastic flow 28, 130, 131, 161, 162, 170, 257, 299
plasticity 325
Poisson's ratio 94, 95, 99, 116, 143, 223, 224, 270
polishing 5, 137, 156, 199, 202, 230, 266
polyamide (nylon) 10, 129, 131, 139
polyether ether ketone (PEEK) 239
polyethylene (PE) 10, 129–131
high-density polyethylene (HDPE) 10, 129, 131
polymer(s) 6–7, 10, 15–17, 34, 39, 78, 79, 99, 105, 129–131, 137, 157, 237, 239, 243, 244, 249
polyoxymethylene (POM) 131, 239, 243–245
polytetrafluoroethylene (PTFE) 10, 129–131, 239, 240, 242–245, 249
powder(s) 8, 22–31, 33–38, 42–49, 63, 64, 134, 188, 196, 216, 227, 228, 267, 280, 297, 313,
starting 24, 45, 54,
precipitation 47, 297
pull-out 86, 287, 300, 302, 304–306, 309 311 *see also* grain, pull-out
crystal 94, 137

r

radiation 29, 110, 114, 146, 269
real contact area 4, 105, 112–114, 123, 127, 141, 142, 145, 149
reinforcement 7, 64, 86, 173, 206, 255, 259, 266
repulsive force 109 *see also* forces
rolling 5, 16, 22, 119, 120, 126, 127, 156
roughness 57, 62, 105–111, 113, 116, 118, 121, 127, 141, 145, 156, 178, 229, 279, 310–313
rule of mixture 270
running-in-period 123, 124, 134, 172, 205, 208

s

secondary phase 188
self-lubricating materials 131
shear
modulus 99
strain 76, 77 *see* strain
strength 15, 173 *see* strength
stress 9, 73, 75, 116, 117, 164, 237 *see also* stress
Si_3N_4 9–13, 64, 65, 169, 174, 175, 195, 239, 242–245, 256, 261, 309, 312
SiAlON 64
SiC 7–12, 29, 64, 129, 174, 175, 187–192, 195–200, 202–210, 227–234, 239, 242–245, 255–262, 266, 267, 270, 280, 281–292, 309
silica 8, 206, 261, 262, 266,
hydrated 190
silicon carbide *see* SiC
silicon nitride *see* Si_3N_4
sintering 12, 22–24
aids 30, 192
field-assisted sintering technique (FAST) 30
hot isostatic pressing (HIP) 22
liquid phase 43, 191, 227, 297
pressure less 22, 27, 29
spark plasma sintering (SPS) 29, 31, 49, 267, 299
techniques, advanced 15

sliding 5, 12, 14–16, 85, 112, 116, 119–121, 123–127, 130, 133–135, 137, 139–143, 145–150, 156–157, 159–166, 170, 171, 174, 177, 178, 180, 181, 187–192, 195–210, 215, 216, 218–225, 239, 266, 319, 320, 323, 325
slip planes 164
solubility 43, 160
spalling 65, 173, 287, 292, 309, 313
S-phase *see* SiALON
steel 7, 9, 10, 13, 15, 16, 37, 53, 55, 56, 57, 59, 63, 64, 91, 111, 112, 116, 127, 128, 163, 178, 179, 196–204, 209, 210, 215, 216, 219–221, 228, 239–249, 280, 281, 292, 298–306, 309, 310
  stainless 24, 82, 298, 300, 301
stiffness 91, 109, 122, 126, 127 *see also* mechanical properties
strain *see also* mechanical properties
  hardening 160, 161
  shear 76, 77
strength 6–9, 11, 15, 26, 64 *see also* mechanical properties
  adhesion 39–41, 160
  bond 37, 39–41, 279
  cohesive 39
  compressive 7, 9, 11, 79, 82, 88, 89, 99, 217, 218
  flexural 89, 90
  fracture 83, 89
  shear 15, 173
  tensile 7, 11, 79, 82, 86, 91, 217, 224
  yield 324
strengthening 6
stress
  compressive 73, 97, 223, 224
  concentrations 80, 82, 83, 206, 305, 306
  field 117, 222, 271
  Hertzian 208
  intensity factor 84, 96
  residual 39, 41, 63, 161, 167, 171, 223, 224, 266, 274
  shear 9, 73, 75, 116, 117, 164, 237
  tensile 73, 86, 89, 116, 161, 222–224
  von Mises 272
Stribeck
  curve 155, 175, 177, 181
  parameter 175, 176
structure
  grain 7, 63, 188
  layered 39, 45, 132
subsurface
  damage 228, 230, 231, 287–291, 319, 322
  region 4, 293
surface
  coating 249, 325
  cracks 164, 170, 280, 288, 291 *see also* crack(s); cracking
  fatigue 123 *see also* fatigue
  finish 39, 62, 297, 313
  layer 63, 140, 238 *see also* layer
  roughness *see* roughness
  stress *see also* stress
  topography 108, 109, 116, 326
surface energy 22, 160, 181, 237, 238–240, 242–244, 246, 248, 249
system-dependent property 320

t

TaC 298–304, 306, 309–313
tangential force 119, 122, 126 *see also* force(s)
tearing 84
temperature 3, 298, 393, 396
  bulk 141, 149
  contact 4, 5, 15, 17, 119, 121, 141, 142, 145–148, 150, 173, 302, 323 *see also* model
  flash 15, 141, 142, 145–150, 174, 320, 322

gradient 145
maps 143
rise 30, 31, 33, 56, 119
theory
Archard's wear 160
delamination 116
fracture mechanics 84, 85
Griffith's 83
Hertzian 115, 118
indentation-fracture 170
thermal
analysis 271
conductivity 10, 11, 42, 57, 64, 65, 141, 187, 195, 265, 270
convection 114, 146
diffusivity 148
energy 59
expansion 11, 206, 223, 224, 272
expansion coefficient 224
properties 10, 14, 150, 298
radiation 269
shock resistance 298, 302, 320
thermoplastics 130
thermosets 131
third body 15, 140, 162 *see also* layer; wear debris
TiB_2 8, 64, 255
TiC 9, 64, 65, 277
TiN 8, 56, 64, 65
$TiSi_2$ 268, 277
Ti_3SiC_2 111, 112
tolerances 59
topography 65, 106, 109, 116, 127, 325, 326 *see also* roughness
toughening mechanisms 15, 86
toughness 7–11, 15, 43–45, 95, 156, 160, 181, 297 *see also* fracture toughness
transfer films, layers 125, *see also* film(s); layer
transformability 216, 218, 219
transformation
hydrothermal 130
martensitic 178
phase 22, 86, 216, 223, 224
stress induced 215
tetragonal-to-monoclinic 224
toughening 15, 80, 85, 86, 219
zirconia phase 223
transition metal 298, 302
tribochemical
layer 121, 130, 172, 173, 180, 181, 322 *see also* layer
reaction(s) 17, 119, 140, 172, 181, 305
wear 15, 155, 159, 165, 171–174, 181, 195, 321, 323
tribochemistry 122, 156, 172, 203, 207–210
tribolayer 172, 202, 222, 300, 305, 306 *see also* tribochemical, layer
tribological
applications 7–11, 14, 30, 57, 71, 79, 84, 99, 187, 195, 255, 319, 325 *see also* applications
contact 9, 79, 85, 119 *see also* contact
tribology 3–6, 15–17, 21, 80, 82, 109, 115, 134, 137, 150, 158, 239, 320–323, 320–323, 325–327
tribomaterialomics 16, 319, 323, 325, 326

u

ultra-high temperature ceramics (UHTCs) 265
ultrasonic method 94

v

vacuum 24, 26, 31, 33, 111, 127, 130, 297, 298
van der Waals force 132
vibration 5, 62, 91, 92, 94, 98, 99, 119, 126, 156, 159, 178

Vickers indentation 87, 97, 162
viscosity 175, 177, 242, 246

w

water 27, 33, 37, 38, 63, 131, 134, 173, 180, 237, 240, 242–246, 249, 320
waviness 106
WC 16, 24, 42, 46–49, 129, 188–191, 195–210, 220, 228, 231, 233, 255–262, 298–306, 309–313
WC-Co 32, 41–49, 130, 197–204, 227, 228, 279–293
wear
  abrasive 9, 158, 159, 161, 162, 168, 170, 220, 224, 227, 228, 230–234 *see also* abrasion
  adhesive 158–160, 177, 202, 220, 305 *see also* adhesion
  computational analysis 16, 265–267, 274, 320, 323, 325
  debris *see* wear debris
  delamination 116, 130, 165, 224, 253, 291
  erosive 14, 16, 32, 157–159, 167–171, 174 *see also* erosion
  fretting 159, 163, 165, 166 *see also* fretting
  loss 131, 157, 259, 281, 282, 291, 293
  mechanism 5, 17, 97, 150, 155, 156, 158, 159, 163, 181, 187, 191, 195, 196, 202, 205, 206, 208, 210, 219, 220, 231, 257, 287, 298, 299, 304, 309, 321, 322, 326, 327
  micromechanisms 14, 313, 319
  mild 5, 178, 323
  model 187, 322, 323 *see also* theory
  oxidative 159, 174, 181 *see also* oxidation
  resistance 6, 7, 9, 11, 12, 15, 16, 54, 56, 71, 79, 95, 151, 158, 164, 173, 187, 191–193, 195, 217, 257, 262, 267, 299, 327
  severe 157, 174, 178, 222, 300
  theoretical analysis 222
  three-body 162
  tribochemical *see also* tribochemical, wear
  tribomechanical 15, 155, 159, 163, 172, 181, 195, 321, 322
  two-body 162
  unlubricated 219
wear debris 5, 131, 160, 165, 172, 178, 201, 206, 208, 209, 306
  polycrystalline 324
wetting 43, 181, 237– 242, 246, 248–250
whisker 7, 86
  reinforced composites 12
WO_3 47, 201, 202, 261, 262
work hardening 105
WS_2 134

y

Young's modulus 79, 91, 94, 95, *see also* elastic modulus
yttria (Y_2O_3) 187, 189, 191–193, 196
  additive 188 189, 191
  content 189, 192

z

zirconia 8, 12, 13, 64, 215–224, 239, 266
  zirconia-toughened alumina (ZTA) 215–226
ZrB_2 16, 265–270, 272–275, 323
ZrO_2 10, 16, 24, 29, 195, 215–225, 242–245, 309, *see also* zirconia
  tetragonal 215, 216

Printed and bound by CPI Group (UK) Ltd, Croydon, CR0 4YY